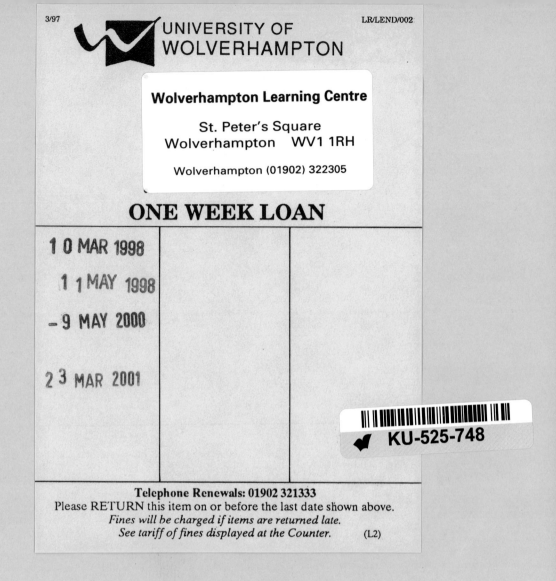

THE ECONOMIC IMPORTANCE OF INSECTS

THE INSTITUTE OF BIOLOGY AND CHAPMAN & HALL

Biology is a disparate science, embracing a spectrum of study from the molecular level to groups of whole organisms. It is increasingly difficult for both professional biologists and students to keep abreast of developments. To help address this problem the Institute of Biology is complementing its own in-house publications with those produced by professional publishing houses. The Institute is pleased to be publishing *The Economic Importance of Insects* with Chapman & Hall. The Institute's Books Committee welcomes proposals for other advanced-level texts.

THE INSTITUTE OF BIOLOGY

The Institute of Biology is the professional body for UK biologists. It is a charitable organization, charged by Royal Charter, to represent UK biology and biologists. Many of its 15 000 members are Chartered Biologists (CBiol), a qualification conferred by the Institute on professional biologists and which is recognized throughout the European Communities under Directive 89/48/EEC.

The Institute's activities include providing evidence on biological matters to government, industry and other bodies; publishing books and journals; organizing symposia; producing specialized registers and coordinating regional branches. The Institute is a prominent member of the European Communities Biologists Association (ECBA) and coordinates liaison between UK biologists and the International Union of Biological Sciences (IUBS).

For further details please write to:

The Institute of Biology
20–22 Queensberry Place
London
SW7 2DZ

THE ECONOMIC IMPORTANCE OF INSECTS

Dennis S. Hill

Animal Resources Program
Faculty of Resource Science and Technology
Universiti Malaysia Sarawak
East Malaysia

The Institute of Biology

Incorporated by Royal Charter

CHAPMAN & HALL
London · Weinheim · New York · Tokyo · Melbourne · Madras

Published by Chapman & Hall, 2–6 Boundary Row, London SE1 8HN

Chapman & Hall, 2–6 Boundary Row, London SE1 8HN, UK

Blackie Academic & Professional, Wester Cleddens Road, Bishopbriggs, Glasgow G64 2NZ, UK

Chapman & Hall GmbH, Pappelallee 3, 69469 Weinheim, Germany

Chapman & Hall USA, 115 Fifth Avenue, New York, NY 10003, USA

Chapman & Hall Japan, ITP-Japan, Kyowa Building, 3F, 2-2-1 Hirakawacho, Chiyoda-ku, Tokyo 102, Japan

Chapman & Hall Australia, 102 Dodds Street, South Melbourne, Victoria 3205, Australia

Chapman & Hall India, R. Seshadri, 32 Second Main Road, CIT East, Madras 600 035, India

First edition 1997

© 1997 Dennis S. Hill

Typeset in 10/12pt Palatino by Keyset Composition, Colchester
Printed in Great Britain at the Alden Press, Osney Mead, Oxford

ISBN 0 412 49800 6

A catalogue record for this book is available from the British Library

Library of Congress Catalog Card Number: 96-84902

∞ Printed on permanent acid-free text paper, manufactured in accordance with ANSI/NISO Z39.48-1992 and ANSI/NISO Z39.48-1984 (Permanence of Paper).

To Ellie, with love

CONTENTS

PREFACE

In the last few decades there has been an ever-increasing component in most BSc Zoology degree courses of cell biology, physiology and genetics, for spectacular developments have taken place in these fields. Some aspects of biotechnology are now also being included. In order to accommodate the new material, the old zoology courses were altered and the traditional two-year basis of systematics of the animal kingdom, comparative anatomy (and physiology) and evolution, was either severely trimmed or reduced and presented in an abridged form under another title. Soon after these course alterations came the swing to modular teaching in the form of a series of shorter, separate courses, some of which were optional. The entire BSc degree course took on a different appearance and several different basic themes became possible. One major result was that in the great majority of cases taxonomy and systematics were no longer taught and biology students graduated without this basic training.

We field biologists did appreciate the rising interest in ecology and environmental studies, but at the same time lamented the shortage of taxonomic skills, so that often field work was based on incorrect identifications. For years many of us with taxonomic inclinations have been bedevilled by the problem of teaching systematics to undergraduates. At a guess, maybe only 5% of students find systematics interesting. It is, however, the very basis of all studies in biology – the correct identification of the organism concerned and its relationships to others in the community.

The present situation is that the British university system offers many aspects of environmental biology in a wide range of programmes and a few pest management courses where a broad integrated approach is taken. Very recently it has been encouraging to note that taxonomy and systematics have returned as topics of scientific importance under the guise of 'biodiversity'. However, in schools and most universities the emphasis in biology teaching has been placed more and more on laboratory work, and computers are assuming an ever-increasing importance. Field studies are being minimized and even abandoned, and now many 'biologists' graduate with a BSc but have never spent any time in the field and cannot tell a meadow pipit from an osprey or a mealybug from a moth. Much environmental study has become theoretical.

There has been professional lamentation about a general decline in student interest in (field) biology, and more specifically the lack of interest in entomology. This is in part rather strange, for there is now a proliferation of books on natural history, birds, mammals and insects, and many are extensively illustrated. I feel that part of the problem may lie with the teachers, many of whom are not naturalists or field-oriented, and sad to say some biology teachers are not interested in biology at all.

Many of we practising applied entomologists are aspiring to encourage student interest in insects and also in field biology generally, and it is to be hoped that mid-level university texts such as these will be used by both BSc undergraduates and also practising biology teachers in secondary schools. After all, the importance of the Insecta, both beneficially and as pests of humanity, cannot be over-emphasized, and the interrelationships between insects and their host plants and other animals are a fascinating source of study.

The illustrations and most of the data used in this book have come from a recently published text, *Agricultural Entomology* (Hill, 1994). That book was based on 40 years of field work and photography in many parts of the world but mainly in the Far East, East Africa and eastern England. It is very striking how similar are the insect faunas of Europe, East Asia and North America, but there are equally interesting differences. Similarly, East Africa and tropical South East Asia have both similarities and differences of importance to humans and agriculture. The bulk of the 1994 book was a systematic review of the class Insecta at family level with reference to the important species that are either beneficial to humans or pests in all parts of the world.

Over the years, I have received much help from colleagues and friends, and these were thanked in the Preface of *Agricultural Entomology*. The insect pest drawings were made by several different artists who were likewise acknowledged. The photographs were taken by me.

An alternative title for this book could be *An Introductiion to Applied Entomology*.

Dennis S. Hill

1

INTRODUCTION

The human animal clearly dominates the world in terms of its impact on the environment and its general effect upon the Earth. Biologically, however, the class Insecta can claim to be the more important. Insects are humanity's greatest rival for the world's food resources, both directly by eating the plants cultivated for food, and indirectly as vectors of disease-causing organisms. In terms of evolution, insect pests kept humanity at a subsistence level for thousands of years. But as we evolved, we developed skills in agriculture, chemistry and medicine, and now we have many weapons with which to attack the insect hordes and the parasites they transmit. However, human progress in the battle against the insects is somewhat like climbing a steep sand dune – for every metre climbed, one slips back half a metre; at any point a quick landslide can precipitate one down to the base again, and each step requires a major effort. A constant striving, and a vast financial expenditure on a day-to-day basis, does keep the insect pests at bay – just!

The dominance of the Insecta in the world fauna and their general biological success can be shown in several different ways. The three usual ways are to consider the number of species concerned, the numbers of individuals, and the range of habitats colonized and niches occupied. The number of animal species named, and thus validated, is about 1 million in total. Of these, some 800 000 are insects. To put the figures into better perspective, there is a total of about 10 000 species of Vertebrata (mammals, birds, reptiles, amphibians and fish). Thus 80% of known animal species in the world are insects. Recent survey work in parts of the tropical rain forests (South East Asia and South America) has disclosed many hitherto unknown species of insect that are as yet unnamed. This has led to some discussion in the scientific press as to just how many undescribed insect species remain to be found in the tropical rain forests of the world. Some proponents suggest that there could be a grand total of up to 4–5 million different species of insect inhabiting the world; some believe the figure to be even higher It would seem that some of these estimates are rather fanciful, but an overall figure of up to 2 million species – that is, double the existing figure – is not unreasonable to assume. Thus out of a total of maybe 2 200 000 species of animals in the world, some 90% are insects.

Consider the numbers of individuals present: in 1994 the estimate of the human population of the world was in the region of 3000 million (or 3×10^9); an estimate of the number of insects alive at any one time in the world was recently published as being 10^{18}. If this estimate is accurate, then there would be 300 million insects for every human being in the world at the present time; a daunting thought! Social insects often live in large communities. Some ant and termite nests may contain up to a million individuals, and many wasps and bees and other ants live in colonies of 20–100 000 per nest. A large locust swarm may contain 1000 million insects, either hopping or flying. In an infested agricultural crop there can be 1000 Cabbage Aphids or Maize Aphids per plant, thus in a field of 10 ha the total insect population can be enormous. The reproductive potential of insect pest species can be very great. In general the number of eggs laid per female insect is correlated to the chances of survival of the offspring. Thus the larviparous/pupiparous tsetse (flies) and Hippoboscidae 'lay' only about 10–20 larvae per year each; each larva pupates immediately after being deposited, so the usual egg and larval mortality are avoided. Insects that lay eggs in open locations where mortality rates are high may lay up to 1–2000 eggs per female, for it is usual for 90–95% of the eggs to be destroyed by natural enemies (predators and parasites) and a similar proportion of larvae to be killed during development. Larvipary is generally rare and most insects lay eggs. Most of the pest species have a great reproductive potential, with the viviparous parthenogenetic female aphids at the top of the list. A single female aphid bearing between six and ten daughter offspring at a time could, in theory, at the end of one year have produced a total of 6 billion aphids. Fortunately for humans, the vast majority of these insects are eaten by other insects and birds, etc., and so the world does not become inundated by aphids.

The third approach is to consider the range of habitats colonized by insects and the total number of ecological niches occupied. Basically, all terrestrial and freshwater habitats are colonized by insects, from the poles to the equator, from below ground level to mountain tops, with anemoplanktonic (aerial) forms found up to altitudes of more than 10 000 feet. Insects survive in the Arctic at −50°C; others are active in hot springs at +40°C, and some beetles survive in

deserts with a noon temperature of +60°C. Vast numbers of insects are phytophagous and feed on flowering plants. Many are scavengers and feed omnivorously on a wide range of dead organic material; many are predacious and prey on or parasitize other insects. Most common plants are host to a large number of different insect species, for example the two species of oak (*Quercus* spp.) in Britain are recorded as hosts for 1000 different species of insect. *Quercus* is the dominant tree throughout the temperate forest belt of the Holarctic region, throughout Europe, Asia and North America, and so the genus covers a vast geographical area. Agriculturally the insect pest load record is held more or less jointly by cotton and cocoa, from which worldwide are recorded some 1400 species of insect and mite.

As nicely summarized by Davies (1988), the reasons for the success of the Insecta are several, not least of which is their adaptability. Insects occupy all imaginable niches in all major terrestrial and fresh water habitats in all parts of the world. The only major habitat uncolonized is the sea, where their relatives, the Crustacea, are dominant. (Actually, a few insect species can be described as 'marine', but not many.) Insects' capacity for flight confers a unique advantage for dispersal, finding food and mates etc., and to escape from enemies. And remarkably, at times when flight power would be a disadvantage, many species have evolved periodic winglessness for greater effectiveness (as with many aphids), and some ectoparasitic forms have lost their wings completely.

Small body size is biologically advantageous as little food is required by one individual and only a short time is required for development – thus tropical aphids can have up to 30 generations in one year and the Diamond-back Moth in Malaysia has 15 generations per year. This means that the evolution of some insect pests in the tropics can be rapid, shown especially in the development of new biotypes resistant to common insecticides. Very small insects often have problems in rain and with water films, which can trap them by surface tension effects. The largest insects tend to be sluggish because their tracheal system of respiration, which relies on gaseous diffusion, is limiting and inefficient above about 2 cm. Thus most insects are classed as 'moderate' in size and have a body length of 0.5–2.0 cm.

The skeleton of insects is an exoskeleton made up of a series of hard sclerites joined by flexible membranes. Primitive arthropods are thought to have been worm-like (Protoannelida), and with the development of a rigid skeleton tagmosis occurred for greater functional efficiency. Thus the head tagma was formed from the six anterior segments, and the thoracic tagma from the following three. The exoskeleton consists of a series of hollow tubes joined by flexible membranes, and is both light and very strong. Moulting (ecdysis) is a problem: although it is necessary to permit body growth, it causes great physiological stress and physical vulnerability. Sick, wounded or otherwise strained insects invariably die at moulting because of the stresses involved. However, ecdysis does permit insect metamorphosis with the great ecological diversification and separation between larval and adult forms. This ecological separation is most developed in the Diptera and Lepidoptera, and also in aquatic forms.

The final reason for the success of the Insecta is their resistance to desiccation, bearing in mind that they have evolved from a worm-like protoannelid. The main features are the development of a waxy (waterproof) cuticle; spiracular closure mechanisms; dry excretory products (these may be crystals of uric acid); and the ability to use metabolic water. The development of an impervious egg shell (chorion) represents the first occasion of this important step to becoming truly terrestrial in the evolution of the Invertebrata.

Insect classification differs somewhat according to the different authorities in different parts of the world. Some groups (orders) are very sharply defined, but in others the precise limits are less clear, and the affinities of some groups may not be obvious. However, in the UK, and a large part of the English-speaking world, for teaching purposes we tend to follow the standard work on entomology based on the book by Imms (1967) – first published in 1925, this is now in its 10th edition and was retitled *Imms' General Textbook of Entomology* by Richards and Davies (1977). In its previous editions, the book was very widely distributed throughout the Third World, as inexpensive editions were available through the English Language Book Society. The 29 orders within the class Insecta are listed in Table 1.1.

The figures to the right of Table 1.1 represent the numbers of families in each order, as per Richards and Davies (1977), and the estimated number of named species. The species number is a minimal approximation, for several thousand new species are being described each year. It is thought that in tropical regions there are many thousands of new species yet to be named.

Taxonomy is the science (or art) of recognizing crucial differences in morphology and anatomy (and sometimes behaviour, etc.) between individuals of different species. New species have to be described, illustrated and named, and all this has to be published in an internationally recognized journal for validation. The original specimen(s) described is then deposited in the Natural History Museum (London) or another major international museum, and is referred to as the 'type specimen'. The scientific name is usually in Latin (but may be Greek) and following the Linnean Binomial System it consists of a generic name followed by the specific. Thus *Aphis pomi* is the Green Apple Aphid; in the UK there are some 80 recorded species of *Aphis*, but only one is called *A. pomi*. Other aphids are placed in other genera, and there are some 90 recorded

Table 1.1   Classification of the Insecta (after Richards and Davies, 1977)

		Class Insecta		*Number of:* *families:species*[a]
		Subclass (1) Apterygota		
Order	1.	Thysanura	(Bristletails)	5:550
	2.	Diplura		6:600
	3.	Protura		4:200
	4.	Collembola	(Springtails)	11:1500
		Subclass (2) Pterygota		
Division		**A. Exopterygota (= Hemimetabola)**		
	5.	Ephemeroptera	(Mayflies)	19:2000
	6.	Odonata	(Dragonflies and Damselflies)	26:5000
	7.	Plecoptera	(Stoneflies)	14:1700
	8.	Grylloblattodea		1:16
	9.	Orthoptera	(Crickets, Grasshoppers, etc.)	17:17 000
	10.	Phasmida	(Stick and Leaf insects)	2:2500
	11.	Dermaptera	(Earwigs)	8:1200
	12.	Embioptera	(Web-spinners)	8:300+
	13.	Dictyoptera	(Cockroaches and Mantids)	9:6000
	14.	Isoptera	(Termites)	7:1900
	15.	Zoraptera		1:22
	16.	Psocoptera	(Book- and Barklice)	22:2000
	17.	Mallophaga	(Biting Lice)	12:2800
	18.	Siphunculata	(Sucking Lice)	6:300
	19.	Hemiptera	(Bugs)	38:56 000
	20.	Thysanoptera	(Thrips)	6:5000
Division		**B. Endopterygota**	**(= Holometabola)**	
	21.	Neuroptera	(Lacewings, etc.)	19:5000
	22.	Coleoptera	(Beetles)	95:330 000
	23.	Strepsiptera		9:370
	24.	Mecoptera	(Scorpion-flies, etc.)	9:400
	25.	Siphonaptera	(Fleas)	6:1400
	26.	Diptera	(Flies)	*c.* 130:85 000
	27.	Lepidoptera	(Moths and Butterflies)	97:120 000
	28.	Trichoptera	(Caddisflies)	11:5000
	29.	Hymenoptera	(Sawflies, Wasps, Ants, Bees, etc.)	60+:100 000+

[a]There is some disagreement internationally over the details of insect classification and estimated numbers vary greatly. In most groups there are undoubtedly many more species to be described and named, so final numbers of species in most groups will be far greater.

genera of Aphididae in addition to *Aphis*. A widely distributed Holarctic species might have been described and named several times, for example in England, France, Japan and the USA. When this happens, the first (oldest) description and name are accepted as valid and the other later ones are regarded as synonyms and are invalid. Historically, most important pest species have a number of synonyms recorded in the literature, and in older publications the precise identity of some insect pests may not be clear. The common name of insect species in the UK is in English. Only certain groups of insects are given local common names – those of sufficient importance include the Macro-lepidoptera (butterflies and larger moths); some larger bugs (Hemiptera) and larger beetles (Coleoptera); some dragonflies (Odonata); some mayflies (Ephemeroptera), and species of importance as pests (Seymour, 1979). A good common name is useful, especially as the scientific name can be subject to change, but it should be meaningful and and not just a translation of the Latin name. Thus 'Green Apple Aphid' is a good name for *A. pomi*.

Identification of insects is based usually on a description and illustration published in a key or book. Failing this, either reference to the original description of a species or comparison with named specimens in a reputable insect museum collection can be successful. As a very last resort, it is necessary to make a direct comparison with the original type specimen(s), but this is usually only possible for professional insect taxonomists. At order level, it is usually very easy to place an insect specimen – thus a beetle is usually distinctly different from a moth or bug. Ecologically the family group is very important – all members of the same family have a similar appearance (i.e. facies), and have similar habits and ecological requirements. Generally, anywhere in the world the members of the same family are easily recognizable – thus an aphid is recognized as an aphid (F. Aphididae) anywhere. However, one has to beware, because for all biological rules/generalizations, there are (rare) exceptions. Every entomologist has encountered insect specimens, especially in the tropics, that at first sight resemble nothing that they have ever seen before, and even to place it in an order can be difficult. Male fig wasps (Hym., Agaonidae), Strepsiptera, male Coccoidea (Hemiptera), Lycidae, Nycteribiidae, Sheep Ked are all at first sight quite difficult to place, and these are adults; there are many larval stages that are very difficult to place taxonomically without extensive study.

It should be stressed in this Introduction that although a great deal of the time a majority of entomologists are concerned with the damage caused by insects to humans and their crops, livestock and possessions, in point of fact in several major ways insects are of vital beneficial importance to human society. There is a built-in imbalance in our general observance of insects, for the damaging aspects are only too obvious whereas most of the beneficial aspects are unobtrusive and tend to be taken for granted.

The monetary value of insects is quite incalculable, but clearly enormous. On the negative (cost) side, how can the effect of malaria be calculated? It is estimated by the World Health Organization that this mosquito-borne disease now attacks 100 million people per year and more than 1 million die. Some 40% of the world population is at risk, not including travellers. We are still losing an estimated 30–40% of crop produce worldwide (valued at US$300 000 million) despite an annual expenditure of $20 000 million on insecticides. On the positive side, the plants pollinated by insects constitute probably half of the world vegetative biomass. And if natural enemies (i.e. insect predators or parasites) regularly destroy some 90% of insect pest populations, without their controlling effect we would be inundated by insect pests. Scattered in the literature are some estimates of the monetary value of some insects and their actions, and these will be quoted later.

2

BENEFICIAL INSECTS

The useful aspects of insects as defined by humans fall mostly into two categories: direct and indirect. Direct beneficiality is less important in these modern times, and it applies when the insect itself is directly utilized. During the early evolution of humanity, particularly at the hunter-gatherer stage, insects must have been a major food item, but with the development of agriculture their importance in the human diet lessened. Insect products such as honey, wax and silk are widely used but of limited economic importance, as also is the now widespread farming (rearing) of various insect species. The most important beneficial aspects of insects are indirect – which is largely why they are relatively unobtrusive and seldom fully appreciated.

The ways in which insects are beneficial to humans are presented below.

2.1 POLLINATION

Most higher plants need pollination in order to produce seed for propagation. Plant propagation occurs in two main ways: vegetative (or asexual) and sexual. Some plants use one method to the exclusion of the other, but many species rely on a combination of the two types for maximum effectiveness. Vegetative reproduction relies upon the use of rhizomes, stolons, tubers, corms, bulbs, suckers, etc., and the young plants are

genetically identical to the parent. These are ideal methods of propagation for ruderals and other pioneer species, for they enable the plants to cover a large area of bare ground rapidly. Many noxious weed species have extensive powers of vegetative reproduction, which in part accounts for their destructive importance. In evolutionary terms, however, this genetic stagnation is a disadvantage, and sexual reproduction involving cross-fertilization (cross-pollination) is needed in order to maintain genetic diversity.

In terms of world flora most plant species rely on a combination of vegetative propagation and sexual reproduction for the production of new individuals and their dispersal. A relatively few species use only seeds as their diaspores, and for these pollination is clearly of paramount importance. So far as is known, no wild plant relies only on vegetative reproduction, but a few cultivated species of antiquity are now only propagated vegetatively. Thus sugarcane is now only grown from stem cuttings, while cultivated bananas set fruit parthenocarpically and are propagated by stem-suckers (although the seed is obviously produced by the wild plants in Malaysia).

Pollination involves the depositing of pollen grains from the anthers (male flowers) on to the receptive stigma of the female flower; a pollen tube extends down the style to an ovule in the ovary, and the fusing of the two nuclei results in the formation of an embryo within a seed. Some plants are self-fertile and their own pollen is used to fertilize the ovules. In most flowers, self-pollination is either impossible or is exceptional. The stamens and pistil may be in separate flowers, either on the same plant (monoecious) or on different plants (dioecious). In normal hermaphroditic flowers the pollen may mature before the stigma is receptive (the plant is then termed 'protandrous') or the stigma may have passed its receptive stage before the anthers ripen (so the flower is 'protogynous'). In addition to these physiological (temporal) mechanisms, most flowers have anatomical developments that more or less prevent the pollen from the one flower reaching the stigma of the same flower – this is termed self-incompatibility. In extreme cases, the flowers are self-fertile. Clearly, in the evolution of the higher flowering plants cross-pollination has been the rule and self-pollination the exception, for obvious genetic reasons. Most self-fertile flowers require regular cross-pollination to maintain seed viability. Thus *Rubus* species (raspberry, blackberry, etc.) when self-fertilizing set smaller fruit with fewer seeds than when insect-pollinated.

The two main pollinating agents are insects and the wind. Flowers pollinated by insects are termed 'entomophilous' and those using wind are 'anemophilous'. Other pollination agents include water currents (for some aquatic plants), some species of bird, bats (in the tropics) and also some slugs. Entomophilous flowers are usually brightly coloured, produce nectar and have an odour that will attract insects. The pollen is not particularly abundant and is often sticky, and the anthers are often not very conspicuous. The stigma is typically inconspicuous and seldom feathery in shape. Anemophilous flowers are open to the wind, often in catkins, with little nectar or scent, and produce an abundance of dry, dust-like pollen. The stamens are borne on long filaments and are easily shaken by the wind. The light pollen floats easily on air currents, and synchronized pollen release controlled by regional weather conditions often results in an entire local area having a yellow hue as a result of *Pinus* pollen or that from grasses. The stigmas are long and feathery and protrude from the flowers; anemophilous trees typically flower well before the leaves appear.

Insects that visit flowers do so for two different food sources: the sugary nectar and the proteinaceous pollen. Some take both but others take only one of the two, although such selectivity may only be seasonal. Clearly with social insects there will be times when sugar is most needed and other times when protein is more important. Most of the insects that visit flowers and effect pollination are living in a mutualistic relationship with the plants; the plants are their only source of food and in return they pollinate the flowers. The flowers produce the nectar and scent purely to attract the pollinating insects, and they usually produce a considerable excess of pollen to allow for that taken as insect food. It has been estimated for apples that if 5% of the flowers develop into fruit, a good crop results. The trees produce a large quantity of pollen, of which only 10% is thought to be needed for pollination (Taylor, 1948) – presumably the rest is available as insect food, although wastage must be extensive. Nectar is required by most bees early in the season (early spring) after the winter 'hibernation' as a source of energy. Thus in some apple orchards in early May, foraging Honey Bees can be diverted from the abundant apple pollen to the dandelion weeds below, which are rich in nectar.

Some pollen-seeking insects may open the anther sacs and are beneficial in causing more effective pollen release – it is thought that the pollen beetles (*Meligethes* spp.) on rape often have this effect. All plants produce pollen (for fertilization) but a number do not produce nectar and the insects visit the flowers solely for pollen.

The Gramineae are all anemophilous and produce bisexual flowers (although maize has separate female and male flowers on the same plant). Cross-pollination is usual, but some species are self-pollinated. Usually when the reproductive organs are mature, parts of the flower undergo turgor pressure. This causes both exposure of the stigmas and the thrusting out of the anthers by rapid elongation of the stamen sac filaments; the anther sacs burst and the abundant pollen is shed. Most grasses are protandrous. Flowering is sequential along the inflorescence spike, and for each flower the process lasts one to two hours. Most cereals open their flowers in the morning from 04.00–07.00 hours, when the air temperature rises to about 24°C. In wheat, the first flowers to open are situated about one-third of the way from the apex of the ear; the rest follow in succession both upwards and downwards from this point. Each flower remains open for 8–30 minutes, and the whole ear completes its flowering in eight to nine days. Most European grasses flower between 05.00 and 09.00 hours, but *Lolium* and *Festuca* flower more towards midday. Some grasses and barley have closed flowers so that self-pollination and fertilization is the rule; cross-pollination is normally impossible. Similarly with entomophilous flowers, the release of scent and anther sac rupture is usually controlled by weather factors to coincide with the activity of the main pollinators.

The overall pollination picture for the flowering plants (Phanaerophyta) is complicated in that there is often a mixture of self- and cross-pollination involving both wind and insects as the agents. For a surprisingly large number of plant species there is uncertainty as to the precise method of natural pollination and fertilization.

Anemophilous plants include the Gramineae (grasses and cereal crops) and many trees, particularly all of the Gymnospermae, oak, beech, birch, elm, poplar, some willows, Polygonaceae,

Cannabanaceae, and hazel, elder, etc. Major groups that appear to be partly anemophilous and partly entomophilous under natural conditions include the Cruciferae, Umbelliferae and Chenopodiaceae. Entomophilous plants include many trees, many shrubs and a large majority of vegetables and flower crops; most of the Dicotyledoneae in fact (Table 2.1).

A large proportion of these plants are cultivated as crops, for shade or as ornamentals. The proportion of non-cultivated (i.e. wild) flowering plants, excluding Gramineae, that are insect-pollinated is not clear at the present time, but would for obvious reasons seem to be large.

The value of the insect pollination of plants to human society is vast, and where wild plants are concerned cannot be calculated. For cultivated plants, however, the value is clearly equally great. It has been estimated that in the USA in the year 1986 the value of insect-pollinated crops was of the order of US $17 000 million for a total of about 90 crops, excluding forage legumes. Modern cattle-rearing relies heavily on the use of forage legumes (alfalfa, clover, etc.), partly as 'hay' or silage, and partly in short-term grazing leys. If seed production of forage legumes or silage grown for livestock food is included, then about one-third of the US food supply depends directly or indirectly on the pollination of plants by insects.

Proctor and Yeo (1973) record that there are some species in most of the 29 insect orders that are involved in pollination, at least to the extent of being anthophilous. Most insects visit flowers either to collect nectar (for its sugary carbohydrates) or pollen (for its proteins). The nectar collectors are the more prevalent, and they usually carry pollen grains, more or less accidentally, from flower to flower and so effect pollination. During the course of evolution this relationship between flowers and insects has developed in complexity until a state of interdependence has become complete mutualism. Many of the pollen-eaters do actually pollinate while

Table 2.1 Entomophilous plants

Trees	Acacia	*Ficus*	Peach
	Almond	Hawthorn	Pear
	Apple	Lime	Plum
	Avocado	Litchi	Rubber
	Cashew	Mango	Spindle
	Coconut	Maple	Sycamore
	Date Palm	Nutmeg	Tung
	Eucalyptus	Oil Palm	some Willows
Shrubs	Cassava	Coffee	Guava
	Cinnamon	Cotton[a]	Lantana
	Citrus[a]	Currants (*Ribes*)	Tea
	Cocoa	Gardenia	
Vegetables	Asparagus	Cucurbits	Sweet Potato
	Beans	Onions	Tomato[a]
	Beets	Peas	
	Cabbage	some Solanaceae[a]	
Miscellaneous	Clovers	Ivy	Poppy
	Flax	Lavender	Sunflower
	many Flowers	Mint	Ranunculus
	Geranium	some Orchids	Rape[a]
	Heathers	*Oxalis*	Tobacco[a]
	Honeysuckle	Passion fruit	*Viola*

Major plant families that are Entomophilous

Amaryllidae	Leguminosae
Araceae	Liliaceae
Boraginaceae	Linaceae
Campanulaceae	Lythrarieae
Caryophyllaceae	Malvaceae
Cruciferae	Myrtaceae
Compositae	Onagraceae
Convolvulaceae	Plumbaginaceae
Ericaceae	Rosaceae
Euphorbiaceae	Rubiaceae
Hypericaceae	Scrophulariaceae
Iridaceae	Umbelliferae
Labiatae	Verbenaceae

[a]Part self-pollinated.

they are collecting their food. The other anthophilous insects actually eat part (sometimes all) of the flower (Meloidae, etc.) but they can pollinate too; however, if numbers are high then total flower destruction can result.

Quite recently agriculturalists and en-tomologists became particularly aware of the importance of insect pollination of crops; this resulted in part from the destruction of bees by the careless widespread use of the organochlorine insecticides. At this time, it was widely thought that Honey Bees were the main pollinating agents. However, recent studies (such as Corbet, 1987) indicate that this is not the case. Where Honey Bees are managed in association with crop monocultures, they are usually the most important pollinators. However, on a worldwide basis it is now thought that the solitary bee species collectively are the most important pollinators, with the Bumble Bees and other social bees next followed by the Honey Bee. Out of a total of about 20 000 bee species (Apoidea) worldwide, more than 85% are solitary (Batra, 1984). In addition to the bees, the other insects of importance as plant pollinators include some flies (Diptera), moths (Lepidoptera), beetles (Coleoptera) and thrips (Thysanoptera) (Table 2.2).

Agriculturally, pollination efficiency is important so as to maintain maximum crop yields, and it is also of importance ecologically. Pollination efficiency depends upon the following factors:

1. Degree of host plant-specificity – polylectic insects visit flowers of many different plants; oligolectic only a few (closely related) plants and monolectic insects visit only one species of flower. Selective foragers will deliberately avoid visiting certain species of flower.
2. 'Tongue' length – flowers with a long corolla are only visited by insects with a long 'tongue' (i.e. suctorial proboscis). Thus spider lilies are only pollinated by Sphingidae (Lep.) with their proboscis up to 25 cm long. Flies (Muscoidea) can only pollinate open-flowered species such as Cruciferae. The floral nectaries are situated at the base of the petals/corolla.
3. Foraging distance – some bees will travel for greater distances (from the nets) than others, up to a distance of 10 km. Dispersal distances for non-social species are of importance.
4. Foraging period – the daily cycle of activity varies considerably with the insect species.

5. All-weather foraging – Honey Bees typically do not fly at low air temperature or in rain, although some species of *Bombus* will do so.
6. Speed of foraging – this is measured as the number of flowers visited per minute.
7. Pollen transport mechanism – the mutualistic species have generally evolved special structures (corbiculae) on either the legs or the body, and thus can carry larger quantities of pollen. In some cases more pollen will arrive on the stigmas.

Thus for pollination of agricultural crops some species of insect are more effective than others: in point of fact, some species are actually useless for some crops. For example, red clover can only be pollinated by 'long-tongued' species of *Bombus*, for the body weight is needed to open the flower and the long proboscis to reach the nectaries at the base of the corolla tube. The open-flowered Cruciferae are probably best pollinated by Muscoid flies, and in practice farmers wishing to preserve a particular variety of cabbage, etc., will use Blow Flies (from fishing maggots) inside a netting cage to effect pollination.

Worldwide there is a general loss of natural habitat with the ever-increasing spread of agriculture and urbanization, together with such practices as hedgerow destruction to produce larger fields. With this loss of natural habitat there is an associated reduction in the numbers of many insect species. At the same time there has been a sudden increase in the acreage of some crops and vast monocultures are being established. The largest monocultures are of maize, wheat and other cereals, and fortunately insects are not needed for their pollination. However, worldwide at the turn of the century, apples were extensively internationally dispersed in many large orchards, and recently in the UK oilseed rape has become a dominant crop. In South East Asia the recent development of oil palm plantations has been spectacular. These ex-

Table 2.2 Insect pollinators

(a) Hymenoptera	**Apoidea**

Colletidae (Plasterer Bees; Membrane Bees)
   Colletes spp. ⎫
   Hylaeus spp. ⎬ Holarctic originally; now widespread
Halictidae (Mining Bees; Sweat Bees)
   Halictus spp. temperate
   Nomia spp. more tropical, especially in New World
Andrenidae (Burrowing Bees; Digger Bees)
   Andrena spp. Holarctic, etc.
Megachilidae (Leaf-cutting Bees and Mason Bees)
   Anthidium spp. tropics
   Chalicodoma spp. (Mason Bees, etc.)
   Megachile spp. (Leaf-cutting Bees) Worldwide
   Osmia spp. (Horn-faced Bees; Mason Bees) Worldwide
Anthophoridae (Cuckoo Bees; Mining Bees; etc.)
   Anthophora spp. Most in Europe and Asia
   Hemesia spp. Important in USA
Xylocopidae (Carpenter Bees)
   Allodape spp. Small; nest in plant stems in tropics
   Ceratina spp. Small; nest in stem pith; Africa, Europe, etc.
   Xylocopa spp. Large; tropical, timber-tunnellers
Apidae (Social Bees)
Bombinae (Bumble Bees)
   Bombus spp. Holarctic
Meliponinae (Mosquito Bees; Stingless Bees)
   Melipona spp. South America; largish in size
   Trigona spp. Africa, Asia; tiny
Apinae (Honey Bees; Social Bees)
   Apis mellifera (Honey Bee) Worldwide
   Apis cerana (Eastern Honey Bee) India, South East Asia
   Apis dorsata (Giant/Rock Honey Bee) India; wild
   Apis florea (Little Honey Bee) India

(b) Diptera
Members of most families are recorded as anthophilous, but the more important groups include:

Midges	Pollinate cocoa in South America
Syrphidae	(Hover Flies)
Muscidae	(House Flies, etc.)
Calliphoridae	(Blow Flies; Bluebottles, etc.)

(c) Lepidoptera
Butterflies are diurnal; most moths nocturnal; nearly 100 families known and most adults are anthophilous and involved in plant pollination

(d) Coleoptera
Some 95 families and more than 330 000 species are named. Many adults anthophilous – usually eating pollen – and some are regular pollinators of certain plants

Table 2.2 *(cont.)*

(e) Thysanoptera
Thripidae (Thrips)
  Thrips spp.
  Frankliniella spp. (100+ spp.) (Flower Thrips)

(f) Other orders
As mentioned in Proctor and Yeo (1973) some members of most other orders are involved with pollination of plants, but none is especially important

tensive new monocultures have usually en-countered pollination problems.

It is now known that probably all synthetic insecticides are harmful to bees, although some 20 years ago it was thought that some were harmless and safe to use when bees were present. The most toxic chemicals are quickly fatal, but the least toxic still up-set insect physiology and behaviour, even though they are not lethal. Nowadays all new synthetic pesticides are routinely tested against Honey Bees. The worst damage levels to Honey Bee populations nationally and internationally were in the period 1950–1970 during the widespread use of the new organochlorine compounds. However, the drastic declines in pollination levels were not appreciated for some time; in Japan by 1980 some 25% of labour time in fruit orchards was being spent on hand-pollination of the flowers. Many fruit and forage legume crops showed poor yields which indicated low levels of pollination. It is now thought that in most parts of the world many crops are being under-pollinated, and yields are lower than expected. The lack of pollinating insects can be remedied to some extent by the increase in the numbers of Honey Bees, but the solitary wild bees are less readily manipulated and much of their natural habitat has been totally destroyed. However, in recent years there has been great effort made in the field of assisted pollination. Some of the first experiments were made with red clover grown for seed; in the USA it was shown that maximizing pollination could produce a ten-fold seed yield increase, and in New Zealand red clover seed crops failed totally until 'long-tongued' Bumble Bees (*Bombus* spp.) were imported and established locally as 'wild' colonies. In 1980 in Japan small colonies of *Osmia cornifrons* (Japanese Horn-faced Bee) were installed in apple orchards in artificial nest sites con-structed from a small bundle of hollow reeds or plastic tubes. The results were spectacular and the experiment a great success, and now this species has been imported into the USA for orchard fruit pollination. In the USA the solitary (but social in small colonies) *Nomia melandes* (Alkali Bee), a native species, is successfully managed in some regions for pollination purposes. Also in the USA the imported *Megachile rotundata* (European Al-falfa Leaf-cutter) is managed for alfalfa pollination. A fairly recent spectacular crop pollination success was achieved by CAB International (1982). Oil palm (native to West Africa) was some time ago introduced into South East Asia as a major plantation crop, but in Malaysia and Papua New Guinea natural pollination was inadequate and the annual bill for hand-pollination amounted to US$11 million. In Peninsular Malaya pollina-tion was better (although not really adequate) and the main pollinators were thrips which were absent from Borneo. In West Africa natural pollination was always good, but the method of pollination was unknown. In 1978, the Commonwealth Institute of Biological Control sent entomologists to West Africa, and they discovered that out of a complex of pollinators (most were beetles) there were several species of *Elaeidobius* (Col., Curculionidae). *E. kamerunicus* was selected for importation into Peninsular Malaya. The species is now well established and a consistent 20% increase in yield combined with the lack of need for hand-pollination results in an estimated annual saving of US$115 million. The latest development in assisted pollination in the UK has been the rearing of Bumble Bees for pollination use in greenhouses. It is now possible to buy a nest of several species of *Bombus* to be kept in a greenhouse. Tomatoes are partly self-pollinating and partly entomophilous, but vibration of the plant is needed to facilitate pollen dislodgement onto the stigmas. Bumble Bee visiting will therefore aid self-pollination as well as effecting cross-pollination.

In the UK the British Bee Keepers Association started an informal pollination service, under which any farmer or grower needing pollination assistance for their crops could be helped. Under the scheme the required number of beehives can be delivered to the orchard or field and will remain there for one to two weeks while pollination is completed. Fruit orchards and rape seed crops most often need assistance, and one hive per hectare and two per hectare, respectively, are usually sufficient. With rape the flowering period is early, and cool weather will inhibit Honey Bee foraging, but the honey is good. Usually the bees do not increase the yield appreciably but they cause the crop to mature more uniformly, and often the seed quality is improved.

In the tropics bees are kept mainly for honey, but it is now apparent that some tropical crops are often under-pollinated, and several international conferences on pollination in the tropics have been held. The International Bee Research Association (IBRA), which started in the UK primarily to help apiarists in their quest for honey, is now devoting much of its attention to the pollination of crops, both in temperate and tropical regions, and is a major source of bibliographical information. IBRA is the proposed world centre for the advancement of apiculture in developing countries. One of its major projects is the compilation of a pollination directory for the crops grown in developing countries (Crane and Walker, 1983).

2.2 APICULTURE

Apiculture refers to the rearing of Honey Bees (*Apis mellifera*), originally for honey, wax and other products, but now ever-increasingly for assisted pollination of crops. As already mentioned, the main source of information in the UK is IBRA which was established in 1949. IBRA now has a completely worldwide remit, publishing journals, books, pamphlets and bibliographies, etc. and convening international conferences on both apiculture and pollination. It was formerly housed at Gerrards Cross, Buckinghamshire, but is now located in Bristol. However, every country in the world has within its Ministry or Department of Agriculture a section concerned with apiculture.

The genus *Apis* has four basic species:

- *Apis mellifera* (Honey Bee) – wild in Eastern Europe, Asia Minor and Africa.
- *A. cerana* (Eastern Honey Bee) – native to India and South East Asia; some rearing in India and China. *A. c. indica* (Indian Honey Bee) is reared in India.

Fig. 2.1 Honey Bee workers (*Apis mellifera*), right-hand bee with filled pollen baskets; south China.

- *A. dorsata* (Giant/Rock Honey Bee) – makes large open combs in the Himalayas; harvested but not reared.
- *A. florea* (Little Honey Bee) – possibly two subspecies; native to India and South East Asia.

Apis occurs as a series of geographical subspecies (sometimes called races) within which a number of races or strains are found locally, each with important differences in their biology. The different forms show basic differences in foraging activity, range of plants visited, docility and aggressiveness, honey yield, etc., and most apiarists have particular preferences. The African Honey Bee, *Apis mellifera scutellata*, is native to East Africa, although several other 'races' are found in Africa. In Uganda and Kenya this bee is responsible for many cases of unprovoked attacks on people (and sometimes livestock): people have been stung to death. It is a superior forager and gives high yields of honey, but is aggressive to other bees as well as occasionally to humans. In the 1950s a few African queens were taken to South America to interbreed; inevitably some of the offspring escaped from 'captivity' and this aggressive race is now replacing local Honey Bees – they started in Brazil and have now spread widely, recently entering the southern states of the USA. The alarm caused by the entry of African bees into the USA has resulted in a couple of novels and at least one film ('Killer Bees'). However, this race is essentially one used to warm conditions and so is not expected to spread far into the USA.

Apiculture consists of the careful management of bee colonies – the insects are not domesticated or tamed in any way, but clearly most commercial races have been selected for docility as well as foraging efficiency. The nest consists of a series of wax combs containing rows of hexagonal cells. The breeding combs (brood combs) typically contain eggs, larvae and pupae at different stages of development, and the other combs are primarily for the storage of honey (a thickened mixture of plant nectars). The bees collect nectar and pollen (and water) as their basic foodstuffs and any excess to their daily requirement is stored in the cells for later use. The colony is long-lived (i.e. permanent) and so in temperate regions the bees collect nectar (and pollen) throughout the spring and summer and store the excess in combs for consumption during the winter when the colony is dormant. In primitive apiculture the practice is to remove part of the honey store, but to leave sufficient for the colony nucleus to survive. Modern practice in temperate regions involves the removal of most (but not all) of the stored honey from the storage combs in late summer and then to provide the surviving bees with sugar solution at regular intervals from late summer and throughout the winter as their main food.

A colony is based on a single, large, fertile female bee termed the 'queen', who is responsible for all egg production: all she does is to lay eggs, at the height of the season some 2000 per day. The worker bees are sterile females with undeveloped gonads; in the event of the death of the queen, some workers can become reproductive and lay eggs. Within the colony there is a division of labour based upon the age of the individuals, with the youngest workers having nest duties and the older ones foraging for nectar, pollen and water. Foraging is very strenuous and at the height of the season (early summer) the workers only live for a few weeks. At this time the bee colony contains up to 50 000 workers. In the UK the overwintering colony is usually 10–20 000 strong. Early in the summer, males (drones) are produced from unfertilized eggs, and young queens are also reared. Queens are females produced from fertilized eggs and which were fed as larvae on a special diet (royal jelly). When a young queen is pupating, it is usual for the old queen to be driven from the hive by the

Fig. 2.2 Traditional early European beehive made of woven straw; Friskney, UK (T. Dunn).

workers. She leaves surrounded by about half the complement of workers and some drones to seek a new nest site and to start another colony. Good bee management involves the apiarist being alert to this development: he can then capture the emerging swarm and entice them to inhabit a new hive.

In the tropics there is no winter period of inactivity but there may be a dry season with a shortage of flowers and water, and generally the bee colony economy is based on the construction of combs and their provisioning for population build-up and swarm development. The original colony usually fragments into several swarms and the old nest may be deserted. In the tropics there is therefore a high yield of wax and a lower yield of honey, with most of the colony's energy going into swarm production. In the tropics most bee colonies swarm every 12–13 weeks on average.

In Europe, in olden times, the first recorded bee hives were dome-shaped, about half a metre tall and made of woven straw (Fig. 2.2). In Africa traditional beehives are either a hollow log or a cylinder of bark, held horizontally on two forked branches, often with a rain shelter over the top or tucked under the eaves of a native hut (Fig. 2.3). There has however been a tendency to change to more efficient modern beehives, sectional structures which originated in the UK in about 1850. In the UK there is now a National Beehive (BS no. 1300; 1960) (Fig. 2.4), but many apiarists favour somewhat modified versions. Basically, the hive is composed of a series of square sections placed one on top of the other. As the colony grows and requires more comb space, additional sections are added, so what started with a single section in the spring can end up with four or five comb sections. The top of the hive is known as the 'crown board' and is covered with a weatherproof metal sheet. The base has a landing platform (Fig. 2.5), resting on the stand, with a narrow entrance so that bees have access but predators such as field

Fig. 2.3 Traditional African beehive – a cylinder of woven grass on two forked tree branches; modern hive in foreground; Alemaya, Ethiopia.

Fig. 2.4 National Beehive (with modified base); Friskney, UK.

Fig. 2.5 Beehive landing platform with foraging workers; south China.

mice are kept out. The first or lowest chamber of the hive is the brood chamber, and above this are placed the more shallow honey storage combs and the crown board.

Traditionally, the main end product of apiculture has been honey, but in recent years the other nest products have become increasingly important. Commercially the use of bees for pollination has become big business. 'Bee-farmers' refers to apiarists with 100 or more hives, and hive transportation has become a regular practice. A major problem with bee-keeping is to ensure a continuous supply of flower nectar and pollen within foraging distance for the entire bee season. In the past most keepers had fewer hives, and they were stationary. Now the commercial apiarists specialize in taking their hives to the sources of nectar. In the UK there is a succession of entomophilous flowers, both wild and cultivated, that are available as sources of nectar and pollen.

Most crop pollination is done early in the season: rape usually in April, apple and other fruits in May and early June. In the summer the larger-scale apiarists have to find other sources of nectar for their bees, and many transport their hives to moorland areas for the heather, a well-known late flowerer. Recent surveys show that there are about 33 000 small-time bee-keepers, averaging four hives each, and 900 bee-farmers with an average of about 100 hives each. Major publications on this topic include: Free (1970), Free and Williams (1977), Martin and McGregor (1973), Crane and Walker (1983) and Kevan and Baker (1983).

Good bee management aims at a steady sustainable yield of honey and other products, while keeping the bee colony healthy to suppress the swarming inclination of the colony.

Apiculture has several basic products: honey, wax, propolis and 'royal jelly', and swarms are sold to establish new colonies. Originally, honey was produced for local consumption but now several countries have areas of high production and are regular honey exporters. It is difficult to estimate the total value of honey production in the world, but it was calculated that in 1988 the value of honey and wax produced in the USA was about US $100 million. The wax forms the comb in which the honey is stored, and for every 6–7 kg of honey produced there is about 1 kg of wax. The annual yield of honey from a beehive (not including that left as winter food) varies tremendously: in Africa it may be only 6–8 kg per hive; in the UK and southern USA, 10–20 kg; and in Australia from *Eucalyptus* trees it may be up to 100 kg per hive.

Honey is a sweet viscous liquid made from plant nectar by the removal of excess water and the inversion of most of the nectar sucrose into fructose and glucose. Its precise constitution varies according to the plants from which the nectar was collected and the prevailing environmental conditions (humidity, etc.), but includes small amounts of pollen, some minerals, vitamins, amino acids, plant oils and many other substances in tiny amounts. Honey's main commercial use is for sweetening a wide variety of foods. It has slight antiseptic properties and has a long medical history of being used for dressing wounds. Honey sold in many African and Asian countries traditionally comes with bits of broken comb and wax fragments and propolis, and sometimes even a few bee larvae. But in the Western world most commercial honeys have been carefully refined by sieving and cleaning. Sad to say there were times in the UK when some of the 'pure' honeys on sale were totally artificial in being just a thick sugar solution. However, in recent years, both in Europe and North America, there has been a public reversion towards so-called 'natural' foodstuffs, and as part of this trend there is interest in unrefined honey. Some honey is now sold as raw honey with bits of wax and propolis included. It is also possible to buy pieces of honeycomb with the honey still sealed inside the individual wax cells. These 'contaminants' are now regarded as being highly beneficial to the human diet, and there is considerable accumulated evidence to this effect (see, for example, Crane and Walker (1983) and Kevan and Baker (1983)).

Fermented honey makes a delicious pale yellow wine, which can be very strong; this is called mead in Europe and the earliest records of its production go back to various monasteries in about AD 500. In Ethiopia, honey wine is called 'tej'. This has a much older history, and has played an important role in local customs for centuries.

Wax (or 'beeswax') is secreted as tiny flakes from four pairs of abdominal glands on the ventral body surface of the adult bee. The flakes are transferred by the legs to the mouth, where they are chewed by the mandibles and mixed with saliva. The chewed wax is then added to the comb. Wax is a yellowish solid, insoluble in water but soluble in the usual organic solvents. Worker bees eat some 7–15 kg of honey to produce 1 kg of wax. Beeswax has a number of traditional uses, for some of which it has been replaced by modern synthetic waxes, although in some instances it is still superior. Its main uses are for making religious candles, in furniture waxes and polishes, and in some cosmetics, ointments, wax paper and lithographic inks. Because of its antiseptic and inert nature, it is also used in some medical (orthopaedic) operations for filling cavities in bones and joints. Most commercial beeswax in Europe, Africa and North America comes from different subspecies of *A. mellifera*, but a lot of wax exported from India comes from the large nest combs of *A. dorsata*.

It is now customary to fit wax comb bases into the wooden frames in the storage comb sections of hives so as to save the bees time in constructing new combs.

Propolis is a dark resinous material which foraging bees collect from buds or the bark of trees, especially poplars. It is carried back to the nest in the pollen baskets on the hind legs of the workers and chewed to make a type of cement to seal holes in the nest cover of the hive and to join the combs to the roof. It is important in temperate regions that the hive or nest cover is draught-proof if the colony is to survive the winter. Propolis is also used to embalm the bodies of dead predators in the hive that are too large to be evicted, such as mice. Formerly, modern apiarists regarded propolis as a nuisance, but in ancient times it was used as a medicine and now, with a revival of interest in its medicinal properties, it has become a very valuable byproduct of apiculture. Propolis has definite antibiotic properties and there are many claims that it has been used successfully to treat a wide range of human ailments. The earliest recorded writings about propolis and its medicinal properties were by Pliny and Aristotle in about AD 70.

Pollen is stored in certain cells on the brood combs and is used as the source of protein for older worker (and drone) larvae. It is basically proteinaceous, comprising a mixture of many different (up to 20) amino acids and vitamins, and many different enzymes. The total list of chemicals analysed is very extensive. Pollen can be collected from the storage cells, or else from a 'pollen trap' which can be fitted to the hive entrance and which causes the workers to drop most of the pollen they are carrying. In recent studies it has been claimed that pollen has therapeutic properties and that it can alleviate symptoms of a wide range of disorders including colds, influenza, asthma, hay fever, arthritis, rheumatism, prostate disorders, menstrual and menopause problems, and virility problems. However, it should be stressed that most treatments using pollens have actually involved the eating of unrefined honey or else a pollen/propolis mixture.

Royal jelly is secreted by the mandibular glands of young worker bees. It is fed to all young larvae for their first three days and to queen larvae for their entire five-and-a-half days. The jelly is highly proteinaceous (40–43% protein) and contains many vitamins, particularly those of the B-complex. That fed to young queens contains about 35% hexose sugars, but that fed to worker larvae contains only 10% sugar: it appears that the jelly has to be supplemented with sugar for the development of larvae into queens. As the worker bee ages the mandibular glands atrophy and the wax glands start to function.

Recent studies on human longevity have shown that in parts of Eastern Europe and Western Asia (around the Balkan states and at the edge of the Himalayas) people often live to a great age (up to 120 years). These are areas of ancient and intensive apiculture, and unrefined honey has long been a regular part of the local diet. The regional strain of Honey Bee (the Caucasian 'race') is a good nectar collector, and local apiarists have long been aware of the beneficial properties of propolis and unrefined honey. The precise value of the separate components in human diet is not clear, but there is an accumulation of evidence that there are definite beneficial effects in having raw, unrefined honey (including bits of wax, pollen and propolis, etc.) as a regular part of the diet. There is great interest in honey, pollen and propolis in many parts of the world, and as demand increases many proprietary formulations containing propolis and pollen, as well as royal jelly, are being sold in most countries.

A parasitic mite called *Varroa jacobsoni* (Acarina : Varroidae) has long been known to attack bee colonies, but has usually been regarded as an insignificant pest. However, in the last couple of years, apiarists in the UK and Western Europe have reported an ever-increasing number of hive infestations (though they can be controlled by insecticide use). But a recent Reuters report (*Sarawak Tribune*, 4 May 1996) describes a serious outbreak of *Varroa* in the USA, and both wild honey bee colonies and domesticated ones are being destroyed. Estimates of damage are of 50% bee losses, or even more, with drastic effects on honey production and also on crop plant pollination.

2.3 SERICULTURE

Silk is secreted by special glands possessed by the larvae of Lepidoptera, and the adults and young of spiders and some mites. However, the commercial production of silk, (i.e. sericulture) is based mostly on the collection of the silk used to form the protective pupal cocoon of one species of Bombycidae (Lepidoptera), *Bombyx mori. B. mori* is known as the 'Oriental Silkworm', the 'Mulberry Silkworm' or just as the 'Silkworm'.

Many caterpillars (the larvae of Lepidoptera) produce silk. The tiny first instar larvae of some Geometridae are dispersed aerially using long strands of silk; tortricid larvae use silk to hold two adjacent leaves of leaf surfaces together; and some Pyralidae roll leaves longitudinally and fasten the roll with strands of silk. In fact, some members of most families in the order Lepidoptera have larvae that produce silk for some purpose. The silk is produced from modified salivary glands in the larval Lepidoptera. However, only a few species produce silk with the properties that make it suitable for commercial application. That is a pupal cocoon made of a single long strand of silk, which can be unwound from the cocoon and rewound on to a spool prior to the weaving of fabric.

Spider silk is produced from posteriorly positioned spinarettes, and is generally too sticky and thin for any possible use in fabric. It has, however, been used for a few obscure purposes industrially. Ecologically, spider silk is important as a means of capturing insect prey, and both spiders and mites (Tetranychidae) use silken parachutes for aerial dispersal.

Many different Lepidoptera have been tried as sources of silk, including the notorious pest the Gypsy Moth (*Lymantria dispar*) which was introduced into North America to be tested as a source. However, only a few have been successful. Widespread commercial silk production has centred around just one species, *B. mori*, which is native to China, but in parts of China, India and Africa local species occur that are of some value (see Table 2.3). These wild species can be used locally; the pupal cocoons can be collected from the wild and the silk unravelled. In these cases no rearing is involved, although suitable larval food trees may be planted so as to encourage the build-up of the moth population. Such local industries can provide a very useful extra source of income for peasant communities. Table 2.3 lists the moth species that produce larval silk of commercial quality – *B. mori* is entirely cultivated, but the other species may either be reared or may be sources of 'wild' silk. In India about a dozen species of Lepidoptera are used for silk production.

The origin of sericulture is obscure, but the first records come from China and date from about 2600 BC. Raw silk was one of the first trade commodities between China and Europe, and was of tremendous value. Historically the greatest sales of silk were from China to Japan where it was used to make silk clothing for the hot, humid summer weather. Today Japan is still the largest silk consumer (for the manufacture of kimonos) but now has its own silk industry with a surplus to export. For centuries the Chinese managed to safeguard their silk industry (by use of the death penalty), but in about 150 BC eggs were smuggled to India, and this started the Indian sericulture industry. Sericulture is now practised worldwide, for mulberry trees are easy to grow: the main production areas are still China and Japan, but these have been joined by Korea, the USSR, India and Thailand. There are even five or six silk farms in the UK. Nowadays in Japan, most caterpillar-rearing is done using a synthetic diet, and

Table 2.3 Moth species used for silk production

S/F Bombycoidea		
Lasiocampidae (Tent Caterpillars; Lappets, etc.)		
Gonometra postica	Africa	*Acacia* spp.
Trabala vishnou (Oriental Lappet Moth)	South East Asia	Polyphagous
Bombycidae (Silkworm Moths: Silkworms)		
Bombyx mori (Silkworm; Oriental Silkmoth)	South East Asia, etc.	Mulberry
Bombyx spp. (6)	India	Mulberry
Saturnidae (Emperor Moths; Giant Silkmoths)		
Actias selene (Moon Moth)	India to China	Many hosts
Antheraea assama (Muga Silkworm)	India	Polyphagous
Antheraea mylitta	India	*Syzygium* spp.
Antheraea paphia (Tussor Silkworm)	South East Asia	Polyphagous
Antheraea pernyi (Chinese Oak Silkworm)	China	Oaks
Antheraea yamamai (Japanese Oak Silkworm)	Japan	*Quercus* spp.
Attacus atlas (Atlas Moth)	Pantropical	Polyphagous
Eriogyna pyretorum (Giant Silkworm Moth)	South East Asia	Camphor
Naudaurelia spp. (Silkworms)	Africa	Several hosts
Samia cynthia (Lesser Atlas Moth; Wild Silkmoth)	India to China	*Lantana*, etc.
Samia ricini (Evi Silkworm)	India, South East Asia	Castor

S/F Noctuoidea		
Notodontidae (Processionary Caterpillars, etc.)		
Anaphe infracta (African Wild Silkworm)	Uganda	*Bridelia*

requires no mulberry leaves. In 1978, world production of raw silk amounted to 408 000 tons of cocoons yielding 819 000 bales of raw silk. The present annual value of world silk production is estimated at being more than US $1000 million.

B. mori (Fig. 2.6) occurs as several distinct races, which is not surprising considering the age of the art of sericulture and the hundreds or thousands of years that some moth populations have been isolated from each other. The four major races are known as the Chinese (both univoltine and bivoltine forms), the Japanese (also uni- and bivoltine forms), the European (univoltine) and the Tropical race (multivoltine).

Each female moth lays about 500 eggs. Hibernation (diapause) takes place in the egg, which is convenient for commercial production. At 25°C larval development takes 20–25 days, and pupation a further ten days.

The caterpillars are usually whitish in colour, and bear a small posterior 'horn' characteristic of the family Bombycidae. The larvae of

Fig. 2.6 Male Silkmoth (*Bombyx mori* – Lep.; Bombycidae) from China; wingspan 25 mm.

Fig. 2.7 Silkworm (*Bombyx mori* – Lep.; Bombycidae) on mulberry leaves; from China; length 30 mm.

the different races show various differences both in body colour and aspects of physiology such as growth rate and disease susceptibility. Their diet has traditionally been fresh leaves of mulberry (usually *Morus alba*), but ever-increasingly a synthetic food material is being used, and in a recent survey it was shown that more than half the silkworms reared in Japan have been fed a synthetic diet. A caterpillar with normal coloration is shown in Fig. 2.7. The cocoon (Fig. 2.8) is white, pink or yellow in colour and is made up of a single silk thread some 500–1200 m in length. The thickness and length of the thread varies with the race of the insect; generally shorter fibres are thicker in the middle region. The polyvoltine races of Thailand produce an uneven fibre with a brilliant lustre. This has to be hand-woven and produces the slightly uneven but very attractive 'Thai silk'. The uni- and bivoltine races of China and Japan produce a large cocoon with a longer and thinner thread. This more uniform thread is smoother and can be mechanically unwound. Each thread of commercial silk is formed of two fibres as two cocoons are reeled simultaneously.

Other species of Bombycoidea produce silk differing in colour, texture and quality, which usually has to be hand-reeled, but the fabric produced can be attractive and its production is a valuable local cottage industry in some parts of Africa and Asia. China, Japan and Thailand have well-established sericulture research institutes. Several silk farms have now been established in England with others scattered throughout Western Europe and the USA.

2.4 INSECT FARMING

For many years there has been widespread acquisition of insects, plants and animals (including birds and their eggs) and trophy hunting of game animals, both for museums and for personal collections. However, with the ever-increasing human population combined with destruction of natural habitats there is now greater pressure to restrict the collection of natural history specimens – in some cases to prohibit the practice completely as far as some (endangered) species are concerned, and in particular for those areas that have been designated as nature reserves. There does, however, remain a genuine commercial demand for a range of species. Some can be supplied by careful, sustainable habitat management, and others are being provided by newly established farming and breeding programmes. More and more animal and plant species are being successfully reared in captivity.

Fig. 2.8 Cocoons of Silkworm (*Bombyx mori* – Lep.; Bombycidae) on mulberry leaf; from China.

The earliest forms of insect farming are clearly apiculture and sericulture, which have already been described. For most of the uses listed below, the original demand was met by field collection, and then typically by field collection of eggs or larvae for rearing in captivity. Today most species are reared successfully and continuously in captivity. In most cases, natural food sources had previously to be collected from the wild, but again recent research and development has produced synthetic foodstuffs which can be commercially manufactured as the diet for many species.

2.4.1 Insect products

Two ancient practices are the production of shellac and cochineal. Shellac is a sticky, brown resinous substance produced from the thick scale of the Lac Insect (Coccoidea; Lacciferidae; *Laccifer lacca*) in India. The scales

Fig. 2.9 A selection of tropical butterflies in a case: six (nos 2, 4, 5, 6, 7 and 9, left to right) are Birdwings from South East Asia (Lep.; Papilionidae).

encrust young twigs of certain trees (*Acacia*, soapberry and *Ficus religiosa*) in India and South East Asia, usually forming a communal scale crust. Twice a year the twigs are collected and the scale mass is scraped off. The body of the female insect yields a bright red dye (lac dye), which is extracted with hot water. The resin (known as 'seed-lac') is melted and filtered, and cools to the flaky texture of commercial shellac. It is used mainly as a high quality wood polish (French polish) and in the manufacture of insulators, buttons, sealing wax and hairsprays among others. Annual export production in India is some 23 million kg, sold mainly to Europe and the USA.

Cochineal is a red dye produced from the dried and powdered bodies of another scale insect – the Cactus Mealybug (Coccoidea; Dactylopiidae; *Dactylopius coccus*), also known as the Cochineal Insect. A native of Mexico and Peru, this insect feeds on Prickly Pear Cactus (*Opuntia* spp.). The insects are swept off the cacti and collected, killed by either immersion in hot water or roasting in an oven, and the application of various extraction techniques then yields a dye of any shade of red to orange, used for both dyeing of cloth and in foodstuff coloration (Cloudsley-Thompson, 1986). In 1860 the annual export of cochineal from Mexico to France was worth 12 million francs. Today the Canary Islands is an area of commercial production, as is Honduras. The commercial uses of cochineal have largely been replaced by the synthetic aniline dyes, but in recent years there has been a resurgence of interest in it as a natural product without the side effects associated with the cheaper artificial dyes. It is now being used again in foodstuff coloration, as well as in cosmetics and homoeopathic medicine.

2.4.2 Butterfly farming

In England, the company Worldwide Butterflies has for a long time offered species of

Lepidoptera for sale, both as pinned dead specimens and live as eggs, larvae or pupae for rearing. It now offers many other insect species – stick insects (Phasmidae) are particularly popular with children as pets. These species can all be reared at home using locally available food plants. The beauty of flying butterflies and the success of captive rearing has led to the establishment of butterfly houses in zoos and leisure centres, and many butterfly farms in many parts of the UK.

In Britain on several important nature reserves there is captive breeding of some butterfly species to augment local populations which have declined considerably in recent years. This population supplementation appears to be working successfully in a number of locations for several species. It is very likely that such captive breeding programmes will become more widespread.

In South East Asia the Birdwing butterflies (Papilionidae; *Ornithoptera*, etc.) have long been sought after by collectors for their spectacular size and beauty (Fig. 2.9). But over-collection has led to some species becoming rare, and some 15 years ago the government of Papua New Guinea declared them a protected species, making their capture and/or export illegal. However, a few years later, enterprising locals started a new venture as a cottage industry in the captive rearing of several species of *Ornithoptera* and others. Thus the reared specimens can be legally exported without endangering the local wild populations.

2.4.3 Fly maggots

According to a recent survey, freshwater coarse fishing is probably the most popular hobby in the UK. The most commonly used bait is Blow Fly maggots (Diptera; Calliphoridae). The flies are bred and maggots reared in dozens of small premises, and a few larger commercial establishments, throughout the British Isles. The final instar (third) larvae are used as fish bait; some are coloured red or yellow using vegetable dyes (Fig. 2.10).

Some vegetable farmers who wish to pollinate their own varieties of *Brassica* crops to produce seed, often buy Blow Fly puparia from fish bait dealers. The emerging adult flies will pollinate the crops very effectively, although the plants do of course have to be enclosed inside fly-proof netting.

Open-flowered plants can be successfully pollinated by Blow Flies, so the flies are sometimes used on certain flowers grown under glass for seed production.

2.4.4 Insects as animal food

The majority of domesticated livestock are herbivorous and relatively easy to feed, but the few carnivorous species can present problems. Many fish, amphibians and reptiles reared in captivity need insects and other live invertebrates in their diet, as do some Insectivora (Mammalia) and some birds. Some are fed entirely on insects and others only partially. In some countries where labour is cheap the insects may be field collected, but the ever-increasing trend is towards captive rearing on either a small scale or as a large commercial enterprise.

Several species of locusts (especially the Desert Locust) and grasshoppers (Acrididae) can easily be bred in captivity so long as provision is made for oviposition (into sand). These are used as food for reptiles, larger amphibians and some Insectivora, some small primates and larger captive birds (such as the Chinese magpie and egrets). Crickets can also be reared though they are more omnivorous and thus more difficult to feed. In parts of Europe, the House Cricket is well established in many large refuse dumps and will breed throughout the cold winter in, for example, Germany, because the heat of fermentation keeps the refuse warm. Some local populations of Hobby (*Falco subbuteo*) rely almost entirely on these refuse crickets as their food supply. Insects for cage bird food are a

Fig. 2.10 Fly maggots sold as fishing bait; the ones which appear darker have been stained red (Diptera; Calliphoridae); Skegness, UK.

considerable commercial item in countries such as China where bird-keeping is extensively practised and many insectivorous birds are kept.

Mealworms are the larvae of *Tenebrio molitor* (Col., Tenebrionidae) (Fig. 2.11), easily reared and used to feed many birds and other vertebrates in captivity. This is a temperate species found throughout Europe, Northern Asia and North America.

Bloodworms are the aquatic larvae of Lake Midges (*Chironomous* spp. – Diptera; Chironomidae), to be found in standing water (ponds, ditches, lakes) and also along the edges of slow-flowing rivers. Most species have red larvae, but some stream-dwellers are green. The larvae are found in dense aggregations on the substrate surface living in soft tubes, especially in shallow water with a high organic content (particularly that polluted with sewage or animal faeces). Their main use is as live food for tropical fish in aquaria (Fig. 2.12). There is a small demand for such food worldwide as fish-keeping is a popular hobby, but in China and parts of South East Asia the practice is more extensive and the demand for live fish food (bloodworms and *Tubifex* worms) is great. In China, the larvae are harvested from the shallow edges of fish ponds, cleaned of faecal material, and sold as live food for the fish kept in household aquaria. A recent development in south China has been to construct shallow concrete ponds of about 10 m^2 and to fill them with water and animal faecal sludge: the local Chironomids breed in large numbers in the pools, which are periodically sieved for the bloodworms. This represents a small but useful source of income for peasant farmers.

Fly maggots of the type used for fishing bait are also used as food for some small Vertebrata as well as some fish. One possible problem with Muscid larvae is that they can survive in the alimentary tract, so causing intestinal myiasis (the invasion of living tissues by larvae). There have been recent experiments in China to try using fly maggots as a source of protein for human diet.

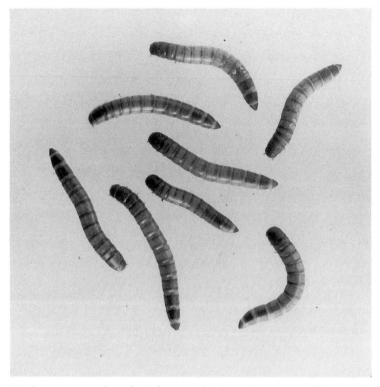

Fig. 2.11 Yellow Mealworms reared as food for cage birds and other small livestock (*Tenebrio molitor* – Col., Tenebrionidae); South China; body length 12 mm.

2.4.5 Insects for teaching and research

Because of the importance of the Insecta to human society, there is a need for different types of insect specimens to be used in teaching at all levels from secondary schools to MSc degree courses. Specimens for dissection (such as cockroaches and locusts) are required in large numbers each year, and are usually either bred locally or bought from a biological supplier. In the UK, USA and most European countries, there are professional biological suppliers which breed the insect species required for schools, universities and laboratories. Very recently the UK Central Science Laboratory of the Ministry of Agriculture, Fisheries and Food at Harpenden has advertised that it can supply cultures of 50 different species of insects.

Fruit Flies of the genus *Drosophila* have been used in genetic research for many decades, as they have relatively simple chromosomes with easily recognized anatomical features such as eye colour and wing length.

Agrochemical companies nowadays have to test their candidate pesticide chemicals against a wide range of insect pest species to assess their destructive effectiveness, as well as against Honey Bees, Ladybird Beetles and other beneficial species (both pollinators and natural enemies of pests). Most of the major agrochemical companies have their own laboratories and rearing insectaries so as to ensure a constant supply of test insects at the correct ages.

Certain insect species have particular qualities that make them useful for some types

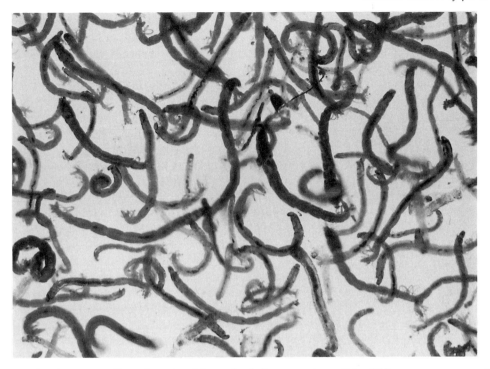

Fig. 2.12 Bloodworms collected and sold as food for aquarium fish (*Chironomous* spp. – Dipt.; Chironomidae); south China.

of research. Thus some mosquitoes are important in malaria research as they are major vectors. *Rhodnius* bugs (Reduviidae) were used for many years at Cambridge University (UK) to study different aspects of insect anatomy and physiology.

2.4.6 Biocontrol agents

A number of predacious and parasitic insect species are now reared commercially in large numbers for biological control programmes (section 2.5). Most are for use in commercial greenhouses and other forms of protected cultivation, but some species are used in open orchards and plantations, particularly in the tropical regions of the world. Most of these species are used to control insect pests of crops and cultivated plants, but some destroy medical and veterinary pests. Other,

phytophagous species are being reared for weed control.

2.5 NATURAL CONTROL OF PESTS

This is insect population reduction achieved by a complex of naturally occurring predators, parasites and pathogens. Pathogens cause disease in the insect population; predators kill and eat the insect pests; and parasites attack the insect pests at all stages of the life cycle. All the different species belong to the same ecological community and share the same habitat, having evolved together over thousands of years. Generally, natural enemies regulate the pest population and suppress or prevent violent population increases most of the time.

Agriculture is the artificial production of a few plant species over a wide area under uniform conditions. The crop presents an almost unlimited supply of food for phytophagous insects. Similarly, the rearing of livestock encourages the build-up of numbers of parasitic and blood-sucking insects. Many crop pests increase in numbers under these conditions, and may cause extensive damage to the crop (in some cases then moving over to other crop plants and causing further damage). At the same time their natural enemies also increase in number, but typically there is a time lag between the pest population build-up and that of the enemies. Thus with short-lived crop plants such as vegetables and some cereals, the natural enemies seldom have time to exert any control over the crop pests. Most pathogens and parasites are host- (prey-) specific and hence the population growth of these natural enemies will be completely dependent upon the pest population. However, most predators are polyphagous (i.e. non-specific in their choice of prey) and thus can have a more immediate population-controlling effect on an insect pest population. In the context of modern agriculture, it is important to encourage both pest-specific parasites and non-specific predators. Table 2.4 lists the more important families of insect predators and parasites.

In perennial orchards (apple, citrus, etc.) and on plantation crops (coffee, coconut, oil palm, cocoa, etc.) natural control is of tremendous importance. On these long-term crops, predacious and parasitic insects have time to build up their populations, and they usually reach levels at which they effectively suppress the pest population. The most common result is that insect pest populations and those of the natural predators and parasites co-exist at moderate levels at which plant damage is not too serious and can often be economically tolerated.

Natural control occurs virtually all the time in all pest populations, to different degrees, and as such it is often difficult to appreciate the extent of the pest population-controlling effect. Often we only see the pest population after natural reduction has taken place. Most insect pests display spectacular fecundity, and in agriculture the main reason why we are not swamped with insect pests is due to the high rate of natural mortality inflicted by the predators, parasites and pathogens. Agricultural research has shown that in many pest situations, the eggs of Lepidoptera (stem-borers, armyworms, leaf-eaters, etc.) regularly suffer predation levels of up to 95%; with most insect pest species, unprotected eggs typically show about 90% mortality (Davies, 1988). Young larval stages are also eaten by a wide range of predators. Size is usually a major critical factor in animal predation, and also, to a lesser extent, in parasitism. Basically, most predators eat prey smaller than themselves. On a field trip in south China, within about five minutes a party including the author saw a large Robber Fly (Asilidae) clutching a Damselfly (Odonata) and then a large Dragonfly eating a small Robber Fly – who ate whom was decided purely by relative size!

Normally, natural enemies will not destroy all members of an insect pest population because with a very small population the prey insects become difficult to find. Biologically, if a host-specific parasite or predator exterminates its host/prey insect then it will itself die of starvation. Thus in any natural (including agricultural) situation, the prey/pest population will fluctuate – following heavy predation, the population will be reduced to a low level, and then prey scarcity will in turn deplete the predator population. Occasionally a disease outbreak (epizootic) will sweep through an insect population and may kill up to 98% – although there will be a few natural survivors that are resistant to the pathogen. However such extreme population reductions are rare.

In addition to the effects of natural enemies, there are innate factors that regulate an insect population. At high population densities, there is a crowding effect, and this often

Table 2.4 Insect natural enemies of insects

A. Predacious insects and arachnids

Odonata   Prey on Diptera and many other insects
   (adults aerial; nymphs in freshwater – all polyphagous)

Dictyoptera – Mantidae   Orthoptera, Diptera and many other groups
   (adults and nymphs both in vegetation – polyphagous on all insect groups)

Hemiptera – Heteroptera

Reduviidae	}	Prey on caterpillars, other bugs and
Anthocoridae		most other insect groups – terrestrial
Pentatomidae		

Aquatic predacious bugs   –   Aquatic insects of all types

Neuroptera

Hemerobiidae	}	Both adults and nymphs eat all types of
Chrysopidae		small insects in foliage

Coleoptera

Carabidae	}	Both adults and larvae feed on many different insects
Staphylinidae		and snails; in litter and soil
Coccinellidae		
Histeridae	}	Some species predacious
Cleridae		
Meloidae	–	Larvae in soil eat grasshopper egg-pods

Members of some other families

Diptera

Cecidomyiidae (some)	}	Larvae only, especially on aphids
Syrphidae (some)		
Asilidae	–	Adults catch flying prey
Therevidae	–	Some
Conopidae etc.	–	Some

Hymenoptera (larvae in nests, most species social)

Vespidae	–	Adults predacious
Scoliidae	–	Larvae prey on beetle larvae in soil
Formicidae	–	Some species predacious

Acarina

Phytoseidae etc.	–	Eat insect eggs and other mites in foliage

Araneida

All families	–	Most use webs to trap flying insects; now known to be very important predators

B. Parasitic insects

Strepsiptera   –   Homoptera and some aculeate Hymenoptera

Diptera (only larval stages are parasitic)

Tachinidae	}	All parasitic on Lepidoptera, etc.
Phoridae		
Pipunculidae		
Bombylinidae		

Hymenoptera (only larval stages involved in parasitism)

Braconidae	}	Homoptera, Lepidoptera, Coleoptera and others
Ichneumonidae		
Chalcidoidea (20 families)	–	All groups of insects attacked, at all stages.

---

results in a reduction of fecundity. Fewer eggs are laid and there is more infertility. Larval development can be impaired and death often occurs at ecdysis, especially at metamorphosis. Adult longevity is often reduced. At low population densities the situation is reversed. More eggs are laid and there is less infertility; larval development is unimpaired and adult longevity is increased. Should the insect population reach a very low level, then typically it will die out as a result of sexual isolation, but this is a rare occurrence in natural situations because of immigration from adjacent populations.

Most insect pest species are biologically aggressive and physically robust, making them difficult to poison with insecticides. On the other hand, however, most parasites (and some predators) are more delicate and more sensitive to chemical poisons. There are now many well-documented cases where the careless use of chemical pesticides in orchards and plantations has resulted in the killing of the natural enemies while destroying few of the pests. Invariably the immediate result has been a pest population upsurge, often of alarming proportions. Classic cases include apple orchards in Europe and North America in the early 1970s: DDT was used to kill Codling Moth (which it usually did) but the predatory mites in the apple foliage were also killed, leading to an outbreak of Red Spider Mite. In South East Asia, oil palm plantations were defoliated by the combined effect of bagworms and slug caterpillars: sprays of organochlorine insecticides not only failed to kill the pest caterpillars but damage levels to the foliage usually increased (Conway, 1972b). The main control recommendation was to reduce the level of insecticide spraying and to allow the natural enemies to control the caterpillar population. If spraying was needed to curb a local population explosion, then only carefully selected chemicals were used, and even then with great care. The broad-spectrum, persistent organochlorine compounds (DDT, dieldrin,

etc.) that were so effective against so many different pests in the 1950s and 1960s are now known to have too many undesirable side effects for their use to be continued. In the fruit orchards of Europe and North America, various Coccoidea became serious pests when organochlorine insecticides killed off the chalcid and braconid parasites that usually kept them under control. The pest species mostly concerned was Mussel Scale, and in many apple and citrus orchards a scale or Red Spider Mite population resurgence could be guaranteed following a single foliar spray of DDT or dieldrin.

The delicate balance between the insect pests, their natural enemies and other insect species in the community is difficult to assess. It is clearly complex, and experimental investigation of the food web is seldom successful. The International Rice Research Institute (IRRI) in Manila has long had a reputation for ecological awareness, and was one of the first major research stations to recognize the importance of spiders in the predation of field crop pests. The IRRI integrated pest management (IPM) team looked at sprayed and unsprayed rice fields in the Philippines, and at the end of the 1992 dry season crop they recorded that the rice fields sprayed with insecticides had four times more pests than the unsprayed fields. The insecticides used on the rice crop did kill the insect pests, reducing the population to the required level (when applied correctly), but the poisons clearly affected the spiders and insect natural enemies more than the pests. Sprayed fields had a total of 56 million more pests per hectare than did the unsprayed (*IRRI Reporter*, 1994). However, in some respects this was a special case, for most of the insect pests concerned were of one species: the Brown Planthopper (*Nilaparvata lugens*). This insect embeds its eggs in the plant tissues and they are not killed by insecticide sprays. So although adults and nymphs will be killed by the insecticide, the hatching eggs will result in a population resurgence, and if the natural enemies (spiders, etc.) are reduced in number, the new Planthopper generation will be very damaging.

A number of plant weeds are eaten by phytophagous insects, but the levels of population control exerted by the insects are low. This is mainly because the insects and plants have evolved together over millennia, and they now co-exist with a certain measure of tolerance to each other. However, translocated weed species can sometimes be controlled in their new location by phytophagous insects introduced as part of a deliberate biological control programme.

It has taken many years for scientists and agriculturalists to appreciate fully the extent of the natural control that is exerted on many insect pest populations in most parts of the world. And nowadays, since the recently developed IPM approach to pest control there is a general concern to perpetuate existing levels of natural control and to enhance this control whenever possible. The importance of the natural control of insect pest populations in most situations in most parts of the world really cannot be over-emphasized.

2.6 BIOLOGICAL CONTROL OF PESTS

This refers to the deliberate introduction of predators, parasites or pathogens to kill the insect pests and reduce the size of their population. The pathogens include fungi, bacteria and viruses that cause diseases in the insect population. In some publications, the term 'biological control' is used very broadly and includes aspects normally referred to under other headings such as 'cultural control', 'use of resistant plant varieties', 'autocide', etc. However, the usual definition is that used above in the strict sense. The term is nowadays often used in the abbreviated form of 'biocontrol'.

In the use of this technique it is vitally important that the existing levels of natural control are not diminished in any way. The basic idea is to enhance and supplement the existing level of natural control.

In most insect pest/crop plant situations, the various relationships are of some antiquity, and very complex, and over the years such host/'parasite' relationships tend to evolve to a level of mutual toleration. The practice of extensive monoculture in modern agriculture does, however, upset this complex balance and encourages the build-up of large pest populations. However, the interrelationships are so complex and so well established that there is seldom room for another predator or parasite to become established. In a few cases it has been possible to augment a local natural predator or parasite population (using the same species), but the effect has usually been short-lived. Seldom is there a vacant niche in the local natural enemy spectrum, as most viable relationships have already developed during the course of evolution. However, the practice of agriculture depends heavily upon the cultivation of 'exotic' crop plants (i.e. plants from other countries) and many of these plants have either brought their own pests with them or the pests have joined them later. These exotic pests, which are transplanted into new environments, seldom have their own natural enemies in the new habitat, and they often produce pest populations. Such pests in their new locations are often vulnerable to control by natural enemies also introduced from their native country. Biocontrol projects are more likely to be successful if carried out in warm tropical regions where insect development continues all the time. In temperate regions with a long, cold winter, insect development is slower and interrupted by the winter, and biocontrol is less likely to work. Some tree-defoliating caterpillars in Canada (such as the Winter Moth) have however been controlled successfully by parasitic Tachinidae (Diptera) and Ichneumonidae (Hymenoptera).

Studies of population dynamics have shown that there is always a time lag between the development of an insect pest population and that of an associated predator or parasite. Thus successful biocontrol projects require time for their establishment. This means that in the field, annual crops would seldom benefit from attempts at biocontrol of their pests. Natural control by predators and parasites occurring in the area will be important in killing pests, but the pest population may well be damagingly large, and deliberate introduction of natural enemies will have little effect because by the time the parasite population is large enough to control the pests the crop will have been harvested. Perennial orchard and plantation crops are suitable for pest biocontrol because of their long-lasting nature: many fruit orchards can be productive for 20–50 years or more. In such situations the predator and parasite populations are maintained at a fairly high level by the abundance of food (pest insects), and the pest population usually does not get large enough to cause serious damage. So by the grower accepting a modest level of crop damage, a fruit orchard can often persist for years without recourse to insecticide spraying, as the natural enemies will kill off the majority of the insect and mite pests.

The advantages of biocontrol of insect pests are numerous:

1. A single introduction can result in long-term (even permanent) control of the pest; precise timing is not required. Financial returns can be enormous.
2. No poisons are used.
3. There is no development of resistance by the pest.
4. There is no destruction of natural enemies or pollinators.
5. There is no environmental contamination or pollution and no interference with natural food webs.

In 1960, Wilson pointed out that it was customary to assess biocontrol projects as either 'successful' or 'unsuccessful', and it is clear that such over-simplification is most undesirable. He suggested that if a biocontrol

project is completely successful, other forms of control can then be dispensed with. As with chemical control, it is usual to have several different levels of control using natural enemies, and on the basis of the work done in Australia, Wilson proposed five different categories:

1. Pests substantially reduced in status.
2. Pests reduced in status.
3. Pests of doubtfully diminished status.
4. Pests of unchanged status.
5. Cases when the introduced enemies failed to become established.

In some agricultural situations, complete control of the pest is the objective, but sometimes partial control by natural enemies is adequate (e.g. a 50% population reduction), especially if carried out as part of an integrated control project.

The original concept of 'integrated control' was conceived at the University of California when it became obvious that the practice of spraying insecticides was inimical to the newly developed system of introducing predators and parasites into fruit orchards. Thus integrated control was originally the careful application of pesticides so as to leave the predators and parasites unharmed.

Most crop/pest situations do not lend themselves to deliberate biocontrol projects, although natural control is always important. As already mentioned, most crops are indigenous to one particular area and have been widely translocated throughout the world to other regions with a suitable climate. Agricultural diversity is being advocated nowadays in virtually every country in the world. These exotic crops in their new locations are being attacked by a mixture of local insect pests, quick to seize upon a new food source, and some of their own native pests which either accompanied the crops or else spread to the new region soon afterwards. Often the exotic pest species flourish in the new location due mainly to the absence of the usual natural enemy complex. These are the pests against which a biocontrol project is likely to be successful, provided that the enemies are brought in from the native country. Local insect pests are termed 'autochthonous' species, and the introduced, exotic ones are known as 'allochthonous' pests.

Probably the earliest recorded practice of biocontrol in agriculture was in south China where natural predators (ants and wasps) were encouraged in citrus orchards in ancient times to prey on various bugs. In the 1950s and 1960s the Chinese expanded their activities in this field, and were probably the first to recognize the importance of spiders as predators of crop pests (see Hill, 1987, Table 2, page 108). This information has only recently been made available to the Western world. The earliest published accounts of biocontrol of crop pests are from California and date from the start of the century. Citrus and peach (and some other fruit and nut plants) had been taken from south China (where they are endemic) in order to establish new orchards in California, Florida and Hawaii (and also south Australia). Unfortunately, the planting material (cuttings, rootstocks, fruit, etc.) was contaminated with scale insects and Fruit Flies, some of which established themselves in the new locations and became serious and damaging to the fruit trees. In California the responsibility for agricultural pest control lay with the University of California Riverside campus, and with commendable foresight the Department of Biological Control was established. At the turn of the century, there were few effective insecticides and the staff of the University realized that natural enemies were the answer to the fruit orchard pests. Expeditions were made to south China to collect predators and parasites of the more important insect pests, and some were successfully established in the USA in the fruit orchards. After World War II, the then Commonwealth Agricultural Bureaux (now CAB International) established the Commonwealth Institute of Biological Control (now the International Institute for Biological Control) and a series of stations was scattered throughout the world, carrying out fundamental work on the biocontrol of insect pests by other insects, and later weeds. Today the importance of the controlling roles played by natural enemies of insect pests is so widely recognized that all major agricultural research stations and centres in most countries are devoting considerable efforts in this field, in both research and practice.

One result of all this effort is that there is now a very extensive literature on biological control. The major publications include the various *Reports* and *Technical Bulletins* published by the CIBC and books by Wilson (1960), Greathead (1971), Huffaker (1971), DeBach (1974) and Delucchi (1976), as well as sections in other more general texts such as that by Kilgore and Doutt (1967).

The cost of a biological control programme can be quite modest (or even cheap) but a large variation has been recorded: however, the better projects have usually been inexpensive and very cost-effective. Some examples are listed in Table 2.5. With many biocontrol projects the actual costs incurred can be accurately assessed, but the value of the damage reduction has to be estimated and this can sometimes be quite subjective. However, this in no way invalidates the procedure. DeBach (1964) estimated that the Department of Biological Control of the University of California, spent over the period 1923–1959 a total of US$3.6 million, which resulted in an annual saving in California of about $100 million. The 'Winter Moth' project in Canada was a very successful temperate example. This species was accidentally introduced into Nova Scotia in the 1930s and became noticed in 1949: as it spread and increased in number, the damage to oak and other deciduous hardwoods (it is polyphagous) soon became serious. Over the period 1955 to 1960, parasites from Europe were imported and successfully established and they brought the Winter Moth under control. The cost of the project was about

Table 2.5   Some successful biocontrol projects against insect pests

Pest	Crop	Predator/parasite	Country	Date	Annual value (US$)
Mythimna separata (Lep.; Noctuidae)	Maize	*Apanteles ruficrus* (Hym.; Braconidae)	New Zealand	1973/75	5m
Phthorimaea operculella (Lep.; Gelechiidae)	Potato	*Apanteles subandrinus* (Hym.; Braconidae)	Zambia	1968	70 000
Diatraea spp. (Lep.; Pyralidae)	Sugarcane	*Glyptomorpha deesae* (Hym.; Braconidae)	Barbados	1965	1m
Operophtera brumata (Lep.; Geometridae)	Oaks, etc.	*Agrypon flaveolatum* (Hym.; Ichneumonidae) *Cysenis albicans* (Dipt.; Tachinidae)	Canada	1955/60	2m
Chrysomphalus ficus (Hem; Diaspididae)	Citrus	*Aphytis holoxanthus* (Hym.; Aphelinidae)	Israel	1965/67	1m
Phenacoccus manihoti (Hem.; Diaspididae)	Cassava	*Epidinocarsis lopezi* (Hym.; Encyrtidae)	Africa	1987	–
Planococcus kenyae (Hem.; Pseudococcidae)	Coffee	*Anagyrus* spp. (Hym.; Encyrtidae)	Kenya	1935–49	10m (total)
Eriosoma lanigerum (Hem.; Pemphigidae)	Apple	*Aphelinus mali* (Hym.; Aphelinidae)	Kenya	1927/28	

US$160 000 and the annual saving to the hardwood industry was estimated at some $2 million. Another good example occurred in New Zealand between 1973 and 1975, when CIBC (Pakistan) established the parasite *Apanteles ruficrus* (Hym.; Braconidae) against the caterpillar of *Mythimna separata* on maize and a range of vegetable crops. The cost of supplying the parasites from Pakistan was US$20 000, and the estimated annual savings about $2 million; the control is still successful. The cases illustrated are regarded as particularly profitable, with very low costs in relation to estimated savings, but even the less successful projects are still quite inexpensive and avoid the problems of pesticide use. With the development of the IPM approach, there is

often scope for the use of limited biocontrol which aims at a modest level of pest population reduction with a concomitant reduced level of pesticide use.

The availability of biocontrol agents was initially pioneered by the CIBC stations and various international research centres, but in recent years many commercial establishments have been set up for bulk production. In 1984, there were relatively few commercial suppliers and so some eight names and addresses were included in Hill (1987), but now the number is so large that it would not be feasible to name specific companies. Some of the biological control agents commercially available in both Europe and North America are listed in Table 2.6.

Table 2.6 Some of the biological control agents commercially available in Europe and North America

Organism	Description	For use against	Supplied as
(a) Predators			
Chrysopa spp.	Lacewings (Neuroptera)	Bugs, small caterpillars	Larvae or pupae
Cryptolaemus spp.	Ladybird Beetles (Col.; Coccinellidae)	Mealybugs, aphids	Active adults
Ambylyseius spp.	Predacious mites (Acarina; Phytoseiidae)	Phytophagous mites and thrips	Active mites
Phytoseilus parsimilis			
(b) Parasites			
Aphidius matricariae	(Hym.; Braconidae)	*Myzus persicae*	Parasitized aphid mummies
Dacnusa sibirica	(Hym.; Braconidae)	Chrysanthemum and Tomato Leaf-miners	Parasitized pupae or active adults
Opius pallipes			
Diglyphus isaea	(Hym.; Eulophidae)		Active adults
Encarsia formosa	Nymphal parasite (Hym.; Aphelinidae)	Glasshouse Whitefly	Pupae in parasitized 'scales'
Trichogramma spp.	Egg parasites (Hym.; Trichogrammatidae)	Lepidoptera	Pupae in parasitized eggs
(c) Pathogens (micro-organisms)			
Verticillium lecanii	Fungus	Aphids and Glasshouse Whitefly	Spore suspension as a wettable powder
Beauvaria bassiana	Fungus	Caterpillars	
Bacillus thuringiensis	Bacterium	Caterpillars	Suspension
Bacillus spp.	Bacteria	Beetle larvae	Suspension
Granulosis viruses and nuclear polyhedrosis viruses	Viruses	Caterpillars; some beetle larvae; sawflies	Viral bodies in a wettable powder

Autocide, the sterile-male (insect) release method (see Chapter 6) is sometimes regarded as a form of biocontrol. There are now several major companies, each usually associated with a nuclear reactor, which can deliver large numbers of sterilized male Fruit Flies, screwworm flies and the like, for local release. Some of these companies also produce insect parasites and predators with their mass-production rearing facilities.

Pioneering work by the Glasshouse Crops Research Institute (GCRI – now called Horticultural Research International) has resulted in the regular use of biocontrol agents in protected cultivation (i.e. glasshouses and polythene tunnels) in the UK. This practice is now widespread throughout temperate regions. The two major pests were Red Spider Mite (*Tetranychus urticae*) and Glasshouse Whitefly (*Trialeurodes vaporariorum*) on tomato, cucumber, chrysanthemum and many other flowers. These are killed by the predatory mite, *Phytoseiulus persimilis* and the parasitic wasp, *Encarsia formosa*, respectively. Later work included the control of *Myzus persicae* and other aphids, mealybugs and various caterpillars, and most recently the leaf-miners (Dipt.; Agromyzidae) that ruin the foliage of chrysanthemums and other plants. The Black Vine Weevil (*Otiorhynchus sulcatus*) and some other beetles have soil-inhabiting larvae that can be very damaging to plant roots, and formulations of pathogenic bacteria and viruses can be used against them. The latest development for the control of Black Vine Weevil larvae is the entomophilus nematode, *Hexamermis*. The end result of all this research is that most growers in the UK with glasshouses, and many of the gardening public, rely on the use of biocontrol agents for the protection of tomato and cucumber crops in glasshouses, so these and other fruits are grown completely pesticide-free.

Biological control (*sensu stricta*) is most applicable to agricultural and forestry insect pests and to date is not being widely used for either medical or veterinary pests. Various fungi, bacteria and entomophagous nematodes are being used successfully in projects to control the larvae of mosquitoes and Tsetse Flies, *Culicoides* and *Simulium*, and some ticks, but generally insects are not being used as biocontrol agents to any extent. Fish, geckos and other lizards, amphibians, birds and other vertebrates are also used for predation of all types of adult and larval insects and Acarina, including domestic, medical and veterinary pests, but the natural insect predators and parasites do not lend themselves to definite biocontrol projects against most medical and veterinary pests.

A major agricultural production constraint is presented by the ubiquitous weed species that so hamper plant cultivation. In many situations, weeds are more serious crop pests than insects. Technically, a weed is any plant proving to be a hindrance to agriculture or plant cultivation generally. A volunteer potato plant growing in a field of wheat is regarded as a 'weed' in the agricultural context. The more serious pest species of widespread and regular occurrence are classed as 'noxious weeds'. These are often ruderals (i.e. secondary pioneers), specially adapted for rapid colonization of bare earth such as roadside verges, agricultural fields and garden plots. Ironically, many of these species are adapted for 'pioneering' and if the resulting community were to be left alone then these transient colonists would often disappear, being succeeded by other species more adapted to crowded conditions. But, of course, agricultural fields are constantly reverted to bare soil by each cultivation, presenting new exposed sites for weed colonization.

Some weed species are local in distribution and presumably indigenous. These are part of the local agroecological community with a place in the local food webs, and have occupied these niches since ancient times. Others are species that have spread during the course of agricultural evolution. They may have been present locally for a very long time and are well adapted into the local community. Both of these types can be classed as 'noxious' if sufficiently biologically aggressive because, agriculturally, seed-bed preparation presents them with bare soil habitats for repeated new colonizations. However, some weeds have been transported from other continents in relatively recent years, and in their new locations have botanically run amok! In the new habitats, some of these very noxious weeds have found ideal climatic conditions and they have bred, spread and colonized very rapidly and extensively. Many of these noxious weeds are South American in origin, and are now widely established throughout Africa and Southern Asia. In these regions the weed species are either tolerated for their useful properties (*Opuntia*, *Lantana*, etc.) or else

Fig. 2.13 Prickly Pear Cactus (*Opuntia* sp. – Cactaceae); Alemaya, Ethiopia.

largely ignored because of a lack of the funds required for their control.

Australia has suffered from a number of major weed species – these are mostly semi-xerophytic weeds, as ruderals, from the Mediterranean Basin, which have become established in dry areas in Australia in livestock pastures. The main species concerned include ironweed/skeleton weed (*Chondrilla juncea* – Compositae), lantana (*Lantana camara* – Verbenaceae), St John's wort (*Hypericum perforatum* – Guttiferae), gorse (*Ulex europaeus* – Leguminosae), Noogoora burr (*Xanthium pungens* – Compositae), ragwort (*Senecio jacobaea* – Compositae), nut-grass (*Cyperus rotundus* – Cyperaceae) and Prickly Pear Cactus (*Opuntia* spp. – Cactaceae). The lantana and prickly pear are of Central American origin. The Commonwealth Scientific and Industrial Research Organization launched a major biocontrol effort against these introduced

weeds in the 1920s, a summary of which was presented by Wilson (1960). Some weeds are very serious pests and very difficult to control – weeds in pastures and aquatic floating weeds do not lend themselves to herbicide treatment even if suitable chemicals could be found. But phytophagous insects can deplete weed populations without any adverse environmental effects. Generally with weed control it is usual to make a multi-pronged attack – typically a leaf-eater (caterpillar or beetle), with a stem-borer (caterpillar or beetle), flower-eater, seed-borer and leaf-miner might all be used simultaneously, depending, of course, on what is available. Sometimes a single insect species may be sufficient.

A classic case in Australia was the Prickly Pear Cactus. Several species were introduced in the 1800s and later escaped from captivity, with two species becoming established as very serious weed pests (Fig. 2.13). By

Fig. 2.14 Caterpillars of the Cinnabar Moth (*Tyria jacobaeae* – Lep.; Arctiidae) eating flowers and leaves of ragwort (*Senecio jacobaea* – Compositae); Gibraltar Point, UK.

Fig. 2.15 Ragwort plant completely defoliated by Cinnabar Moth caterpillars (*Tyria jacobaeae* – Lep.; Arctiidae); sand dunes at Gibraltar Point, NNR, UK.

1900 the Cactus was spreading throughout Queensland and New South Wales, until by about 1925 it covered some 60 million acres which were declared useless for agriculture (DeBach, 1974). Collecting in South America and the southern USA yielded a number of different phytophagous insects and between 1920 and 1935 the Prickly Pear Board introduced more than 50 different species, most of which were confined to Cactaceae as host plants (Wilson, 1960). The most successful species was a pyralid moth called *Cactoblastis cactorum*, imported in the 1920s. Its larvae are host-specific and tunnel up inside stems causing the plant to collapse. Some of the Cactus Mealybugs (*Dactylopius* spp.) were also effective in killing *Opuntia* plants. The overall effect of the insect releases has been to control *Opuntia* spp. quite effectively over a vast area of eastern Australia at a very modest cost. A similar project has been conducted in California (Huffaker, 1971).

A temperate example of interest is the control of ragwort (*Senecio jacobaea* – a Compositae) by the monophagous defoliating caterpillars of the Cinnabar Moth (*Tyria jacobaeae* – Lep.; Arctiidae). The brightly coloured (aposematic) yellow and black caterpillars eat both flowers and leaves (Fig. 2.14) and may defoliate the plant (Fig. 2.15) which might then die. However, the plants are biennial and have great powers of recovery, and a well-established plant in the

UK may suffer complete defoliation and yet sprout anew from the base. This noxious weed is poisonous to cattle and can dominate pastures. It is native to Europe and now widely spread throughout South Africa, Australia, New Zealand and North America. The Cinnabar Moth has been introduced into New Zealand and parts of the USA and a measure of control has been achieved.

Most of the early attempts at biocontrol were aimed at insect and mammal pests (such as the rabbit in Australia), largely using other insects and mammals (including the fox and stoat) as predators. But in recent years there has been a great deal of effort, and success, in the control of weeds, both terrestrial and aquatic, by phytophagous insects, and the literature is becoming quite extensive.

The procedural requirements for successful biological control programmes are several and very important. Nowadays the projects originating from the IIBC are quite sophisticated and typically very effective, but scientists have had to learn the hard way. Important publications such as Wilson (1960) and Greathead (1971) reveal that many of the early attempts at biocontrol in Australia and Africa were unsuccessful. A selection of these failures are included for instruction in Hill (1983, page 89) and Hill (1987, page 117). In the past, many failures were the result of careless handling of the predators and parasites in shipment, together with accidental delays in transit or distribution, rather than being due to intrinsic factors in the host/parasite biology. However, there have been a considerable number of cases where an introduction was made without any real scientific basis for the choice of predator or parasite – it seems often to have been a matter of convenience. The important points are as follows:

1. Delivery time – in the early days intercontinental shipment was by sea and many predators/parasites died in transit. Modern air travel makes this problem negligible, for delivery times can be reduced to days.

2. Numbers – sufficient numbers of individuals have to be released simultaneously in order to establish a population. Commercial rearing facilities are now good enough for large consignments of parasites to be purchased at a time.

3. Identity – it is vital that the identity of the pest be clearly established. Kenya Mealybug was misidentified in the 1930s and Asiatic parasites gave no control. In 1935 it was described as a new species native to East Africa, and parasites which gave good control were imported from Uganda. Also, the identity of the parasite must be certain.

4. Origin – most serious pests are not indigenous to the region in which they are most damaging, so in the search for suitable predators or parasites it is necessary to go back to the country of origin or sometimes to a local area within that country. Thus using parasites from the Philippines and Ceylon (Sri Lanka) for the control of Kenya Mealybug was a waste of time.

5. Host-specificity – it is useless to introduce a parasite which usually attacks another genus of pest, even though the two genera might have very similar biologies. Parasites from a closely related species in a different part of the world might be worth trying, but usually many parasites are species-specific. However, predators are usually less prey-specific than parasites, and so a predator of another (similar) genus in the country of origin may be worth considering as a biocontrol agent.

6. Local enemies – when the pest is indigenous to the region where it is damaging it is usually not worth while to attempt to breed large numbers of a local parasite for release, since the pest/parasite relationship is usually of a long-standing stability

which will not allow changes of this nature. If the parasite complex is extensive there is little chance of inserting another species. However, occasionally there appears to be a vacant niche for either a parasite or predator and then it is worth while attempting an introduction.

7. Predator/parasite survey – when the original parasite survey is made, it is important that all local parasites are found. An introduction of a species already present, but missed because of inadequate surveying, is invariably a waste of effort. This error was made with the Coffee Berry-borer (*Hypothenemus hampei* – Col., Scolytidae) in Kenya.

2.7 INSECTS IN NATURAL FOOD WEBS

The outstanding feature of the insects is their sheer abundance. They occur in all terrestrial and freshwater habitats throughout the world, although they are poorly represented in the sea (they are, however, common on the shore, especially along the strandline). Small in size, insects occur as an enormous number of species, and the numbers of individuals are vast. In terms of species and individuals, they outnumber all other elements of the world fauna, separately and collectively, to an immeasurable extent, and in terms of biomass they easily outweigh the rest of the world fauna on land. All possible habitats (on land and in freshwater) are colonized by insects and all possible niches occupied, but these can be very broadly categorized into soil, terrestrial and freshwater habitats. Insects' importance in all terrestrial and freshwater food webs cannot be over-emphasized.

2.7.1 Soil habitats

There is variation in the soil fauna according to the physical and chemical properties of the soil – the most inert soil contains little life except for some micro-organisms and, at the other extreme, a deep rich, moist loam can be home to thousands of small arthropods. In soil, in the absence of light, food webs are essentially detritus-based, and the primary consumers are scavengers (detritivores). Animal life is mostly confined to the top 10–20 cm of the soil, although some species descend deeper. In temperate regions the soil fauna is dominated by Collembola (Springtails) and soil mites (Oribatae, etc.), woodlice and Annelida. In the tropics, however, ants (Formicidae) and termites (Isoptera) are dominant. An article in *New Scientist* (26.1.91) describes how termites in Southern Africa (mostly *Odontotermes* spp.) build underground nest systems that result in surface ridges and gulleys up to 2 m high and 1 km in length. These 'termite bands' affect soil drainage and often improve growing conditions for crop plants. Certainly the plant materials collected on the surface and taken deep into the soil increase the fertility of most soils. It would seem possible to increase crop yields by making use of the termite activity in the soil. In addition to these dominant groups there are crickets (Orth.; Gryllidae) and a range of beetle larvae and pupae, many fly larvae and pupae, a few caterpillars and many Lepidopterous pupae. Leaf litter is usually associated with soils because of the intimate association, but the fauna here is not fossorial or minute, tending to be cryptozoic and light-avoiding. A dominant group in leaf litter in warm regions is the Blattaria (cockroaches).

2.7.2 Terrestrial habitats

In reality it is almost meaningless to generalize to this extent because of the diversity of terrestrial habitats, but the important point is that the Insecta are a dominant element of the terrestrial fauna and they are the basis of most of the food webs (when viewed broadly). Insects are the small basic herbivores, as well as there being some small carnivores. They feed on the terrestrial

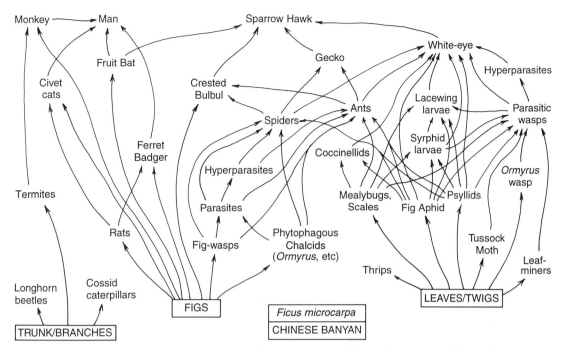

Fig. 2.16 Terrestrial food web based on the Chinese Banyan (*Ficus microcarpa*) in Hong Kong; 1975.

vegetation and are in turn food for amphibians, many reptiles, most birds and some mammals (Insectivora, etc.). Figure 2.16 shows a terrestrial food web for the Chinese Banyan (*Ficus microcarpa*) in Hong Kong which is fairly typical of many webs.

The main groups of insects that occupy terrestrial habitats are the grasshoppers (Orthoptera), some termites (Isoptera), most bugs (Hemiptera), thrips (Thysanoptera), most beetles (Coleoptera), all adult flies (Diptera) and some larvae and pupae, all adult Lepidoptera (moths and butterflies) and most larvae and pupae, and all Hymenoptera except for the fossorial ants. In addition there are, of course, most of the other less abundant groups of insects, such as Psocoptera and stick insects (Phasmida). Most of the parasitic groups could also be included broadly in this habitat, including lice (Mallophaga, Siphunculata) and fleas (Siphonaptera).

2.7.3 Freshwater habitats

The insect groups commonly found here are quite well defined, and include the nymphal stage of Ephemeroptera (Mayflies), Odonata (Dragonflies and Damselflies) and Plecoptera (Stoneflies). The adults are mostly found closely associated with the water body. Some Heteroptera (Hemiptera) are referred to as water bugs – all stages are aquatic. Some Diptera have aquatic larvae (and pupae) and groups such as mosquitoes (Culicidae) and Lake Midges (Chironomidae) are tremendously important as the primary consumers at the base of most freshwater food webs. Caddisflies (Trichoptera) are all aquatic as larvae and can be important in many freshwater habitats. In ponds and lakes small Crustacea (*Daphnia, Cyclops, Gammarus*, etc.) are clearly very important, but the insect larvae are equally so.

Freshwater habitats fall into two separate

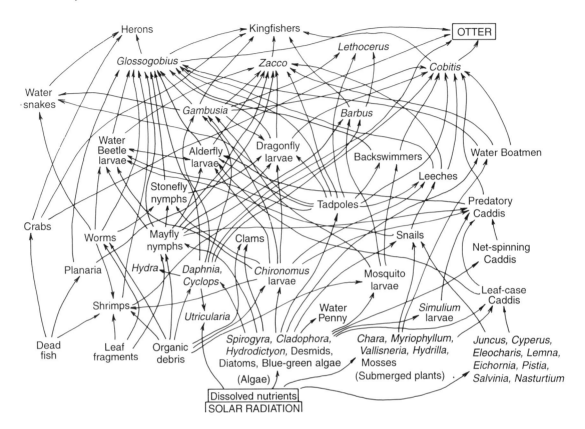

Fig. 2.17 Freshwater stream food web; Tai Po Kau Forest Reserve, New Territories, Hong Kong; 1975.

categories: flowing water (i.e. streams and rivers, or lotic habitats) and standing water (i.e. ponds, lakes and reservoirs, or lentic habitats). There are some important physical and biological differences between these two habitat types, but overall the insect fauna tends to differ mainly at the species level. Thus there are stream species of Mayflies and Dragonflies and separate pond/lake species, but both orders of insect are found in both types of freshwater habitat. In Hong Kong in 1975 I constructed food webs (for teaching purposes) for the freshwater stream in the Tai Po Kau Forest Reserve in the New Territories (Fig. 2.17), and also for the Plover Cove Reservoir, in the New Territories (Fig. 2.18). Both of these food webs, although repre-

senting very specific locations, do illustrate several basic aspects of many typical freshwater food webs. They also show the relative importance of some of the insect groups.

The Mayflies, Stoneflies and Caddisflies (as adults as well as nymphs) are important in the diet of many carnivorous fish. This has led to the use of artificial imitation insects as lures for fishing. In several countries, a flourishing local cottage industry is the manufacture of fishing flies, large ones for salmon and smaller ones for trout and sea trout. A few of the larger 'flies' are totally artificial, but the majority mimic adult Mayflies, Caddisflies, Stoneflies and other insects such as bees, termites and various other flies.

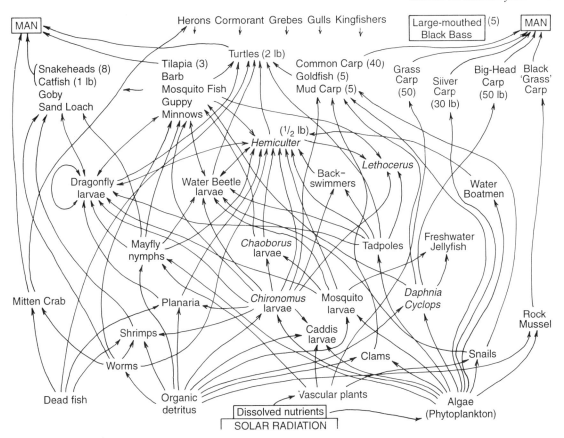

Fig. 2.18 Food web for a freshwater reservoir; Plover Cove Reservoir, New Territories, Hong Kong; 1975.

2.8 INSECTS AS HUMAN FOOD

So far as is known, insects must have played a major role in the feeding of primitive humans, as a source of animal fat and protein. Studies on the diet of primitive peoples in remote communities reveal that many insects are regularly eaten, and it can be assumed that when early humanity was at the hunter-gatherer stage of evolution, insects would have been consumed regularly. In temperate regions at the present time a few insects are eaten, largely out of gastronomic curiosity; some can be found for sale in Japan and parts of China – these include honey-ants and chocolate covered bees and termites.

In parts of the tropics many insects are a regular dietary item. In Uganda the Longhorned Grasshopper (*Homorocoryphus nitidulus*) is unusual in that it feeds on grasses and cereal crops (Gramineae) and it periodically swarms in large numbers. The local name for the grasshopper is 'Nsenene', and the adults are collected and eaten either raw or cooked. Children can be seen taking a small paper bag of Nsenene to school for their lunch. Also in Uganda the large termite *Macrotermes bellicosus* is very abundant around Lake Victoria, and in most villages the various termite mounds are regarded as the property of different families. At the time of swarming the women and children put

Fig. 2.19 Queen Termite (*Macrotermes bellicosus* – Isoptera; Termitidae) in opened 'royal cell' with attendant male, soldier and two small workers; Kampala, Uganda.

Fig. 2.20 Bombay Locust (*Patanga succincta* – Orth., Acrididae) which is eaten as 'Sky Prawn' in Thailand (body length 60–80 mm).

baskets over the mound entrances and collect the young alates.

In many parts of Africa and tropical Asia termite mounds are dug out to collect the queen which is sometimes eaten on the spot. A queen *Macrotermes* has a swollen abdomen filled with eggs and the fat body measures up to 6 cm by 2 cm (Fig. 2.19). Large grasshoppers, especially locusts, are eaten in both Africa and Asia – the Desert Locust is regularly eaten in many parts of Africa on both sides of the Sahara Desert. In Thailand, Bombay Locust (*Patanga succincta*) (Fig. 2.20) is roasted and eaten – in many street foodstalls you can buy ten roasted locusts on a satay stick. A recent development in Bangkok used as a tourist 'attraction' is the sale of roasted locusts under the more enticing name of 'Sky Prawns'! In 1980, a government control project paid villages in north Thailand to hand-catch Bombay Locusts, which they killed and kept for food. An article in the UK *Guardian* newspaper (October, 1991) reported that the Thai Farmers' Bank quoted a rise in the price of fresh locusts in Bangkok from 6 cents (US dollars) per lb in 1983 to $1.40 in 1991.

Beetle larvae, especially chafer grubs

(Scarabaeidae) are eaten in Australia and Papua New Guinea by the Aborigines in the 'Bush' as well as the well-known witchety grub (the larva of *Xyleutes leucomochla* (Lep.; Cossidae)) that feeds in the soil on the roots of some Acacia trees. In Sarawak a common weevil larva (*Rhynchophorus* sp.) larva bores the trunk of sago palm, and in Kuching Market 'Sago-worms' are regularly on sale, along with some red ants and grasshoppers. In some regions, Honey Bee larvae and pupae are eaten together with honey in the wax comb in which they live. As discussed in section 2.2, honey from Honey Bees (*Apis mellifera*) is an important minor item of human diet worldwide. In the health-conscious Western society, honey (especially when unrefined and mixed with pollen, propolis and bits of wax) is highly regarded as a health food with medicinal properties.

The insect abdomen contains an exten-sive fat body, and well-fed insects can be squashed and rendered down to make cook-ing fat. Well-documented cases include the Bombay Locusts in Thailand, alate termites and Lake Midges (Dipt.; Chironomidae) over Lake Victoria. The Lake Midges from the Rift Valley Lakes in East Africa are collected when swarming in vast numbers and can be com-pressed into blocks of 'midge butter'. Two books on the topic of insects as human food are Bodenheimer (1951b) and Holt (1988), and other sources are listed in Gilbert and Hamilton (1990).

2.9 MISCELLANEOUS

There are a few other instances where insects are used by humans and can reasonably be termed 'beneficial' as distinct from being harmful.

(a) (b)

Fig. 2.21 (a) Adult and (b) nymphal Stonefly (? *Perla* sp.) (Plec.; Perlidae) from freshwater stream at Tai Po Kau, New Territories, Hong Kong.

2.9.1 Insects as ecological indicators

In the UK there is growing concern over the continuing destruction of natural habitats and the general loss of our ancient tracts of land. Biological indicators of habitat antiquity are usually floral, but sometimes the insect fauna has been used with apparent success. The total insect spectrum, reflecting the overall biodiversity, can be used in this respect, and in some regions there are special 'indicator species' similar to some of the Arctic/alpine plants found on hilltops throughout the UK that are regarded as post-glacial survivors.

Sometimes there is interest in categorizing habitats for classification purposes, and the insect fauna can sometimes be used as a measure of affinity between similar or related habitats.

Some insects have been used as pollution indicators. The order Plecoptera (Stoneflies) have larvae that live in fast-flowing freshwater streams, and are reputed to be very sensitive to a range of industrial pollutants as well as to oxygen deficiency. Their absence or presence, expressed both qualitatively and quantitively, can give an indication of the extent of any pollution in that stream (Fig. 2.21). Aerial pollution is sometimes a matter of noxious gases and sometimes soot or other forms of particulate carbon in the air. A now classical study carried out around Manchester used the Peppered Moth (*Biston betularia*) with its rare dark morphs (Kettlewell, 1973). The occurrence of the three morphs (colour forms) – pale (typical), dark and intermediate (Fig. 2.22) – was thought to be related to the colour of the bark of local birch trees. With increased air pollution, and a high soot content, the tree bark became discoloured and dark (sooty), and the numbers of the dark form (*carbonaria*) increased. However, in clean areas the pale form still predominated. It was thought that the key factor was bird predation when the nocturnal moths rested during the day on the bark of the birch trees.

Fig. 2.22 Peppered Moth (*Biston betularia* – Lep.; Geometridae) showing the three different colour forms (morphs); Cambridge, UK.

Dark moths on a pale background are conspicuous, and vice versa. After a national and local clean-up campaign (through the Clean Air Acts), the dirty air became cleaner, and after some 30–40 years the trend has been reversed – the trees have become cleaner and the dark form of the moth has became scarce again (Clarke *et al.*, 1994). The proportion of d.f. *carbonaria* recorded by Clarke in 1959 was 93%; this had fallen to 23% by 1993.

2.9.2 Medical uses

Medical uses of insects are several but minor. As mentioned in section 2.2, bee honey has long been known to have mild antiseptic qualities and has been used in the past for

Fig. 2.23 Cricket cages from China; the larger (10 cm cube) is a cage for keeping the male cricket inside; the smaller one is a carrying cage; both are made of rattan.

wound dressings. Beeswax, because of its sterile and inert nature, is still used in some orthopaedic operations to fill spaces in bones. The genetic research based on the Fruit Fly *Drosophila* has important medical connotations in the study of inherited human defects and disorders. There is a long history of the use of some larger fly maggots for the cleaning of open wounds. Some of the fly infestations were accidental, as indeed they must have been initially in historical times, but later some were deliberate. The fly species used had larvae that were purely saprozoic and showed little tendency to practise myiasis. In remote tropical locations it is possible that this practice persists, but records are unclear; however there are recent reports of its use in both Europe and the USA for special cases.

2.9.3 Forensic uses

Forensic entomology has existed for a long time, but in recent years it has received publicity and is probably being used more frequently (Smith, 1986). According to a UK television programme ('The Witness is a Fly'; BBC2; 17.4.94), the succession of saprozoic insects that naturally infests a human corpse or dead animal can be divided into eight distinct waves from a fresh corpse to a dry and desiccated cadaver three years old. The first-wave insects prefer fresh flesh as food; subsequent waves respond to different odours as decay proceeds; and the final two waves feed on dry protein including skin and hair. Some insect species have specialized habitat requirements and yield information on locality that can be important, and others can give seasonal information, helping to pinpoint the time of death.

2.9.4 Entertainment

Entertainment is provided by insects and is usually associated with gambling. One exception is the commercial flea-circus where fleas are harnessed with minute straps and pull tiny carts for human recreation. Apparently the only species that could successfully be harnessed was the Human Flea, and when this became scarce the circuses had to be abandoned. Sailors in the British Royal Navy in former years used to bet on cockroach races, using the small *Blatella germanica*, which was abundant in most ships. In South East Asia and parts of China, beetle-wrestling can still be seen in cafés and on some street corners. Stag Beetles (Lucanidae) and Unicorn Beetles (Scarabaeidae; Dynastinae) are used for wrestling – these are species where the male has greatly elongated mandibles or else horn-like projections on either the head or prothorax, and the rival males interlock their 'horns' and push to dislodge each other from a narrow branch. A somewhat more sophisticated sport is fighting crickets. Male crickets can be quite aggressive, and will 'sing' (stridulate) when displaying to other males. If put together two adult males will actually fight – although I have personally never witnessed this. In China, there is often great interest in these crickets, and one can buy small cricket cages made of carved rattan (Fig. 2.23) – with a volume of about $10\,cm^3$. Separate carrying cages are much smaller but are sufficiently sized for the purpose.

3

PEST DEFINITIONS

As already mentioned, we do tend to be overly preoccupied with insects as pests because of the damage they do and their importance in relation to human affairs. In order to study the subject and to communicate properly, it is necessary to have the correct vocabulary: hence the need for specific definitions in this small chapter. For a detailed 'Glossary of Entomology', the extensive work by Torre-Bueno (1937), now revised and updated by Nichols (1989), is highly recommended.

When confronted with an insect/host situation, it is vitally important to determine that the insect is in fact a pest before contemplating any control measures. In the past, much money and time have been wasted over cases where the insect either was not a pest or else the damage done was so slight as not to warrant any real concern. Sometimes ecological chaos can result from unwarranted pesticide applications.

3.1 DEFINITIONS

3.1.1 Pest

The definition of a pest can vary in detail according to the precise context in which it is considered, but in the widest sense a pest is an insect (or organism) that causes harm to humans, their livestock, crops or possessions. They key word is 'harm' and is usually interpreted as 'damage' which, of course, can often be measured (often quantitatively). Moreover, damage can usually be equated to **economic loss** in terms of actual money. Harm at its lowest level of interpretation includes nuisance and disturbance. Thus a buzzing mosquito at night can prevent sleep, and in tropical Africa Face Fly can be very distracting and will impair working efficiency.

3.1.2 Pest status

Pest status is given when the insect (population) causes a particular level of damage, but the designation really only applies to that particular insect population on that host at that time. On another host or at a lower population level, the insect might well not be of pest status.

3.1.3 Economic pest

The activities of the insect pest result in damage which is often expressed eventually as a loss of money, i.e. a loss of income to a farmer or loss of revenue to a community. When the damage level reaches a certain point at which financial loss is significant, then the insect species (population) concerned is designated an economic pest. The presence of an economic pest requires that control measures are taken.

Clearly, a decision as to when the damage level is significant is very subjective, and will vary according to the pest and the damage. Agriculturally, most perennial horticulture crops (coffee, apple, oil palm, etc.) are regarded as far more valuable than annual agricultural crops like wheat, other cereals and potato. As a general agricultural guide, it has been suggested that an insect becomes an economic pest when it causes a yield loss of 5–10%.

3.1.4 Economic damage

This is the amount of damage done to a crop or host by a pest that will financially justify the cost of taking artificial control measures against the insect pest. In the past, this has usually meant the application of insecticides. This assessment will vary greatly from situation to situation (crop to crop) according to the basic value of the crop (or host), its current market value and the cost of the control measures being considered. The main reason for the overuse of insecticidal seed dressings in the past was because they were an effective measure against certain soil-dwelling insect pests and were relatively very cheap.

One problem in the tropics is that many peasant farmers practising subsistence farming find they cannot justify the cost of pesticide use at all.

3.1.5 Pest species

This is an insect species whose members are regularly (or usually) of pest status and cause damage of economic consequence. Such a species usually requires population-controlling measures. Thus Colorado Beetle in Europe, Tsetse (Fly) in tropical Africa, Carrot Fly in England, and Codling Moth in Europe and North America are all well-known pest species. However, the individuals are not necessarily of pest status, for they might be on hosts of no economic importance (Tsetse Fly on wild game; Carrot Fly on hemlock

plants in the hedgerow, etc.), or they may be present only in very small numbers. Another problem is presented by insects such as male mosquitoes – they do not feed on blood, but do of course mate with the blood-sucking females. Just because an insect belongs to a pest species it should not be assumed that it is itself a pest – evidence is required.

3.1.6 Economic injury level (EIL)

Entomologists are primarily concerned with insect pests and their populations, the size of which can be determined when the status of the pest species is being assessed. The EIL is the lowest pest population that will cause economic damage: most protection activity is aimed at preventing the pest population from reaching this level. Too often, unfortunately, we fail to anticipate pest problems, and then the main concern is to reduce damage levels by reduction of the pest population to a point below the EIL.

Precise assessment of economic damage will vary considerably from crop to crop, place to place and time to time, according to several economic and ecological factors. However, this would entail considerable expenditure of effort and time (and hence money) which is usually just not feasible. Thus official local recommendations, that are typically based on years of empirical observations, will use mean levels of damage, yields, costs, etc. in making definitions of local EIL for major pests on important crops.

3.1.7 Economic threshold

Anticipation of a serious pest problem can lead to complete avoidance of damage, and this could be said to be the ultimate objective of pest control programmes. To this effect, the economic threshold is designated as the population density of an increasing insect population at which control measures should be started in order to prevent the population reaching the EIL (Stern *et al.*, 1959).

3.1.8 Pest complex

Any cultivated plant (or animal) will normally be attacked by a number of different species of insects, mites, nematodes, parasites, pathogens, birds and mammals, and these together form a complicated interacting pest complex. In the field of plant health/protection (and also animal health and human health) it is very necessary to consider the entire pest complex, particularly since some pests interact even to the extent of mutual **synergism** (potentiation). A most important aspect is that of **predisposition**. This is when an organism is attacked by one pest/parasite and thus stressed and weakened, being made more vulnerable to other pest/pathogens/parasites. Unfortunately, living organisms can be rendered vulnerable by many forms of stress, both physical and physiological and, in the case of humans, also mental. In the research laboratory, biologists may be dealing with single pest organisms, but in the field they are invariably confronted with a pest complex which makes the situation far more complicated. Key pests (section 3.1.11) have to be identified for their damage is the most serious and their control the most imperative.

3.1.9 Pest spectrum

This refers to the total range of different types of pest and their species recorded attacking a particular crop plant or host organism. Interpreted widely, it will refer to all the species recorded on that host (plant) worldwide, and at the other extreme all the species recorded locally. Thus cotton has almost 1400 species of insect and mite pests worldwide, but in any one locality (e.g. Uganda) the total is only about 40 species; it is important to know which interpretation is being used.

One aspect of interest is that often any one host will be attacked by the same spectrum of pests in different parts of the world. Thus cotton will have a bollworm complex of three to four species of both Pyralidae and Noctuidae (Lepidoptera) but these will be of different species in Africa, Asia, the USA and South America. Similarly, apple crops worldwide will be attacked by a tortrix complex (Lep.; Tortricidae) that feed on both fruits and leaves, but the species concerned will differ from continent to continent.

3.1.10 Pest load

This is the actual number of species of different pests and the number of individuals to be found on either a crop or a single plant, or other host, at any one time. Usually, there would be a pest complex concerned but in some cases it could refer to a monospecific population.

3.1.11 Key pests

In any local pest complex, there are usually a few major pests that are the most important. These dominate the complex in various ways and usually cause the most important damage, and their control is urgently required. Locally, any crop can be expected to have one or two key pests, and they may vary regionally and seasonally. Because of their importance, the control of key pests will dominate any overall control programme for the crop, and it is important that economic thresholds be established for these pests. Key pests usually have a high reproductive potential, often a good survival mechanism, and they usually inflict serious direct damage to the crop (e.g. Codling Moth).

3.1.12 Major pest (or serious pest)

In previous books (Hill, 1983, 1987) it has been necessary to distinguish between the more important pests and the less important in terms of their treatment. The more important pests are termed 'major pests'. These are species that are either serious pests locally or are economic pests over a wide geographical area. A few species are included in this category because they are both polyphagous and widely distributed and so likely to be regularly encountered, even though on any one host damage may be slight.

Major pests are the species that need to be controlled. In the past a broad-spectrum insecticide (organochlorine) was usually applied to the crop, and it killed most of the major pests. Nowadays, the integrated pest management approach requires that all the major pests be considered individually, and in some cases each species has a separate control approach. Usually in most localities the major pest species remain fairly constant over the years. However, in some tropical regions it is reported that the major pests of rice have completely changed over a 50-year period, so some changes should be expected over long periods of time.

3.1.13 Minor pest

These are the less serious species of insect that are recorded feeding or ovipositing on a crop plant. Usually they inflict only slight damage on the host, and often their effect is indiscernible. Sometimes a major pest of one crop/host will be a minor pest on another – this is quite common agriculturally. Moreover, some species that are major pests on a crop on one continent are minor pests on that crop on another – this difference is often seen in comparisons between the Old World (Eurasia and Africa) and the New World (North and South America).

The crop plants with the largest number of recorded pests on a worldwide basis are cocoa (with 1400), cotton (1360), sugarcane (1300) and coffee (838). Some widely established wild plants and trees also have extensive lists of phytophagous insects feeding on them: oaks (*Quercus* spp.) in the UK have a total of 1000 recorded insect species. Cotton crops have some 1360 species of insects and mites recorded feeding on them worldwide, but in any one country, for example, Uganda or Ethiopia, the list of species is usually about 20–30. Of these between four and eight are major pests and 10–20 minor pests, with two or three being key pests.

3.1.14 Pest species recruitment

Most major crops, forest trees and some types of livestock have a definite centre of origin in one particular part of the Earth, although during historical times they have been transported all over the world to become established in areas of suitable climate. A good agricultural example is seen with *Citrus* crops. *Citrus* is endemic to south China/Indochina, and by the turn of this century it had been established in many subtropical parts of the world (the Mediterranean, South Africa, Australia, Hawaii, California and Florida). Examination of the pest spectrum in all these new regions reveals that most of the insect and mite pests are allochthonous, in that they originated in the south China region where *Citrus* is endemic and have gradually spread into the new areas of cultivation. However, in each region there are also some autochthonous (local) pests that have adapted to the new host plants. Generally, each species of cultivated plant (and animal) will have a pest spectrum composed of a mixture of allochthonous species and autochthonous species, the proportions of which will vary from species to species. The purpose of Plant Quarantine Regulations is to prevent the spread of noxious pest organisms from region to region, for many major crops/hosts in their new locations do not carry the full load of native pest species, some of which could be devastating in the new locations.

3.2 DEVELOPMENT OF PEST STATUS

Because of the basic economic importance of pest insects, there are two main ways in which insects become pests: these are as

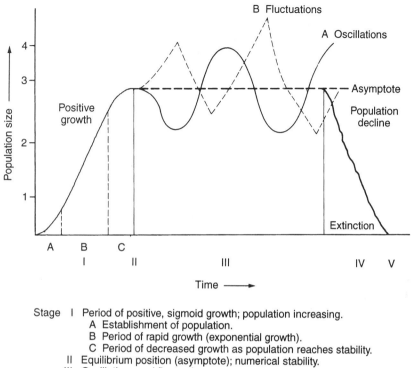

Stage I Period of positive, sigmoid growth; population increasing.
          A Establishment of population.
          B Period of rapid growth (exponential growth).
          C Period of decreased growth as population reaches stability.
    II Equilibrium position (asymptote); numerical stability.
   III Oscillations and fluctuations.
          A Oscillations – symmetrical departures from equilibrium.
          B Fluctuations – asymmetrical departures.
   IV Period of population decline (negative growth).
    V Extinction.
   VI Special cases; accentuated sudden changes in growth rates.
          A Population 'spurts'.
          B Population 'crashes'.

Fig. 3.1 The growth of populations (from Allee *et al.*, 1955).

a result of ecological and also economic changes.

3.2.1 Ecological changes

Under this heading there are three main aspects of importance, and these are broadly:

- The size and state of the pest population.
- The nature of the damage done by the pest to the crop/host.

- The value of the damage done as assessed by human society.

The latter point clearly overlaps with economic changes, but it is also closely related to the population condition. Because of the importance of the damage done by insect pests, this topic is dealt with in a separate chapter (Chapter 5).

(a) The pest population

A most important point is that any insect

species is only an economic pest at or above a particular population size, for at that population density damage of economic proportions will be done. Consequently, control measures are usually designed only to reduce the pest population below this particular level. Only rarely is complete eradication of the pest population aimed at and more rarely is it ever achieved! Allee *et al.* (1955) drew a schematic representation of the growth of a population (Fig. 3.1). The graph is self-explanatory, and applies to all populations of living organisms. Four separate hypothetical population levels are indicated, represented by the numbers 1 to 4. Population level 1 would apply to a very serious pest such as Colorado Beetle in the UK, and Rosy Apple Aphid (*Dysaphis plantaginea*) for control measures are recommended when just a single aphid is found per tree, at bud-burst. Population level 4 represents species that are only occasionally of economic consequence when their population is very large, and would include various locusts (Orth.; Acrididae) in the tropics and cutworms (Lep.; Noctuidae) in continental Europe and the UK. Most of the more common pests are represented by the hypothetical population levels 2 and 3 at which economic damage is done. Level 2 is below the usual population asymptote and damage can be more regularly expected, but level 3 is above the asymptote so damage would be less frequent.

For each pest species there will be a population level at which it becomes an economic pest, and usually the control measures to be applied will aim at lowering the population below this level.

A most important point is that populations are almost never static, and typically they wax and wane, often oscillating or fluctuating about a mean asymptote, but sometimes with extreme fluctuations from very low to very high. In the UK insect population study is made more difficult as some species are univoltine and many are bivoltine, so that

each life-stage (larva, adult, etc.) might only occur once or twice during the year. This constant change in population size is why the study is called 'population dynamics', to stress its ever-changing nature. In the UK the occurrence of the cold (often foodless) winter period results in a uniform and regular development of most insects, so that most develop synchronously, with the result that at any one time almost all of the insects (of that species) will be at the same stage of development. In the tropics, however, insect development and breeding is typically more or less continuous, so that at any one time all instars will be present, and the population constantly changes in structure. Insects have four main stages of development: egg, larva, pupa and adult, although each stage can easily be subdivided into further instars. The age structure of a population can be separated into all the different instars in the construction of a **life-table**, and the relative proportions of the different age groups will reflect the dynamic state of the population, in particular whether it is increasing or decreasing in size. Details of the construction of a life-table are given in Chapters 10 and 11 of Southwood (1978) and also in Davies (1988). The state of the population is important to know, for if it is declining then it is usually too late to apply control measures. Many control recommendations issued by the Ministry of Agriculture, Fisheries and Food in the UK are based on the assumption that the pest population is increasing.

A very simplistic equation summarizing the state of an insect population is as follows:

$$P_2 \rightleftharpoons P_1 + N - M \pm D,$$

where P_2 is the final population, P_1 is the initial population, N is natality (= birth rate), M is mortality (= death rate) and D is dispersal (emigration and immigration). This illustrates the main ways in which a pest population can be attacked, that is by reducing the birth rate, increasing the death rate, preventing immigration and encouraging emigration.

In summary, the simplest way in which many insect species become pests is just by an increase in their numbers.

The term 'population resurgence' is used to describe the rapid increase in numbers of a pest which occurs after the effects of a pesticide treatment have worn off. The pesticide treatment initially suppresses the population by killing a large proportion of the insects, but the survivors rapidly breed in large numbers once the pesticide has gone. This is seen most dramatically with rapidly breeding species like parthenogenetic aphids. It also occurs if the insecticide treatment accidentally kills off a disproportionate number of the natural enemies of the pest, rather than the pest itself.

(b) Monoculture

In the very early days of agriculture, most plants would have been grown in small plots rather like a patchwork, like most smallholder cultivation now. By Roman times, however, relatively large areas were being sown with a single crop. Even then, crop production did not really constitute a 'monoculture', except probably for wheat and barley fields. However, during this century there has been an ever-increasing practice of uprooting hedges and filling ditches to make very large fields for cultivation. British fields were typically in the order of 2–10 ha in size, surrounded usually with a ditch and a hedge; nowadays, many are 10–50 ha and without hedgerows. In North America and parts of Asia virgin prairie has been ploughed to sow cereals in large-scale agriculture. Now that agriculture in temperate regions is fully mechanized, it is necessary that field size is large. New crop varieties have become more specialized in their cultivation conditions, and this has also led to more extensive crop monoculture

and less crop rotation. With an extensive monoculture, the pests that feed on that plant are provided with unlimited food, and they will increase in numbers accordingly.

A classic case is the development of Colorado Beetle as a major pest of potato in the USA. Prior to about 1860, this insect lived on various wild Solanaceae in small numbers, but when large-scale potato cultivation started it moved on to the potato crops, which it clearly preferred as food. The Beetle became a common species present in large numbers and caused extensive damage to plant foliage. It became established in Europe (initially in France) in 1921, and has now spread throughout most of Western Europe on the potato monoculture.

In parts of Asia, the USA and Canada there are vast regions devoted to the monoculture of cereals, referred to as the 'corn belts' (maize) and 'wheat belts', where the crops are measured in terms of square kilometres rather than hectares. Similar areas occur with cotton and sugarcane cultivation, and now in the tropics there is a trend towards large-scale cultivation of oil palm, rubber and coconut on huge commercial estates and plantations. Another example is the extensive increase in rape as a commercial oil crop in the UK and other parts of Europe. The usual cabbage pests are of little importance on rape, the major pests being the pod attackers which hitherto had only been found on seed crops and wild hosts. Now all the pod-feeders have become major pests in UK agriculture.

A recent trend in agriculture that causes concern among crop protectionists is the large-scale cultivation of **cloned** plants. Cloned cultivation represents the extreme condition of monoculture where all plants are genetically identical. This can be very desirable agriculturally for production purposes, as evinced by the success of the new Standard Malaysia Rubber in 1965, which heralded a renewed commercial interest in natural rubber. However, cloned crops are very susceptible to pest attack – disease is the main problem, and a new strain can spread through a cloned crop like wildfire. Apparently the entire maize production of the USA is based on only four to six varieties, and in recent years there have been two major disease epiphytotics which caused great concern. Insect pests do not currently pose such a serious threat in practical terms, but they could.

(c) Insect distribution, dispersal and migration

The two outstanding features of any insect population are its tendency to increase and the subsequent dispersal of the large population. The dispersal may be entirely local and for short distances, but it can also be a regular annual event, and some individuals may travel hundreds (or even thousands) of kilometres.

The distribution of an insect species is governed by several factors, the main ones being climate, availability of food and the evolutionary history of the place. Since insects are poikilothermous, it is likely that temperature is the most important climatic factor, and it is possible to divide insect species into three main groups on this basis:

1. Tropical insects – cold death point about 10–15°C.
2. Temperate insects – cold death point at 0°C by freezing.
3. Boreal (Arctic) insects – death point well below 0°C (−20 to −30°C); body fluids supercool and freeze to glass.

Within these broad divisions of insect distribution there is some variation because of temperature adaptability. **Eurythermal** species can tolerate a wide range of ambient temperature, and will occur over a wide geographical range. **Stenothermal** species only thrive in a narrow range of temperature, and this can be high, low or intermediate. Some of the most important pest species are very adaptable and can tolerate a wide range of environmental conditions.

There appear to be three different zones of abundance for most insect species, and this concept is of importance so far as pest species are concerned:

1. Zone of natural abundance – this endemic zone is where climate conditions are near optimal and the species is always present, regularly breeding, and often in large numbers. The species is regularly a pest of importance.
2. Zone of occasional abundance – the environmental conditions are less favourable, being either cooler, drier, etc. or else having a period of unsuitable climate (e.g. winter or dry season). Some breeding does occur, but the population is kept low by the overall climatic conditions. Only occasionally does the population rise to pest proportions. Sometimes climatic conditions are severe enough to destroy the insect population, which then has to be re-established by dispersal from the endemic zone.
3. Zone of possible abundance – this is the zone into which adult insects spread from the other two zones. They may survive here for a while, occasionally even breeding, but are rarely of pest status. Changing climate usually kills off the population after a short time.

These three zones are not necessarily fixed in their demarcation, for the controlling factor may be either food or climate. If a crop becomes more widely cultivated, then the pest will follow it within the limits of climate. A point of recent interest is that the climate of the UK seems to be becoming warmer, and during the last few years a number of continental insect and other arthropod species (spiders) have become established in southern regions.

The dispersal success of an insect depends on the effectiveness of the method of dispersal and its basic adaptability, including feeding requirements and survival mechanisms for periods of adverse conditions.

COLORADO BEETLE

D. FITCHEW

(Natural size)

THE COLORADO BEETLE* is a dangerous foreign potato pest now rampant in many European countries. It is very likely to invade Great Britain, and potato growers should keep a sharp watch for it. The illustration above shows what the pest looks like—beetles on the left, grubs on the right. Instructions—what to do and what to avoid—are given overleaf.

* *Leptinotarsa decemlineata* Say.

MINISTRY OF AGRICULTURE AND FISHERIES, ST. ANNES, LYTHAM ST. ANNES. LANCS.

PLEASE READ THIS CAREFULLY AND KEEP IT FOR REFERENCE

● **Examine** your crops at least once a fortnight, from the time the shoots appear above ground until the haulm has died off. Look for striped beetles or red grubs eating the potato leaves (*see illustration*).

● **In England and Wales.** Send any beetles or grubs that look like those in the illustration to the Ministry of Agriculture, 28, Milton Road, Harpenden, Herts., with a letter giving your address and stating where the insects were found.

● **In Scotland.** Send your specimens and letter to the Department of Agriculture for Scotland, St. Andrew's House, Edinburgh, 1.

● **Pack specimens in a tin** or other strong box that will not get crushed in the post. Put a potato leaf in the box.

● **Do not spray** or interfere in any way with the crop on which you find specimens until you get instructions from the Ministry. If you do, you may cause the pests to spread and make it more difficult to eradicate them.

● **If you are in any doubt** what to do, report the facts to the nearest Police Station.

● **Caution.** The Colorado Beetle Order of 1933 requires occupiers of land to report to the Ministry any suspected outbreak of the Colorado Beetle. It also prohibits the keeping of any live Colorado Beetles and the spraying or other treatment of any of the crops concerned, except under the authority of the Ministry.

● **Please show this leaflet** to farm workers, allotment-holders, school children, or any other persons who might assist in the discovery of Colorado Beetles.

THOSE WHO HELP ARE DOING WORK OF REAL NATIONAL IMPORTANCE

T51-9103

Fig. 3.2 MAFF advisory pamphlet for Colorado Beetle.

Migration is strictly a double journey – emigration out of an area and then the return (immigration) into the same area later. Birds, mammals and fish regularly practise very spectacular long-distance migrations, but the only insects recorded to migrate are the Milkweed Butterflies of North America which overwinter in New Mexico. Some long-distance dispersals by insects, such as locusts and Lepidoptera (Noctuidae), are different in that several generations of insect may be involved. However, the 1988 invasion of the West Indies by Desert Locusts from North Africa involved the high altitude dispersal of flying locusts sucked up by wind storms and carried across the Atlantic Ocean in only a few days. Usually, such a movement would take perhaps a year and three generations. The annual invasions of Western Europe by Lepidoptera from North Africa and of Canada from the southern USA, as well as Leafhoppers (Hem.; Cicadellidae) into

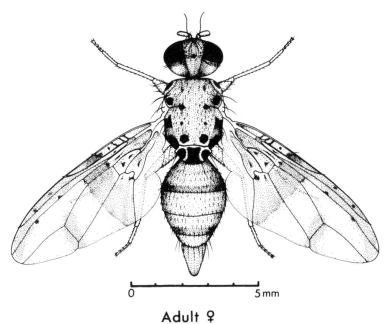

Adult ♀

Fig. 3.3 Adult female Fruit Fly (*Ceratitis* sp. – Dipt.; Tephritidae).

Japan from South East Asia, typically involve several generations of insects which disperse relatively slowly, stopping to feed and breed along the way: the journey takes several weeks or a month or two. The sporadic spectacular pan-Africa dispersal of the Desert Locust can take up to ten years to complete.

Invasion by pest species occurs with some regularity. In the UK there is virtually an annual accidental introduction of Colorado Beetle (Fig. 3.2), and in the USA several Fruit Flies (Dipt.; Tephritidae) annually threaten the fruit industry in California and Florida (Fig. 3.3). A spectacular invasion of the UK by the American Lupin Aphid (*Macrosiphum albifrons*) took place in 1981. It was first recorded in London and, despite the statutory eradication measures that were taken, became established and spread eventually as far north as Yorkshire. It is confined to lupins, so far as is known; the flower spike

(Fig. 3.4) is killed and the entire plant may die.

(d) Spread of insects by humans

Most major crops were originally endemic to one particular region, and during the centuries of world exploration and trade many were carried to other regions where climatic conditions were suitable for cultivation. Often some of the endemic insect pests were also transported along with the crop plants, and many became established in the new locations. They often became serious pests, for they were without the restraint of their endemic natural enemies (Elton, 1958). These introductions were usually by accident and carelessness, and the small insects or eggs were overlooked on the plants, produce or in packing material. In the past, sea freighting was a slow business, but the present practice of air freight means that

Fig. 3.4 Lupin Aphid (*Macrosiphum albifrons* – Hem.; Aphididae) on a lupin flower spike (which later died); Skegness, UK.

insect stowaways are much more likely to survive the journey.

Comparatively recent accidental introductions that became agriculturally disastrous include in 1975 the Greater Grain-borer (Col; *Prostephanus truncatus*) of maize, from Central America into Tanzania (Fig. 3.5). In 1973 the Cassavae Mealybug (Hem.; *Phenacoccus manihoti*) travelled from South America to West Africa; and the Cassava Green Mite (Acarina; *Mononychus tanajoa*) in 1971 also crossed from South America into East Africa. These three pests threatened agricultural production of maize and cassava over a very wide area and caused damage running into

millions of pounds. Their importance led to concerted international control co-operation, which to date has been highly successful. A similar case of carelessness involved the importation into Morocco of goats from South America which were infested with Screw-worm (Dipt.; *Cochliomyia hominivorax*) in 1989. A population of adult flies became established in North Africa and there were fears that this major veterinary pest would spread throughout the continent, but again international co-operation succeeded in eradicating the pest in North Africa.

Deliberate introduction by humans has long been practised – many plants and animals in the UK owe their introduction to either the Romans or later to the Norman conquerors (rabbit, little owl, sycamore, etc.). More recent deliberate introductions include the Gypsy Moth (Lep.; *Lymantria dispar*) from Asia into the USA in 1870 as a possible silk moth, and the importation of African Honey Bee into Brazil in the 1950s. In general, though, most insect introductions have been accidental.

3.2.2 Economic changes

Since economics represents a keystone of pest control, it is not surprising that a change in basic economics can alter the status of a pest completely. This is seen most clearly with agricultural crops and their insect pests, but it does also apply to veterinary and medical pests.

(a) Change in demand

The value of the crop and its quantity are the points of concern here. Damage to the produce (fruit, etc.) that is not important when prices are low can be very serious when prices are high. Conversely, if a particular food crop is in short supply, then some pest damage will be tolerated, whereas if the crop product is very abundant then damaged items will be rejected.

(a)

(b)

Fig. 3.5 (a) Greater Grain-borer (*Prostephanus truncatus* – Col.; Bostrychidae) (ex ICI); and (b) damaged maize grains.

The greater the demand for a crop, the greater its value becomes and also, therefore, the incentive to grow it. With some crops the demand may be simply for increased quantity, but nowadays very often the demand is for increased quality. The recent increase in the sale of vegetables and fruit as prepacked, washed produce in supermarkets has led to general-public rejection of blemished produce. For example, slugs and wireworms do not affect the yield of potatoes, but the holes and tunnels made spoil the appearance and storage qualities of the tubers. On washed potatoes such damage is very evident, but on tubers with soil stuck

to the skin the damage is usually not noticed at the time of purchase.

When new crops come on to the market, their pests suddenly become important. Sometimes a new crop, or new variety, will replace a previous one and there may be a sudden change in the pest spectrum. Some new varieties can be grown over a longer period of time and may then be exposed to a different range of insect pests. Carrots can now be grown continuously in East Anglia with a whole series of different varieties at different times. Similarly, the different species and varieties of *Brassica* that are loosely referred to as 'Chinese cabbages' can now be grown more or less continuously in the area from south China to Malaysia; this gives the opportunity for various pests to keep on breeding and is why the Diamond-back Moth (Lep.; *Plutella xylostella*) is so important as a *Brassica* pest in the region.

(c) Change in production costs

Crops for export, and cash crops in general, are far more valuable than local food crops, and consequently their pests are more important. Many vegetables and fruits grown in the tropics and subtropics are now being sold fresh in Europe after being air-freighted – these products are expensive; a high level of produce hygiene is demanded and the major pests are of high economic importance.

This point is of particular humanitarian concern. In most of the tropical regions of the world there is official concern, and international aid, for the study and suppression of pests of the main export/cash crops such as coffee, tea, cotton and rubber, whereas the local food crops are neglected in this respect.

Some of the new high-yield crop varieties are very successful; the rice varieties produced by the International Rice Research Institute in the Philippines were responsible for some of the successes of the 'Green Revolution' in tropical Asia in the 1970s. However, some of these varieties are agriculturally very demanding and require extensive use of fertilizer. The extra fertilizer produces a lush plant growth and also encourages weeds. The lush foliage in-duced Brown Planthopper (Hem.; *Nilaparvata lugens*) to greater fecundity, and this formerly obscure minor rice pest rapidly became the number one insect pest of rice in tropical Asia. Another reason for its spectacular rise in importance is that it is capable of very rapidly producing new biotypes (biological varieties) which are resistant to regularly used insecticides.

4

DAMAGE CAUSED BY INSECTA AND ACARINA

Ideally with good insect pest management, actual damage to the host should be avoided. In practice, however, it is more usual to find damage at some level as part of the pest assessment process. There are many different ways in which the host organism is damaged by the pest, and sometimes it is convenient to group them under two headings: indirect and direct damage.

4.1 INDIRECT DAMAGE

This is when the insect pest does not cause discernible harm to the body of the host organism; sometimes the pest does not actually touch the host at all. In the agricultural context, considerable damage can be done to cultivated plants by weeds competing for water, minerals, light, etc., and in point of fact this competition is one of the major constraints to crop production worldwide.

4.1.1 Disease vectors

In the zoological world, by far the most important form of indirect damage is when the insect acts as a vector for a parasite or pathogen. In many cases there is a very intimate relationship between the parasite and the vector, and often the vector is the only means of transmission. The simplest form of dispersal is known as **mechanical transmission**; typically the insect picks up the

parasite on its body surface while feeding on the host organism, and it may either deposit the parasite on to a new host body or else may contaminate the food which will later be eaten by the host. But many insect pests are fluid feeders (sap-suckers on plants; blood-suckers on vertebrates), and these can mechanically transmit pathogens and parasites by contamination of the proboscis (like a contaminated hypodermic syringe).

In agriculture almost all virus diseases are spread by feeding insects – aphids (Aphididae), leafhoppers (Cicadellidae), planthoppers (Delphacidae) and some other bugs are the main groups of vector. For fungi and bacteria there are some insect vectors (a few beetles, etc.) of plant diseases, but transmission is usually not by insects. However, in a forestry context, many pathogenic fungi are transmitted to trees by beetles. Veterinarily, biting flies transmit parasitic Protozoa and some Nematoda (filaria), but the greatest damage to livestock comes from ectoparasitic ticks (Acarina) that transmit a wide range of parasitic protozoans and related organisms (Rickettsia, spirochaetes, piroplasms, etc.). In relation to humans, insect vectors include fleas, lice, some ticks and a range of biting flies (mosquitoes, etc.) transmitting Protozoa and viruses, as well as synanthropic flies (*Musca*, etc.) which spread bacteria, etc. by food contamination. A few biting flies

Table 4.1 Some major diseases transmitted by Insecta and Acarina

Vector	Disease	Parasite	Transmission	Distribution
(a) Plants				
Dysdercus spp. (Hemiptera)	Cotton staining	*Nematospora* spp. (fungi)	Feeding puncture; biological	Pantropical
Scolytus spp. (Coleoptera)	Dutch elm disease	*Ceratostomella ulmi* (fungi)	Mechanical; on insect body	Europe; North America
Mysus persicae (Hemiptera)	Sugar beet yellows	Virus	Feeding	Europe; USA
(b) Livestock				
Glossina spp. (Diptera)	Nagana	*Trypanosoma* spp. (protozoa)	Bite	Africa
Musca domestica (Diptera)	Anthrax	*Bacillus anthracis* (bacteria)	Contamination	Worldwide
Argas persicus (Acarina)	Poultry Piroplasma	*Aegyptianella pullorum*	Bite	USA
Rhipicephalus spp. (Acarina)	East coast fever (of cattle)	*Theileria parva*	Bite	Africa
(c) Man				
Xenopsylla cheopis (Siphonaptera)	Plague	*Pasturella pestis*	Bite	Pantropical
Aëdes aegypti (Diptera)	Yellow fever	Virus	Bite	Africa; South America
Anopheles spp. (Diptera)	Malaria	*Plasmodium* spp. (protozoa)	Bite	Pantropical
Glossina spp. (Diptera)	Sleeping sickness	*Trypanosoma* spp. (protozoa)	Bite	Africa
Musca domestica (Diptera)	Cholera	*Vibrio cholorae*	Food contamination	Pantropical

(Tabanidae) in the tropics transmit endoparasitic nematodes and cause onchocerciasis, etc.

On the whole, however, **biological transmission** is even more important. This is when the parasite/pathogen undergoes a vital stage of development inside the insect vector. Often in such cases the insect is more than just a vector and is in fact an intermediate host playing a major role in the parasite's life cycle. A very wide range of parasitic organisms is transmitted this way, and they can be further subdivided according to the precise details of their development in the insect vector body.

Table 4.1 lists a few major disease-causing organisms transmitted by insects and Acarina to some agricultural crops, some livestock and to humans. More examples are given in Chapter 5.

In agriculture most insect pests cause direct damage to plants, but with livestock there is an almost equal distribution of direct and indirect damage. However, with humans major insect pests are mostly acting as parasite vectors and there is little direct damage.

4.1.2 Disturbance

This is not very important but can occasionally be quite irritating. Cattle are often

Fig. 4.1 House Flies on a table in an open-air café in south China.

very sensitive to the sound of adult Warble Flies (*Hypoderma* spp.) and in the UK there are regular reports in the press of herds of cattle stampeding after being frightened by the buzz of a flying Warble Fly. In the stress of this situation the animals can be very upset: gravid cows often abort and dairy cows can have a greatly reduced milk flow. Buzzing mosquitoes can successfully keep people awake for half the night and cause them considerable distress. In parts of Northern Europe, Asia and Canada biting flies harass both livestock and human hosts, and holidays/expeditions to these regions often have to be cut short. Mosquitoes, Biting Midges and Black Flies are the usual culprits and they breed in the Arctic tundra in vast numbers. A recent television programme showed a biologist by a lake in Iceland; his silhouette was completely blurred by the swarm of *Simulium* (Black Flies) around his body. At a less dramatic level, in Africa the Bush Fly (*Musca sorbens*) causes regular irritation by its persistent harassment of both humans and livestock. The distraction caused by its attention results in a greatly reduced level of work efficiency. The presence of skin ectoparasites such as fleas and lice can also be very irritating to both livestock and humans – their movement on the skin surface alone is sufficient to disturb. The presence of insects such as wasps and ants is disturbing to most people, and many an open-air restaurant has had to close temporarily due to the presence of a local wasps' nest. Similarly the sight of a large number of flies (Fig. 4.1) can be equally repelling. A few insect species produce an odour that is offensive, for example several species of cockroach can produce a strong smell in enclosed premises if they are abundant.

4.1.3 Contamination

Foodstuffs of all types can be rendered unsaleable or unusable by the mere presence of an insect pest; there does not need to be any actual damage evident. With ever-increasing standards of public hygiene, all establishments involved in food preparation, storage and sale are very susceptible with regard to insect contamination and could both lose their commercial licence and be legally penalized. Insect excreta contamination of vegetables, such as caterpillar droppings in lettuce and cabbages, can make them unsaleable.

A recent interesting case of contamination was that of opened cotton bolls in Africa and honey-dew from Cotton Whitefly (*Bemisia tabaci*) – the sugar solution made the cotton lint sticky and the fibres could not be adequately separated (ginned) mechanically.

Fig. 4.2 Apple damaged by Codling Moth caterpillar (*Cydia molesta* – Lep.; Tortricidae); Skegness, UK.

4.2 DIRECT DAMAGE

Direct damage occurs when the insect pest causes visible harm to the host organism. There are different degrees of damage in this category. The most serious type of direct damage is when the part to be harvested is damaged by the insect pest. For example, a single Codling Moth caterpillar can ruin a large apple (Fig. 4.2), but a tortricid caterpillar destroying an apple leaf (Fig. 4.3) has no effect at all upon the total yield from that tree, and neither would 50 sap-sucking aphids. Most root crops (potato, sugar beet, etc.) can tolerate partial defoliation without noticeable loss of yield; there have been many experiments to demonstrate this. Many crops can tolerate up to 30% defoliation without discernible yield loss. So often damage to the other parts of the plant body is not nearly as important as that to the part to be harvested. Figure 4.4 shows six different *Brassica* plants in relation to possible damage. If the seedling is nibbled the whole plant is destroyed. Broccoli is grown for the flower heads and Brussels sprouts for the lateral buds (i.e. sprouts); they can tolerate leaf and root damage. Cabbage has to have a large tight 'heart' so leaf damage can be important. Turnip is grown for the swollen root, so only root damage is serious. Rape is grown only for its seeds as a source of oil, so leaf and root damage can usually be ignored. Cabbage Root Fly can destroy a young plant, but on larger plants the damage may be insignificant, whereas a single maggot inside a 'sprout' can be very damaging (especially if in a frozen pack in a supermarket). Thus damage which can be ignored on one plant may be serious on another crop, and sometimes the method of sale is the critical factor. Cabbage Root Fly damage to turnip is usually a surface tunnel on the root, which would be removed by peeling and would be hidden by a little soil. However, on a washed and prepacked root it would be obvious and could prevent a sale. An interesting case in Hong Kong concerned a

Fig. 4.3 Apple leaf eaten by tortricid caterpillar; Skegness, UK.

consignment of frozen broccoli heads bought for airline meals but rendered useless by the presence of Diamond-back Moth pupae inside small, flimsy, silken cocoons inside the flower heads. The small caterpillars would have fed on the leaves but had crawled up the stems to pupate hidden on the flower head stalks.

With veterinary and medical pests generally, the direct damage tends to be blood loss or the removal of tissues by feeding. Clearly a small loss of blood uncomplicated by the absence of disease/parasite transmission, is of little consequence to a healthy animal, but a maggot causing myiasis can be quite damaging.

4.3 HOST VULNERABILITY

Most plants produce far more leaf cover than they actually need, presumably to survive grazing. When grazed, the plant responds by **compensatory growth**, such as tillering in Gramineae and extra shoots in dicots. The **leaf area index** of many crop plants can be as high as 4 or 5, so that much of the lower foliage will be shaded by upper leaves, and with closely spaced crops the lower parts of the plant may be shaded by adjacent plants. It should be remembered that in all ecosystems the members of the plant kingdom are the basic primary producers and so grazing by animals is to be expected. It is to be equally expected that plants will evolve methods of counteraction. Some plants also produce more flowers and fruit than they can actually sustain, so there is a regular natural flower-fall followed by fruit-fall; this is shown especially in apple trees.

Thus many crop plants are not as vulnerable to insect pest attack as might appear at first sight. However, sometimes a host plant is far more vulnerable to a pest than might be thought. This is particularly the case with aphids. Often a heavy infestation can be very unsightly and conspicuous (especially if associated with sooty mould)

Examples of five closely related crops (*Brassica* spp.) which have different parts of the plant body harvested, and also a generalized seedling.
1. *Brassica* seedling
2. Broccoli (*B. oleracea* var. *botrytis*), grown for flower heads
3. Brussels sprouts (*B. oleracea* var. *gemmifera*), grown for lateral buds
4. Cabbage (*B. oleracea* var. *capitata*), grown for 'heart', i.e. telescoped main shoot
5. Turnip (*B. rapa*), grown for swollen root
6. Rape (*B. napus*), grown for seeds as a source of oil

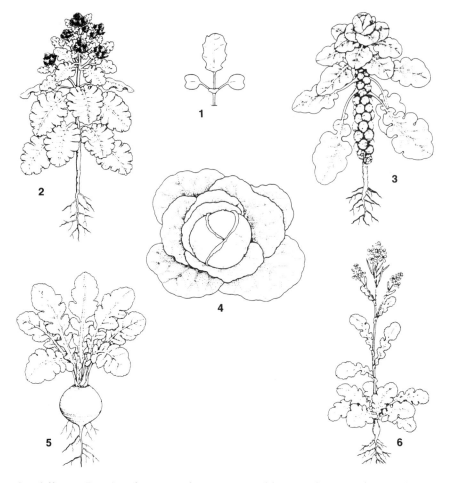

Fig. 4.4 Six different *Brassica* plants in relation to possible pest damage (from Hill, 1994).

but actual damage may be slight. At other times, however, a light infestation can cause serious damage. Clearly small seedlings can be debilitated by aphid attack, but older plants can also be vulnerable.

For agronomic purposes there have been well-defined growth stages for most major crops for some time, and in recent years crop protectionists have used these stages to define periods of crop plant vulnerability to major insect pests. Figure 4.5 shows the accepted growth stages for wheat (and other temperate cereals) (FAO/CABI, 1971). Research has shown that wheat seedlings can

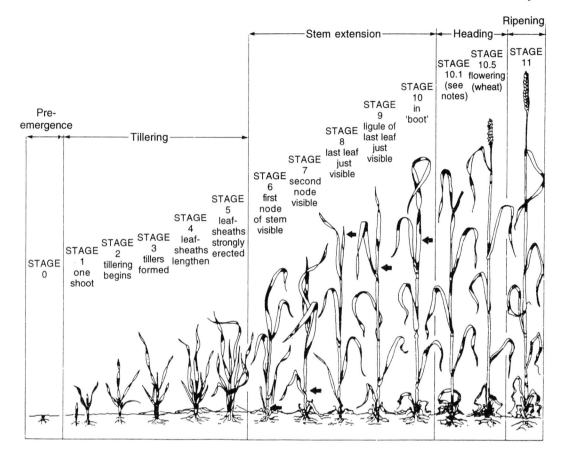

Stage 0 Pre-emergence
1 One sprout (number of leaves can be added) = 'grairding'
2 Beginning of tillering
3 Tillers formed, leaves often twisted spirally; in some varieties of winter wheats, plants may be 'creeping' or prostrate
4 Beginning of the erection of the pseudostem, leaf-sheafs beginning to lengthen
5 Pseudostem (formed by sheaths of leaves) strongly erected

6 First node of stem visible at base of shoot
7 Second node of stem formed, next-to-last leaf visible
8 Last leaf visible, but still rolled up, ear beginning to swell
9 Ligule of last leaf just visible
10 Sheath of last leaf completely grown out, ear swollen but not yet visible

Tillering

Stem extension

10.1 First ears just visible (awns just showing in barley, ear escaping through split of sheath in wheat or oats)
10.2 One-quarter of heading process completed
10.3 One-half of heading process completed
10.4 Three-quarters of heading process completed
10.5 All ears out of sheaths
10.5.1 Beginning of flowering (wheat)
10.5.2 Flowering complete to top of ear
10.5.3 Flowering over at base of ear
10.5.4 Flowering over, kernel watery ripe

11.1 Milky ripe
11.2 Mealy ripe, contents of kernel soft but dry
11.3 Kernel hard (difficult to divide by thumbnail)
11.4 Ripe for cutting; straw dead

Heading

Flowering (wheat)

Ripening

Fig. 4.5 Growth stages in wheat, oats, barley and rye (FAO/CABI, 1971).

Fig. 4.6 Wheat flag-leaf infested by Wheat Aphids (*Schizaphis graminum* – Hem.; Aphididae); Skegness, UK.

be easily damaged at the young seedling stage (establishment) (stage 1), and again at the start of flowering on both the flag-leaf and the head (stage 10). The flag-leaf of cereals is responsible for photosynthesis, providing nutrients for grain formation. In recent years the common practice of autumn sowing of cereals has ensured that vulnerable seedlings avoid aphid infestation – by the time aphids are abundant in the spring the plants are long past the vulnerable stage in their development. Figure 4.6 shows a wheat flag-leaf infested by Wheat Aphids.

When studying crop losses and damage assessment (section 4.4) it is vitally important to be aware of any periods of crop plant vulnerability, so that these can be taken into account. Some crop plants only need protection during their most vulnerable periods, i.e. at the time of plant establishment and also probably at flowering. Precision drilling with pesticide-treated seed helps to protect some young seedlings. Transplanted seedlings (especially Brassicas) are now being planted with a soil ball (Fig. 4.7) rather than bare-rooted; this helps to establish the seedling, so reducing vulnerability.

In general all organisms are likely to be vulnerable to pest/parasite/pathogen attack at times when they are subjected to stress. Natural periods of stress are associated with establishment and reproduction, and in human cases with old age. Plants growing in unsuitable conditions and livestock imported into unfavourable climates (e.g. European cattle into tropical Africa) are very likely to be vulnerable to local pest organisms and will probably perish. Recent work in California has shown that Eucalyptus trees are severely damaged by weevils when they are water-stressed, but the surprising point was that the water-stress appeared minimal.

Fig. 4.7 Brassica seedling with soil ball, for transplantation; Friskney, UK.

The present movement towards the integrated pest management approach to control of pests requires that we know as much as possible about pests and the hosts they attack. Especially important is the identification of any developmental stage where the host is vulnerable.

Predisposition can be a very important stress-related factor in that there can be interaction between parasitic organisms attacking a host, so that their combined effect is more damaging than the sum of their separate effects. This is most noticeable in humans and livestock when sick individuals are prone to a whole range of pathogens and parasites, some of which may be lethal to them in their weakened state.

4.4 DAMAGE ASSESSMENT

It is vitally important that any insect pest infestation be assessed with some level of accuracy. There are innumerable records of pest control efforts that were totally wasted because they were not necessary or were instigated too late. At the same time, some serious infestations have been left unchecked because the damage potential was grossly underestimated. Each host organism needs to be studied in detail in relation to the major pests by which it is attacked. Moreover, this needs to be done in each of the major regions of the world where it is cultivated/reared, so that control measures are only employed when really necessary.

The ultimate aim of agriculture (*sensu lato*) is to produce a sustained yield of worthwhile economic proportions. With a crop in the field it is difficult to estimate the expected yield loss, but a damage assessment can be made and the size of the insect population causing the damage can be determined. The assessment sequence would be:

Insect pest population→
(can be determined)

Damage assessment→Yield reduction
(can be made)   (estimated?)

So many factors interact to determine the level of crop yield that to define the effect of just one factor (i.e. insect population size) on yield is difficult. However, there has been a gradual accumulation of empirical data for most crops over the years that now enables us to make some general correlation between pest population/damage level and expected yield loss. This work was mostly done to establish the economic injury levels for the major crops and pests, and are not actually damage assessments. They can, however, be used for this purpose.

4.4.1 Pest infestation

The infestation of a host by a pest is expressed in two different ways:

1. Incidence – this is the proportion of the host population with the insect pests or damage symptoms. It is usually expressed as a percentage.
2. Severity – the severity of an infestation is a measure of the size of the insect population on the host. This is usually expressed as the number of insects per plant or per ten leaves, etc.

Most of the early work in this field was done by plant pathologists, who suffer the disadvantage that the causal organisms (i.e. pathogens) can usually only be seen under a microscope. However, the host plant has symptoms that are often diagnostic and are always static. Entomologists can often locate the insect pest, but some are mobile and can disappear rapidly when disturbed, leaving only the damage to be recorded.

Thus ideally any infestation assessment should combine both pest incidence and infestation/damage severity. With sufficient collected empirical date, however, a correlation with overall insect numbers (i.e. size of the population) can be made. The pest population can be assessed by either some form of quantitative survey or use of a light trap (or other type of trap); an even more simple method for farmers to use these days is the pheromone trap. However, it should be remembered that usually the simpler the method of assessment the less accurate the results are likely to be.

Population assessment is essential, and it is important that a logical and straightforward system is used. During national sur-

veys of flora in the UK it was decided that any competent biologist could use a four scale assessment as follows, but for general student use it could be argued that the alternative (six scale) is preferable:

Abundant	(A)	Very common	(VC)
Frequent	(F)	Common	(C)
Occasional	(O)	Uncommon	(U)
Rare	(R)	Rare	(R)
(Absent)		Very rare	(VR)
		(Single record)	
		Absent	(A)

These categories can usually be used in most general surveys with a fair level of reliability. In some specific infestation investigations, however, it may be necessary to be more detailed.

4.4.2 Damage assessment

As already mentioned, some insect pest infestations can be quite misleading: some heavy infestations cause little damage, whereas a light infestation at a crucial time can be serious. With most pests, slight damage can be ignored, but if it appears that more serious damage is likely to follow, then after a damage assessment it is necessary to take control action. Several different damage assessment schemes are widely used by plant pathologists for disease symptoms. In terms of numbers of leaves infected and extent of individual leaf lamina damaged, assessment can be straightforward. Often a scale of 0–9 or 1–10 is used by plant pathologists.

With insects on cultivated plants, damage assessment is not quite so easy. With relatively simple damage, such as apples infested by Codling Moth larvae, damage can be counted and expressed as a percentage of apples per tree infested. With leaf-eating, probably the commonest form of plant damage, the lamina area removed can be measured and expressed as a percentage of total leaf area. With sap-sucking bugs such as aphids, damage is far less obvious and the main criterion usually has to be actual numbers infesting the plant or plant part. There is general agreement that it is preferable to take a large number of quick, easily categorizable samples than a small number of large, detailed samples. When dealing with animal pests, it is usual to employ four or six damage categories, as follows:

Very severe (VS) or (4) or $\begin{cases} 6 \text{ (Plant death)} \\ 5 \end{cases}$

Severe (S) (3) 4

Mild (M) (2) $\begin{cases} 3 \\ 2 \end{cases}$

Very mild (VM) (1)

1 (Noticeable but no damage)

0 (None)

There are some advantages in using a numerical scale, for then a field damage assessment can be made as an overall mean of a large number of individual plants. This can then be extended to a regional survey based upon a number of different fields.

It should be stressed that each different type of damage to a crop will require a separate assessment method, although certain types of damage will require similar methods of assessment.

With veterinary and medical pests it is not as easy to quantify damage; often there needs to be a greater reliance on pest numbers, such as the number of lice per animal or the number of fly visits per captive host per hour. Damaged stored produce, such as grain or pulses, is often crudely measured as an overall weight loss, but this is an over-simplification.

5

HARMFUL INSECTS

Harmful insects are the species that cause damage to humans and their livestock, crops and possessions worldwide. Some are direct pests in that they attack the body of the host organism (plant or animal) and either suck sap or blood or eat the tissues. Indirect pests are mostly concerned with the transmission of pathogens or parasites causing disease. In some cases it is the adult insect that causes the damage, in others it is the larval stage (caterpillar, maggot, etc.). Sometimes it is both adult and larva that are pests, and they may act in concert or separately.

5.1 MEDICAL PESTS

With medical pests the ultimate damage is a human death. The less severe result is a period of sickness which may range from a couple of weeks in hospital to a stay in bed at home with appropriate medication. Such medical statistics are readily available in most developed countries and the World Health Organization (WHO) makes estimates worldwide. What is seldom taken into account is the chronic debilitation resulting from non-lethal infestations and infections throughout the tropics. In many cases the diseased people continue to work, but somewhat sluggishly and inefficiently, and this chronic disability may last for years or even a lifetime. The overall effect on the local economy can be quite disastrous. For example, recent estimates (WHO, 1990) say that at present 267 million people in 103 countries are affected by malaria, with 1–2 million dying annually. According to the WHO, the five 'top' tropical diseases at present kill 2 million people a year (Patel, 1993) and it predicts that by the year 2010 this total will be 4 million people a year. This is despite all our recent advances in medicine and pest control. Of these five 'top' diseases four are insect-based: malaria, leishmaniasis, sleeping sickness and lymphatic filariasis (the fifth is schistosomiasis which is not insect-based). The WHO estimates that half a billion people suffer from these diseases today. The main groups of insects that are medical pests worldwide are listed in Table 5.1. Some important diseases of humans that are transmitted by Insecta and Acarina are listed in Table 5.2.

The different groups of insects that are medical pests are reviewed in more detail below.

5.1.1 Order Siphunculata (Sucking Lice)

These are permanent obligatory ectoparasites, found only on the Mammalia. They are small and wingless, and mouthparts are adapted for piercing skin and sucking blood from the host. Legs have terminal claws adapted for clinging to hairs. Their feeding can cause intense irritation. Eggs are laid attached to hairs and are colloquially known as 'nits'.

Table 5.1   Major insect medical pests

Group	Common name
Order Siphunculata (Sucking Lice)	
Family Pediculidae	Human Lice
Order Hemiptera, Suborder Heteroptera	Bugs
Family Reduviidae	Assassin Bugs
Cimicidae	Bed Bugs
Order Siphonaptera (Fleas)	
Family Pulicidae	Human Flea
Ceratophyllidae	Cat and Dog Fleas
Leptopsyllidae	Rodent Fleas
Tungidae	Jigger/Chigoe Fleas
Order Diptera (Flies)	
Family Psychodidae	Sand Flies
Culicidae	Mosquitoes
Simuliidae	Black Flies
Ceratopogonidae	Biting Midges
Tabanidae	Horse Flies
Glossinidae	Tsetse Flies
Muscidae	House Flies, etc.
Cuteribidae	Bot Flies
Calliphoridae	Blow Flies, etc.
Sarcophagidae	Flesh Flies
Order Hymenoptera (Wasps, Bees, Ants, etc.)	
Family Vespidae	Wasps
Formicidae	Ants
Apidae	Bees
Order Acarina (Ticks and Mites)	
Family Ixodidae	Hard Ticks
Trombiculidae	Chiggers
Sarcoptidae	Itch Mites
Acaridae	Domestic Mites, etc.

Pediculus humanus, the Human Louse (Fig. 5.1), occurs as two distinct varieties: var. *corporis* is the Body Louse (which occurs on the body between skin and clothing) and var. *capitis* is the Head Louse (which lives only in the fine hair of the head). The two species are cosmopolitan in distribution, and quite abundant (especially in warmer countries).

Body Lice transmit various diseases under suitably crowded and unhygienic conditions. Trench fever and Epidemic fever have killed many thousands of people in several major epidemics. Head Louse has become quite common again in recent years in many European countries after many years in abeyance, especially among long-haired schoolchildren.

The Human Crab Louse, *Pthirus pubis* (Fig. 5.2), has stouter legs and larger claws – it is found only on the coarser body hair in the

Table 5.2 Some human diseases transmitted by Insecta and Acarina

Vector	Disease	Parasite	Transmission	Distribution
Triatoma spp.	Chagas' disease	*Trypanosoma cruzi*	Bite contamination by faeces	S. America
Pediculus humanus	relapsing fever	*Borrelia recurrentis*	Skin contamination	Europe, Asia, Africa
	epidemic fever	*Rickettsia prowazeki*	Bite contamination	Eurasia
	trench fever	*Rickettsia quintana*	Bite contamination	Eurasia, Mexico
Xenopsylla cheopis	plague	*Pasturella pestis*	Bite	Pantropical
	murine typhus	*Rickettsia typhi*	Bite contamination	Pantropical
Anopheles spp.	malaria	*Plasmodium* spp.	Bite	Pantropical
Aëdes aegypti	yellow fever	Virus	Bite	Africa, S. America
	dengue	Virus	Bite	Tropics
Mosquitoes	encephalitis	Viruses	Bite	Africa, America
Mosquitoes (*Anopheles* and *Culex*)	filariasis (elephantiasis)	*Wuchereria bancrofti*	Bite and through skin	Pantropical
Phlebotomus spp.	leishmaniasis (Kala-azar, etc.)	*Leishmania* spp.	Bite, etc.	Africa, Asia, S. America
Glossina spp.	sleeping sickness	*Trypanosoma* spp.	Bite	Africa
Simulium spp.	onchocerciasis (river blindness)	*Onchocerca volvulus*	Invasion of bite	C. and S. America, Africa
Chrysops spp.	loasis	*Loa loa*	Bite	Africa
Chrysops, etc.	tularaemia	*Pasturella tularensis*	Bite	Holarctic
Cockroaches, *Musca domestica*	dysentery, amoebic	*Entamoeba histolytica*	Food contamination	Pantropical
Musca domestica and *M. sorbens*	dysentery, bacillary	*Bacillus* spp.	Food contamination	Pantropical
also *Ophyra*, *Atherigona*, etc.	conjunctivitis	Bacteria	Eye visits	Pantropical
	typhoid fever	*Salmonella typhi*	Food and water contamination	Pantropical
	cholera	*Vibrio cholerae*	Food contamination	Pantropical
	yaws	*Treponema pertenue*	Wound contamination	Pantropical
	trachoma	Virus	Contamination	Pantropical
Trombicula spp.	scrub typhus	*Rickettsia tsutsugamushi*	Bite	South East Asia
Ornithodorus spp.	African relapsing fever	*Borrelia* spp.	Bite (or contamination)	Africa
Amblyomma and *Dermacentor* spp.	Rocky mountain spotted fever	*Rickettsia*	Bite	USA

Taken from Hill, 1994.

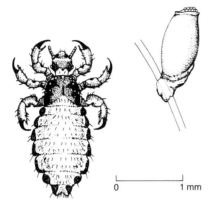

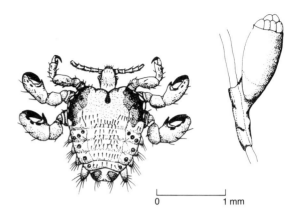

Fig. 5.1 Human Louse (*Pediculus humanus* – Siph.; Pediculidae) and egg attached to hair ('nit'); Uganda.

Fig. 5.2 Human Crab Louse (*Pthirus pubis* – Siph.; Pediculidae) and egg attached to pubic hair; Uganda.

pubic region, armpits and occasionally even on eyebrows and eyelashes! The bites are very irritating and infestations may be difficult to eradicate, but no diseases are transmitted.

5.1.2 Order Hemiptera, Suborder Heteroptera (Bugs)

These bugs are often quite large in size, winged and with a sucking proboscis; they may be brightly coloured. They all have toxins and enzymes in their saliva, which is injected into the host organism during feeding, so their 'bite' is usually quite painful. Many of these bugs are predacious upon other insects, snails, etc. and will often 'bite' if handled – some of the plant-feeders will also insert their proboscis if handled and that can be painful. However, only two families are of regular importance as medical pests:

(a) Family Reduviidae (Assassin Bugs)

This is a large group, but only a few are domestic pests, mostly in warmer countries. They feed mainly on rat blood but they do take blood from sleeping people, especially those sleeping on the floor. *Triatoma* (Fig. 5.3) is a common New World genus

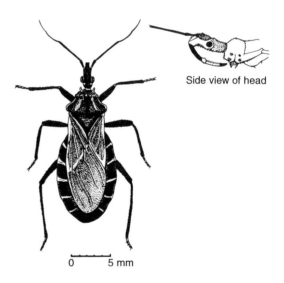

Side view of head

Fig. 5.3 Domestic Assassin Bug (*Triatoma rubrofasciata* – Hem.; Reduviidae); Hong Kong.

found also in Asia, and *Rhodnius* is pantropical – in South America they both transmit *Trypanosoma couzi*, a blood parasite causing Human Trypanosomiasis (Chaga's Disease). The present estimate is that 20 million people in South and Central America are infected, and many of these will die.

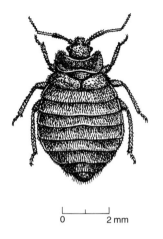

Fig. 5.4 Common Bed Bug (*Cimex lectularis* – Hem.; Cimicidae); Hong Kong.

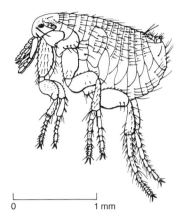

Fig. 5.5 Human Flea (*Pulex irritans* – Siph.; Pulicidae), adult female; Hong Kong.

(b) Family Cimicidae (Bed Bugs)

Bed Bugs are small, wingless, human ectoparasites that suck blood. There are two species that attack humans: *Cimex lectularis* is the Common Bed Bug (Fig. 5.4) and has a more rounded body; it is common throughout Europe, Asia and North America. The Tropical Bed Bug (*C. hemipterus*) is found mostly in Africa and Southern Asia. The bugs are nocturnal and hide in crevices in the bed during the day – at night they emerge to feed on sleeping people. Eggs are laid in crevices in beds as well as the wooden frames. The bites are irritating and unsightly, but no diseases are transmitted.

5.1.3 Order Siphonaptera (Fleas)

These are small, apterous, ectoparasitic insects of characteristic appearance – the body is laterally compressed for squeezing between hairs and the legs are large for jumping on to the host. Mouthparts are adapted for piercing skin and sucking blood from the host animal. The hosts are all mammals or birds, and host-specificity is the rule. Larvae are free-living scavengers that live in the nest or lair of

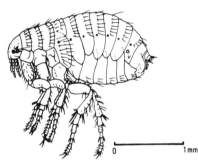

Fig. 5.6 Tropical Rat Flea (*Xenopsylla cheopis* – Siph.; Leptophyllidae), male; Hong Kong.

the host, or in corners and crevices of houses, and pupation takes place here. Both eggs and pupae can remain dormant under adverse conditions. Emergence of the young adult is triggered by vibration from the host.

The Human Flea (*Pulex irritans* – Fig. 5.5) is now quite rare in most parts of the world, and humans are often attacked by animal fleas. The most important flea pest is the Tropical Rat Flea (*Xenopsylla cheopis* – Fig. 5.6) which is the main vector of the plague pathogen (*Pasturella pestis*; now called *Yersinia pestis*). Plague is still endemic in many warmer parts of the world and it remains a major medical hazard.

The epidemic that swept through Europe in the Middle Ages was known as the Black Death – this was the bubonic form transmitted by the Tropical Rat Flea. It finally killed about two-thirds of the population of Europe. As recently as 1994 there was an outbreak of plague in western India, where the urban rat population is enormous, and several hundred people died before vaccines and antibiotics brought the outbreak under control. Unfortunately, this was the pneumonic form which attacks the lungs and is spread aerially by droplet infection.

An unusual human parasite is the Chigoe Flea (*Tunga penetrans*); the young female flea burrows into the skin at the side of a toenail, and once inside she swells to about 3–4 mm by enlargement of the abdomen. Infestation of the foot is painful and irritating.

5.1.4 Order Diptera (Flies)

These are medical pests in three different ways. Most are blood-sucking adults (and some transmit parasitic organisms); other adult flies such as the House Fly contaminate wounds and foodstuffs with pathogenic bacteria, fungi, etc., and the larvae (maggots) of some species live in human flesh causing myiasis. Adults are winged and fly well (some diurnally; some nocturnally). The blood-suckers have enzymes in their saliva that cause reaction by the human body, causing irritation (itching) and inflammation that can be quite serious. The more important families that are medical pests are listed below:

(a) Family Psychodidae (Sandflies and Mothflies)

Members of the subfamily Phlebotominae are Sandflies. Females of the genus *Phlebotomus* are blood-suckers that feed on humans (and other vertebrates), and they transmit *Leishmania* parasites, causing kal-azar and other forms of leishmaniasis. Throughout the tropics different species and subspecies of the parasite are involved and they are transmitted by different species of Sandfly in the different regions.

There is a group of viruses known as the Phlebotomus Fever Group – more than 27 different viruses are concerned. Sandfly fever is the most important – this occurs in the Mediterranean region through to Pakistan and is spread by *P. papatasi*. In humans it is a short, sharp, non-fatal fever.

(b) Family Culicidae (Mosquitoes)

Feeding adults take blood but larvae are aquatic in freshwater bodies. Only females take blood, some at night and some during the daytime, while others are crepuscular. The 'bites' are irritating, and in some locations a person may be attacked by dozens of mosquitoes at a time; in some Arctic areas one can be attacked by thousands at a time. Major diseases transmitted by mosquitoes include malaria, yellow fever, dengue, encephalitis and filariasis. They are worldwide in distribution and breed in the Arctic in vast numbers during the short summer season; they are abundant in all parts of the world, dependent mainly on available water for breeding purposes. A useful publication on medical entomology is by Kettle (1990), and there are now two editions of the Natural History Museum (London) book on this topic, Smith (1973) and Lane and Crosskey (1993). A typical mosquito, together with larva and pupa, is shown in Fig. 5.7.

The main pest genera are as follows:

- *Aëdes* – there are many species (900+) worldwide: many are banded black and white (Fig. 5.8); most are urban- or forest-dwellers for they breed in tiny water bodies in containers such as tin cans, old tyres, broken bottles, leaf axils and tree holes. Most countries have a species of *Aëdes* as a common urban pest. *A. aegypti* is the vector of yellow fever and dengue in the tropics. The WHO reported that in 1994 cases of yellow fever reached their highest levels

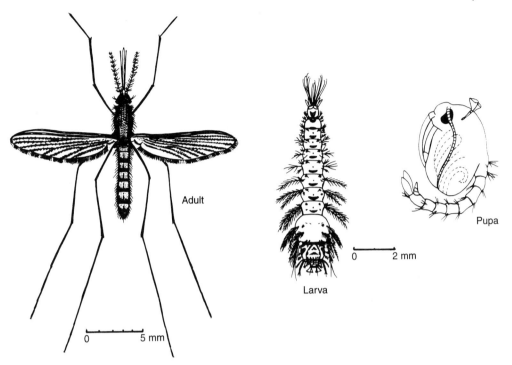

Fig. 5.7 Malarial Mosquito, adult, larva and pupa (*Anopheles* sp. – Dipt.; Culicidae); Hong Kong.

since 1948, both in West Africa and South America. The death rate varies, but it has been as high as 80%. The WHO records show that 17 728 cases of yellow fever were reported between 1986 and 1990, with 4710 deaths.

- *Anopheles* – these are the malarial mosquitoes, worldwide in distribution but most abundant in the warmer parts of the world. Kettle (1990) records 377 species worldwide. As mentioned previously, the WHO estimates that at present 267 million people in 103 countries are suffering from malaria, with an annual death rate of 1–2 million. The disease is debilitating if not always fatal, and some strains reoccur regularly, even if there is no re-infection. A generation ago it was thought that malaria was no longer a disease of consequence since DDT controlled both adults in

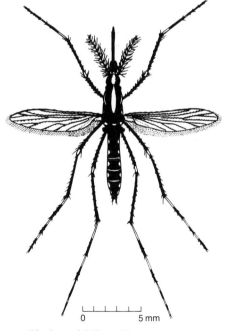

Fig. 5.8 Black and White House Mosquito (*Aëdes albopictus* – Dipt.; Culicidae); Hong Kong.

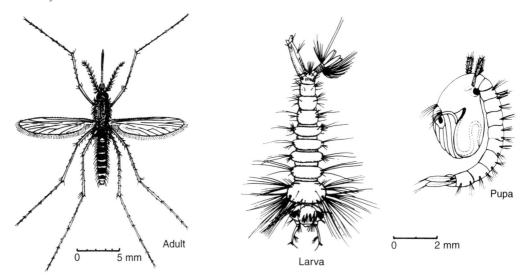

Fig. 5.9 Brown House Mosquito (*Culex quinquefasciatus* – Dipt.; Culicidae); Hong Kong.

buildings and larvae in water bodies, and chemotherapy was very effective against the parasite. Now, however, the situation is reversed in that mosquitoes have developed resistance to the insecticide (which also had to be withdrawn from sale because of its damaging effects on the environment), and the *Plasmodium* parasite developed resistance to the usual drugs. The implications of all this are that malaria is once again a major killing disease in the tropics with millions of sufferers, many of whom have chronic conditions.

- *Culex* – a very large genus, worldwide, with many different subgenera and sub-species; 744 species are recorded. *Culex quinquefasciatus* (=*C. pipiens fatigans*) (Fig. 5.9) is almost always found in the tropics, and is the vector of many pathogenic organisms. In some countries it is known as the Brown House Mosquito; it breeds in standing water (such as ponds and pools) but also smaller bodies of water in rain-butts. *Culex* and some other closely related genera are the major vectors of arboviruses and filarial worms. *C. quinquefasciatus* trans-mits the filarial worm, *Wuchereria bancrofti* (Nematoda) causing filariasis. Together with a close relative, *Brugia malayi*, it attacks people in Africa, tropical Asia and the Pacific region, and in 1974 the WHO estimated there were more than 250 million cases. The disease is a chronic debilitating one, and 1–2% of cases may result in elephantiasis, where blocked lymphatic vessels cause gross swelling of body extremities.

One mosquito of special interest is the large blue and white *Toxorhynchites* with its odd bent proboscis (Fig. 5.10). This distinctive insect lives in tropical rain forests and is occasionally found on domestic premises where it causes unnecessary alarm. The adults are harmless nectar-feeders, but the larvae are predacious and live in tree holes, etc. in the forest. They feed on the larvae of *Aëdes*. In parts of the wet tropics, several species of *Toxorhynchites* are being used in biological control programmes to reduce the numbers of *Aëdes* breeding in inaccessible container habitats.

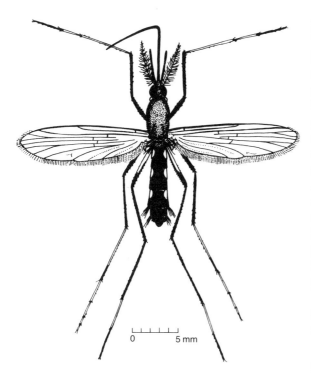

Fig. 5.10 The Predatory Tropical Mosquito (*Toxo-rhynchites splendens* – Dipt.; Culicidae); Hong Kong.

(c) Family Simuliidae (Black Flies) (1450 species)

These small, black, stout-bodied flies are active and fly well – most females are diurnal but some males fly at night. They are world-wide in distribution, and breed mostly in fast-flowing freshwater streams. Black Flies are abundant in the tropics, where they are vectors of several filarial worms, and very abundant in Arctic regions and moun-tain streams, where their main pest effect is nuisance due to their blood-sucking harass-ment. Within the family the preferred feeding hosts for the blood-sucking females are birds (hence their abundance in the Arctic) but mammals (including humans) may also be attacked.

In Africa *Simulium damnosum* transmits the filarial nematode, *Onchocerca* in humans by its blood-feeding. The resulting disease is called 'onchocerciasis', and the worm usually settles in the eye and causes 'river blindness'. Typi-cally, the victim may not be blinded until after some years of infection. In Africa, especially West Africa, it is estimated that 1 million people are blind as a result of this parasite; in all 17 million in 34 countries are infected and at risk of losing their sight. Other species trans-mit *Onchocerca* worms in Central and South America. Apparently a cheap and effective drug ('Ivermectin') has been developed and one dose will give protection for a year. However, in West Africa the most affected people live in remote areas and the 'Sight Savers' project is facing serious distribution problems.

In parts of the UK, biting *Simulium* flies are a nuisance for people on holiday in these regions.

(d) Family Ceratopogonidae (Biting Midges)

Some of these tiny midges are blood-sucking; most female *Culicoides* feed on vertebrate blood, while some feed on other insects and others feed on fluids from plants. Larval development takes place in damp or wet soil, and in the UK the larvae overwinter. In the tropics some species are found in forests, but most are in mangrove swamps where they may be very numerous. The bites are usually very irritating, but the flies do not transmit parasites in humans. On certain beaches in the Seychelles a species of *Culicoides* breeds in the damp sand at the top of the shore and the biting females drive sunbathers from the beach. In parts of the UK the tourist industry is decidedly hampered by midge problems when the adults emerge in July.

(e) Family Tabanidae (Horse Flies; Clegs)

These are large stout-bodied diurnal flies, many of which are grey or brown in colour with large eyes. Male flies feed on nectar and plant juices, but the females also are blood-sucking and prefer vertebrate blood (mostly mammals, reptiles and amphibians). They occur worldwide, but are most abundant in South America and Africa. The larvae live in soil or water (wet soil is preferred) and some are predacious, others feeding on vegetable debris. The 'bite' of the female fly is not really painful, but its attack can be disturbing and people are alarmed.

More than 3500 species of Tabanidae are recorded; the three best-known genera are:

- *Tabanus* (Horse Flies, etc.) – large, often dark coloured flies with clear wings; found worldwide; most typical of open woodland and grassland (savanna) (Fig. 5.11).
- *Chrysops* (Deer Flies) – largish flies with beautiful emerald-coloured eyes and dark-banded wings; found worldwide but mostly Holarctic and Oriental. In North America typically in open woodland (deer country) (Fig. 5.12).
- *Haematopota* (Clegs) – slightly smaller, with speckled wings, and banded eyes. They are most abundant in Africa and Asia, rare in the New World and often seem to prefer the edges of forest.

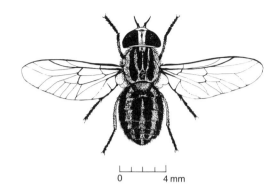

Fig. 5.11 Horse Fly (*Tabanus* sp. – Dipt.; Tabanidae), ex cow; Hong Kong.

Tabanids are very effective mechanical transmitters of blood-dwelling parasites from host to host, and they also cause harassment and discomfort to humans when bitten. The diseases involved include tularaemia (caused by the bacterium, *Pasturella tularensis*); this is a Holarctic disease of rodents and lagomorphs. It is transmitted to humans either by contact with the raw flesh or mechanical transmission by ticks and Tabanidae. In the USA the main vector is *Chrysops discalis*. *Loa loa*

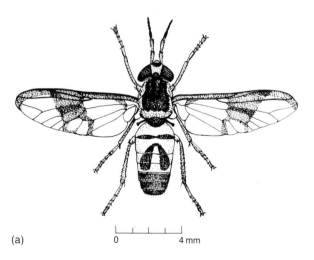

(a)

(b)

Fig. 5.12 (a) Deer Fly (*Chrysops* sp. – Dipt.; Tabanidae); Hong Kong; and (b) adult feeding on the arm of the author; Skegness, UK.

is a nematode transmitted biologically from monkeys to humans by several species of *Chrysops* in tropical Africa. The worms cause painful swellings of the wrists and ankles.

(f) Family Muscidae (House Flies, etc.)

This is quite a large group comprising some 3900 species, but dominating because some species are very abundant and many are synanthropic. The larvae are mostly saprophagous scavengers, often in the soil or refuse heaps, but some are phytophagous and a few are carnivorous. Adults typically have a proboscis ending in a swollen sponge-like labellum used for soaking up liquids, but the Stomoxyinae have developed hardened prestomal teeth with which skin can be ruptured for blood-sucking. Most muscids regurgitate fluid to digest solid food externally before sucking up the liquid mass, and they thus transmit mechanically several important bacteria and other pathogens via contaminated food.

House Flies (*Musca domestica*; Fig. 5.13) are recorded as carrying some 100 different pathogens and as transmitting 65 of them (Kettle, 1990). These include helminths, viruses, bacteria and protozoans. The most important pathogens are as follows: eggs of some Cestoda and Nematoda, including the threadworm (*Trichuris*) and hookworm (*Ancylostoma*); cysts of Protozoa, including *Entamoeba* which causes amoebic dysentery; viruses causing polio, infectious hepatitis and trachoma; and bacteria etc. which cause dysentery, diphtheria, typhoid, tuberculosis, leprosy, cholera and yaws. In a newspaper report ('*Guardian*', 1994) it was stated that in the UK since 1990 there has been a six-fold increase in dysentery, but most of this is probably due to water contamination. The tropical Bush Fly (*Musca sorbens* complex) is a serious nuisance in Africa and Asia (with a very close relative in Australia), as also is the House Fly. A few other species of *Musca* and some similar genera are also synanthropic and

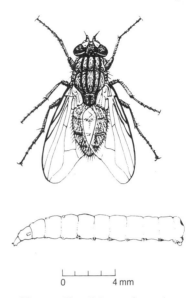

Fig. 5.13 House Fly (*Musca domestica* – Dipt.; Muscidae), adult and larva (maggot); Hong Kong.

to a lesser degree involved as minor medical pests, either by feeding adults spreading pathogens or larvae causing myiasis.

(g) Family Fannidae

These are now generally separated from the Muscidae as a small family in their own right. The genus *Fannia* has some 220 species, ten of which are reported as medical pests, albeit of a minor nature. *F. canicularis* is the cosmopolitan Lesser House Fly (Fig. 5.14) whose larva is adapted for living semi-aquatically in decaying organic material. *Fannia* larvae are recorded as causing cases of human vaginal myiasis.

(h) Family Cuterebridae (Rodent Bot Flies)

Species of *Cuterebra* have been recorded causing human myiasis. *Dermatobia hominis* is the Human Bot Fly of South America which regularly causes human myiasis.

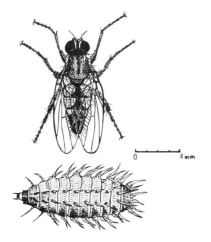

Fig. 5.14 Lesser House Fly (*Fannia canicularis* – Dipt.; Fannidae), adult and larva; Hong Kong.

(i) Family Calliphoridae (Blow Flies; Bluebottles, etc.)

Several species are synanthropic and their large size makes them very conspicuous and domestically annoying. The larvae are mostly saprozoic on carrion or dung, but some have developed parasitic habits. Some larvae have been used as surgical maggots – sterile larvae have been placed in human wounds where they feed on the necrotic tissues and clean the wound completely without invading living tissue. *Lucilia* larvae apparently also secrete allantoin which suppresses bacterial development so that the cleaned wounds resist infection. Figure 5.15 shows a typical Bluebottle. The most important species known as medical pests are:

- *Auchmeromyia lateola* (Congo Floor Maggot) – Africa south of the Sahara; larvae lurk in forest litter and suck blood at night from humans sleeping on the ground.
- *Calliphora* (Blow Flies; Bluebottles) – synanthropic in Holarctic region and cooler parts of Australia. Adults may spread food and faecal pathogens. Larvae sometimes cause myiasis in humans.
- *Chrysomya* (Tropical Blow Flies, etc.) – some species are the tropical equivalent of *Calliphora* and *Lucilia*; others are termed 'Old World Screw-worms' and cause myiasis in humans as well as attacking livestock.

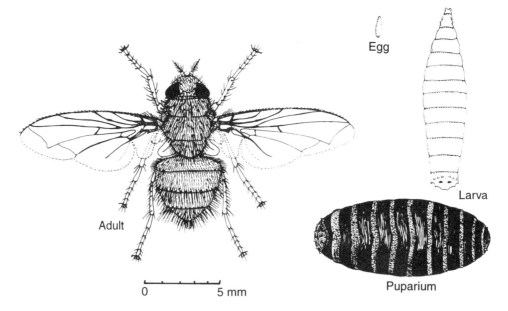

Fig. 5.15 Bluebottle adult and immature stages (*Calliphora* sp. – Dipt.; Calliphoridae); Hong Kong.

- *Cochliomyia* (=*Callitroga*) – these are the New World Screw-worms that devastate livestock and sometimes infest humans.
- *Cordylobia anthropophaga* (Tumbu/Mango Fly) – found in tropical Africa; eggs are laid on drying clothing and the larvae bore into the flesh of the wearer. Infants can be seen with literally hundreds of small boil-like pustules, each containing a single maggot.
- *Lucilia* (Greenbottles) – most adults are metallic green in colour and the larvae are basically carrion-feeders, some of which invade living flesh causing 'sheep strike'. Others have been found in various human body cavities.

(j) Family Sarcophagidae (Flesh Flies)

These are larviparous large grey flies with a black-striped thorax (Fig. 5.16), some of which are associated with human food contamination. The larvae can survive passage through the human intestine and cause intestinal myiasis with its symptoms of pain and vomiting. *Wohlfartia* has several species which are commonly involved with human wound myiasis in Asia and Africa.

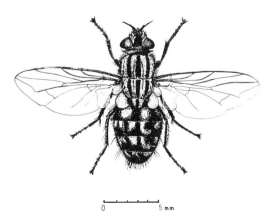

Fig. 5.16 Flesh Fly (*Sarcophaga* sp. – Dipt.; Sarcophagidae); Hong Kong.

(k) Family Glossinidae (Tsetse)

A tiny family with only 22 species in one genus (*Glossina*), confined to tropical Africa, south of the Sahara. They are of great importance in Africa as the vectors of the *Trypanosoma* species which causes sleeping sickness in humans and similar diseases in animals. Their presence is the reason for the underpopulated regions of Africa (see below). They are all blood-suckers and feed on Vertebrata, and most show a degree of host preference. Some mechanical transmission takes place but usually the trypanosomes undergo development in the insect body and so the Tsetse (Fly) is typically both vector and intermediate host. The two most important species (groups) are:

- *Glossina morsitans* (group) – these transmit *T. rhodesiense* (Rhodesian sleeping sickness) – more localized, and regarded as basically game tsetse, in open savanna woodland.
- *Glossina palpalis* (group) – these are vectors of *T. gambiense*, the main cause of sleeping sickness – more of a waterside forest species.

Sleeping sickness is also called African (human) trypanosomiasis, and is widespread in 36 African countries. The two protozoans are identical in appearance but produce different symptoms. Rhodesian sleeping sickness is a rapid disease and usually kills in three to nine months, but the Gambian form is more of a chronic wasting disease which often takes several years to be fatal. The precise number of people suffering from sleeping sickness at the present time is not known, but at the turn of the century it killed a quarter of a million people around Lake Victoria (East Africa), and, according to the WHO, some 10 000 new cases are recorded in Africa each year. The reason for there not being more cases is that humans have learned the hard way to avoid settling within the areas of tsetse dominance known as the 'fly-belts'. Large areas of tropical Africa therefore remain unpopulated.

5.1.5 Order Lepidoptera (Moths and Butterflies)

Generally Lepidoptera are neither medical nor veterinary pests, but a few species should be mentioned:

(a) Family Pyralidae (Snout Moths)

A dozen species in several different genera are recorded as 'Eye Moths'; at night adults fly to the face of cattle and other mammals (sometimes humans) to suck lachrymal fluid from the eyes. This occurs in parts of India and South East Asia (Banziger, 1968), but there is no evidence that disease organisms are carried.

(b) Family Geometridae (Loopers, etc.)

In this family there is another species of Eye Moth, also found in South East Asia, in the genus *Problepsis*.

(c) Family Noctuidae (Owlet Moths)

Also recorded by Banziger (1968) were a dozen species of Noctuidae that behaved as 'Eye Moths' and took tears from sleeping humans and other large mammals at night.

Another group of closely related species have their haustellum (proboscis) modified into a short, stout and heavily barbed piercing organ – some use this proboscis to suck sap from thick-skinned fruits, while a few species have become blood-suckers, piercing the skin of sleeping mammals (mostly cattle and buffalo) at night in South East Asia.

(d) Family Lymantriidae (Tussock Moths, etc.)

These moths are of moderate size and typically covered with long setae; the ovipositing female will usually cover the eggs with a mass of setae from the end of her abdomen for protection and concealment. The caterpillars are quite large and very bristly, and in some cases (especially species of *Euproctis*) the setae are painfully urticating; they have barbed spicules and easily penetrate human skin where they cause intense irritation. This lasts for up to a couple of days, and the eggs, larvae and pupae of these species should be avoided – they can be particularly upsetting to children, especially as the setae are so fine as to be virtually invisible.

5.1.6 Order Hymenoptera (Wasps, Ants, Bees, etc.)

The main feature of the Aculeate Hymenoptera is the possession of a sting (modified ovipositor) – the injection of venom into the human body is painful. Wasp and bee stings regularly kill around a dozen people in the UK each year by anaphylactic shock. This is a hypersensitivity to venom following a sting after an initial sensitizing sting; death typically occurs in a matter of minutes.

Common Wasps (*Vespa* spp., etc.) (Fig. 5.17) have a typical aposematic (warning) coloration of yellow and black (or in some cases red and black) and people often react quite dramatically to their presence. The sting is used for killing insect prey and is very efficient. The author was once attacked in the Seychelles by Paper Wasps (*Polistes* sp.) and two workers delivered 18 and 17 stings, respectively, in about as many seconds, the last stings being just as painful as the first.

Honey Bees use their stings for defensive purposes and if a person is attacked, the barb of the sting lodges in the epidermis and the entire stinging apparatus is pulled out of the insect abdomen – the bee then dies. But Honey Bees will attack without hesitation intruders who threaten their nest colony. The African Honey Bee (*Apis mellifera scutellata*) is renowned for occasional outbursts of ferocity. In East Africa, people are regularly stung to death by a swarm of Honey Bees. Around the outskirts of Kampala, Uganda, it was usual for at least a dozen people a year to die from these

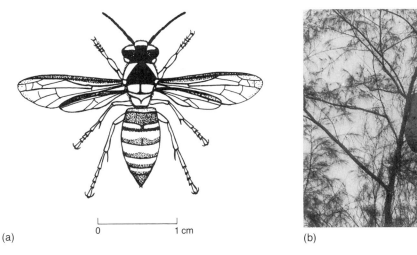

(a)

(b)

Fig. 5.17 (a) Common Wasp (*Vespa bicolor* – Hym.; Vespidae) worker; Hong Kong; and (b) aerial nest high in tree.

unprovoked bee attacks. Both wasps and bees are most likely to attack people if their nest is disturbed or threatened and this can be dangerous since each nest will contain some thousands of workers.

Ants show considerable evolutionary development, some 'advanced' forms having lost the sting as a penetrating structure, it being replaced with a system for squirting formic acid. Many species have large, effective mandibles and they bite savagely if provoked. Ants have a 'soldier' caste whose function is to defend the nest and the foraging workers and also to kill large prey, so they tend to be naturally aggressive. There are a few species, known as 'Biting Ants', which nest in foliage and sometimes workers in coffee and citrus plantations will refuse to tend bushes or trees that contain ant nests (Fig. 5.18). However, often these ants can be handled without fear of attack. On the other hand, the Army/Driver Ants of tropical Africa and South America (which live without any fixed abode) justifiably spread fear at their appearance in any domestic situation, for they are fierce, aggressive and very numerous, and bite as well as sting.

5.1.7 Order Acarina (Class Arachnida) (Mites and Ticks)

These are not insects but are sufficiently closely related as to be normally included in the study of entomology. The main difference between ticks and mites is that ticks are larger and their mouthparts are modified into a 'toothed' hypostome which is used for penetration of the host skin for sucking blood and also as a means of attachment to the host. Many of the hard ticks that attack livestock will attack humans if given the opportunity; sometimes the pathogens they carry will infect the human hosts causing various types of 'tick fever'. The encouragement of wild deer and rabbits in rural regions results in campers being used as hosts for several different ticks, and in parts of North America tick-borne diseases are becoming quite common. Many ticks are not very host-specific – more likely they tend to be habitat-specific.

(a) Family Trombiculidae

Chiggers are larval forms of *Trombicula* (Fig. 5.19), tiny in size, that attach to the skin of

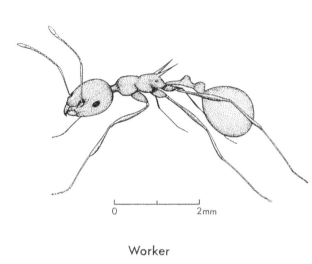

Worker

(a)

(b)

Fig. 5.18 (a) Biting Ant; and (b) nest between two coffee leaves; Kampala, Uganda (photo D. McNutt).

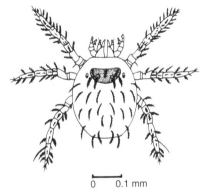

Fig. 5.19 'Chigger' – larval stage of *Trombicula* sp. (Acarina; Trombiculidae); Hong Kong.

humans and various animals in the warmer parts of the world. They feed on serous exudates from sub-epidermal cells (Harwood and James, 1979). There are two groups of species; one group produces an intense itching and causes severe dermatitis, and the other transmits the *Rickettsia* that causes 'scrub typhus'.

(b) Family Demodicidae (Hair Follicle Mites)

These are minute, worm-like mites, quite host-specific, that live on the skin and in the hair follicles of various mammals. *Demodex folliculorum* is the human Hair Follicle Mite (Fig. 5.20), which is usually found in the hair follicles of the face; it is very common and ubiquitous in distribution.

(c) Family Sarcoptidae (Itch Mites)

These are tiny, hemispherical mites with reduced legs and often without claws. They make tunnels in the skin up to an inch or more in length, where the eggs are laid. *Sarcoptes scabei* is the ubiquitous Human Itch Mite or Scabies Mite (Fig. 5.21) which burrows in the soft skin between the fingers, the bend of elbow and knee, shoulder blades and around the penis or under the breasts. The infestation is irritating and requires careful chemical application for eradication.

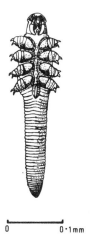

0   0·1mm

Fig. 5.20 Hair Follicle Mite (*Demodex folliculorum* – Acarina; Demodicidae); ventral view; Hong Kong.

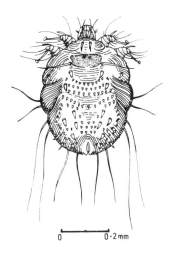

0   0·2mm

Fig. 5.21 Human Itch (Scabies) Mite (*Sarcoptes scabei* – Acarina; Sarcoptidae); Hong Kong.

(d) Family Acaridae

The classification of some groups of mites is still contentious and different taxa are used with different names by different authors.

The Acaridae is a diverse group, but a number are to be found in domestic situations, both in houses and food stores. The mites are small in size, may infest skin and clothing and can be found on the scalp.

Dermatitis is a common result of their infestation, and they can also invade lungs and intestines. The dermatitis caused is sometimes known as 'Grocer's Itch' and can be caused by several common house mites that are found on foodstuffs, including the Flour Mite.

The House Dust Mite (*Dermatophagoides pteronyssinus*) is sometimes placed in the related family, Pyroglyphidae. This mite has become a major international pest in recent years as more houses are centrally heated and insulated. They live in beds, mattresses and other secluded domestic sites and feed on organic debris, especially flakes of skin. In the UK in recent years there has been a dramatic increase in the number of cases of asthma. This is apparently due to several different factors, but a major cause is the increase in House Dust Mites. People become allergic to chemicals in the faecal droppings, not usually to the mites themselves, and this dust when inhaled causes asthma, rhinitis and can also cause eczema. Recent estimates suggest a total of 11 million sufferers of these problems in the UK. An interesting recent case in the UK concerned feather-filled pillows that were infested, causing the sleepers severe respiratory problems. The mites were killed by placing the pillows in a domestic deep freeze (at −20°C) for six hours.

5.2 VETERINARY PESTS

Most livestock are reared as a direct source of meat for human food, but others are kept for their various products such as dairy cattle for milk, sheep for wool and poultry for eggs. Some animals are kept as pets or even for ornamental purposes.

In general, if an animal is slaughtered then all its constituent parts are commercially utilized (for example, cow hides for leather, hooves and horns to make glue and bones for bonemeal). Draught animals are still important in the Third World and their health can be vital to local farmers.

The main food (meat) animals reared extensively worldwide are cattle, sheep, goats, pigs, chickens, ducks and marine and freshwater fish. Lesser numbers of turkey, quail, pigeon, guinea fowl, shrimps and lobsters are also regularly produced. Game-ranching is becoming popular in many countries – in the UK, to utilize land that is regarded as marginal or too poor for cultivation but whose foliage can be used as food for native wildlife, such as Red Deer. In the tropics, the native ungulates are adapted to local conditions and can tolerate a difficult climate and/or insect pests and local pathogens, whereas exotic cattle generally succumb to these hazards. Draught animals include horses, donkeys, mules, camels, water buffalo and oxen. In some desert areas, camels are also used for milk and occasionally for meat. On the whole, most animals reared by humans are multipurpose in terms of use.

With many types of livestock, health and commercial production are measured in two different ways. For example, beef cattle are assessed by their growth weight (body weight over time) whereas dairy cattle are assessed by their rate of reproduction and milk yield, both daily and annually. Similarly with sheep: both growth rate and wool yield are measured. Veterinary pests in extreme cases cause the death of the animal; at lower damage levels they affect health and vigour, causing a reduction in weight gain, abortion of embryos and reduction of meat/wool/egg production.

Table 5.3 lists the main groups of veterinary pests. These fall into two categories: the direct pests that damage the host animal by feeding on its blood or tissues, and the indirect pests that act as vectors or intermediate hosts (or both) of parasites or pathogens, causing disease.

Damage to livestock varies in extent, usually according to the age and health of the host. Young animals, old animals in poor health and animals kept in insanitary conditions are more likely to suffer heavy infestations, and these will have an adverse effect on the animal. Low levels of infestation will still cause irritation and will upset the host. To relieve itching, the animal will scratch and rub against hard objects, which can damage fur and hide. In Australia sheep lice (*Damalinia ovis*) are regarded as serious pests as infested wool has a lowered value. Some heavy lice infestations appear to be the result of ill-health or stress.

Some of the major diseases of livestock transmitted by Insecta and Acarina are listed in Table 5.4. The more important species to be regarded as veterinary pests are reviewed below.

5.2.1 Order Mallophaga (Biting Lice; 'Bird' Lice) (2800 species)

The name 'Bird Lice' is not strictly accurate, for some species are found on mammals although most are on birds. These are small, apterous, obligate, permanent ectoparasites that live in the feathers of birds and the fur of mammals. With their biting mouthparts they feed on fragments of skin, feathers, hair and scabs, and blood from scratches or wounds will also be taken. Harm to the host consists mostly of skin irritation. When laid the eggs stick to hairs or feathers.

Host-specificity is marked, and some species show preference to certain body regions so that a single bird host can have several different species of Mallophaga on different parts of the body (wings, head, flanks, etc.).

Species of agricultural interest include:

- *Bovicola bovis* – worldwide on cattle.
- *Columbicola columbae* (Pigeon Louse) – worldwide on pigeon.
- *Damalinia* spp. – several species are found on sheep, cattle, horses and goats.
- *Felicola subrostrata* (Cat Louse) – worldwide on cats but not common.
- *Lipurus* – a typical bird-biting louse found on poultry (Fig. 5.22).
- *Menopon gallinae* (Chicken Shaft Louse) – worldwide on chickens.

Table 5.3 Major insect veterinary pests

Group	Common name
Order Mallophaga	Biting Lice; 'Bird' Lice
Several families	
Order Siphunculata	Sucking Lice; Animal Lice
Family Haematopinidae	Cattle Lice, etc.
Linognathidae	Cattle Lice, etc.
Order Hemiptera, Suborder Heteroptera	Bugs
Family Cimicidae	Chicken Bugs, etc.
Belostomatidae	Giant Water Bugs
Order Diptera	Flies
Family Culicidae	Mosquitoes
Simuliidae	Black Flies
Tabanidae	Horse Flies; Clegs, etc.
Muscidae	Sweat Flies; Stable Flies, etc.
Gasterophilidae	Bot Flies
Oestridae	Warble Flies
Cuterebridae	Rodent Bots
Calliphoridae	Blow Flies; Greenbottles
Glossinidae	Tsetse
Hippoboscidae	Keds; Louse Flies
Order Siphonaptera	Fleas
Several families	Animal Fleas
Order Acarina	Ticks and Mites
Family Dermanyssidae	Red Mites
Argasidae	Soft Ticks
Ixodidae	Hard Ticks
Demodicidae	Hair Follicle Mites
Sarcoptidae	Mange Mites
Psoroptidae	Scab Mites

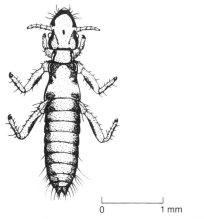

Fig. 5.22 Typical Bird Louse (*Lipurus* sp. – Mall.; Lipuridae); Hong Kong.

- *Menacanthus stramineus* (Chicken Body Louse) – the commonest and most damaging chicken louse.
- *Trichodectes canis* (Dog-biting Louse) – worldwide on dogs; can act as a vector for Dog Tapeworm (*Dipylidium caninun*) (Fig. 5.23).

Harm to the host depends on the density of the lice – small numbers cause no problem but large numbers are very irritating. In general, there are conflicting opinions as to the effect of Mallophaga on animal production; it is likely that the Siphunculata (section 5.2.2) are the more damaging.

Table 5.4   Some animal diseases transmitted by Insecta and Acarina

Vector	Disease	Parasite	Transmission	Distribution
Musca domestica	anthrax	*Bacillus anthracis*	Contamination	Worldwide
Culicoides spp.	blue tongue of sheep	Virus	Bite	S. Africa
Simulium ornatum, etc.	cattle filariasis	*Onchocerca gutturosa,* etc.	Bite	Europe, N. America
Simulium venustum	(ducks)	*Leucocytozoon anatis*	Bite	N. America
Glossina spp.	nagana	*Trypanosoma* spp.	Bite	Africa
Tabanidae	surra	*Trypanosoma evansi*	Bite	India, South East Asia
Tabanidae	(cattle)	*Trypanosoma theileri*	Bite	Cosmopolitan
Stomoxys calcitrans	mal de Caderas	*Trypanosoma equinum*	Bite	S. America
Ctenocephalides felis	dog tapeworm	*Dipylidium caninum*	Eating flea	Cosmopolitan
Dermanyssus gallinae	fowl spirochaetosis	*Borrelia anserina*	Bite	Australia
Argas persicus	duck tick paralysis	*Anaplasma marginale*	Bite	USA
	poultry piroplasm	*Aegyptianella pullorum*	Bite	USA
	fowl spirochaetosis	*Borrelia anserina*	Bite	Europe, Asia
Ixodes ricinus	pyaemia of lambs	*Staphylococcus aureus*	Bite	Britain
	redwater of cattle	*Babesia bovis*	Bite	Africa
Boöphilus annulatus	Texas cattle fever	*Babesia bigemina*	Bite	USA
Rhipicephalus spp.	East Coast fever (of cattle)	*Theileria parva*	Bite	E. and S. Africa
	redwater of cattle	*Babesia bigemina*	Bite	Africa
Amblyomma hebraeum	heartwater of cattle	*Rickettsia ruminantium*	Bite	Africa

Taken from Hill, 1994.

5.2.2 Order Siphunculata (Sucking Lice; Animal Lice)

These are closely related to the previous group, and are sometimes classified together with the Mallophaga as the order Phthiraptera (lice). All are also apterous, permanent, obligatory, ectoparasites to be found on the bodies of Mammalia, but their mouthparts are adapted for piercing and sucking and they feed on the blood of the host. Eggs are laid very firmly attached to the hairs and the whole life cycle is completed on the body of the host. The blood-sucking is particularly irritating, and there are two genera that are of importance as veterinary pests:

- *Haematopinus* – occurs as several different

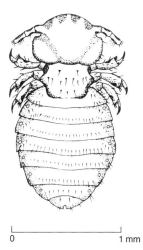

Fig. 5.23 Dog-biting Louse (*Trichodectes canis* – Mall.; Trichodectidae); Hong Kong.

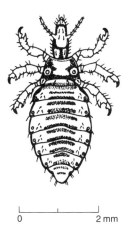

Fig. 5.24 Cattle (Sucking) Louse (*Haematopinus* sp. – Siph.; Haematopinidae) from a cow in south China.

Fig. 5.25 Giant Water Bug (*Lethocerus indicus* – Hem.; Belostomatidae); Hong Kong; body length 90 mm.

species on horses, cattle, pigs and buffalo. They are large lice, some 4 mm long. A total of 22 species are known and all parasitize ungulates. They cause serious irritation and upset the host animal (Fig. 5.24).

- *Linognathus* – occurs as 50 species, mostly on Artiodactyla but a few on Carnivora; six occur on domestic animals. Species parasitize dogs, sheep, goats and cattle,

and there is some host overlap so that specificity is not absolute.

5.2.3 Order Hemiptera, Suborder Heteroptera (Bugs)

This group is really of little veterinary importance, although there are two species of Cimicidae that attack poultry in North and Central America. The Belostomatidae are the huge, fiercely predacious Giant Water Bugs, up to 8–10 cm long (Fig. 5.25), found throughout the tropics. The rostrum is short, sharp and curved, and is used to spear small fish, frogs and insects. The toxic saliva is used to kill and partly digest the prey, which is held immobile by the raptorial forelegs. In fish ponds these bugs can be serious pests.

5.2.4 Order Siphonaptera (Fleas) (1400 species)

These small, wingless ectoparasites are found worldwide, are usually host-specific and occur on most warm-blooded vertebrates from which they suck blood. Their life cycle is geared to the nest or lair of the host animal, for the eggs fall off the host body and larval development takes place in the nest. The scavenging larvae feed on organic debris; only the adults are blood-sucking. Most species of birds and many larger mammals have their flea parasites; about 100 species are confined to birds and the other 1300 species are found on larger mammals. However, only mammals that have a nest or lair (den) are used as hosts, so Rodentia are major hosts. Cats, dogs, foxes, badgers and the hedgehog are also important hosts. Among the usual livestock reared by humans, only goats and pigs are recorded as hosts. Certain flea species will pass from host to host in domestic situations,

which makes them important as disease vectors – especially for the plague pathogen. In the USA 20 species of flea are recorded as biting humans, and most of these are from cats, dogs and rodents. The Human Flea (*Pulex irritans*) is actually the normal flea of pigs, and is now scarce in households. The common domestic species are now the Cat Flea (*Ctenocephalides felis*) (Fig. 5.26) and Dog Flea (*C. canis*), both of which readily feed on humans. Several species on poultry can be of some importance, causing the birds to become anaemic and emaciated, and young birds can be killed. Both Cat and Dog Flea transmit tapeworms, especially *Dipylidium caninum*.

5.2.5 Order Diptera (Flies)

Veterinary pests (as with the medical ones) are sometimes the biting flies that suck blood, and also the larvae that practise myiasis and invade body cavities and even the living flesh. The blood-sucking adults may also act as vectors (and sometimes intermediate hosts) for different parasites and pathogens causing diseases in the host animals. There has clearly

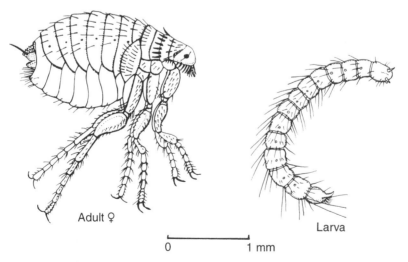

Fig. 5.26 Cat Flea, adult female and larva (*Ctenocephalides felis* – Siph.; Ceratophyllidae); Makerere, Uganda.

been a long evolutionary history of inter-dependence between various biting Diptera and the parasites they carry. Many of these biting flies are equally important as medical and veterinary pests; usually different species of fly are concerned but in some cases the same species of fly will feed on humans or livestock – whichever is readily available. The main insect species concerned follow:

(a) Family Culicidae (Mosquitoes)

There are 3000 species worldwide and the females of most species need to take blood meals for the development of eggs. All groups of Vertebrata are used as blood sources, but the Mammalia and birds are most popular. However, veterinarily mosquitoes are not important pests. The feeding adults will take some blood and cause some irritation, but not usually a great deal. The role of mosquitoes as vectors affecting livestock worldwide is not serious either (Harwood and James, 1979; Kettle, 1990) although some virus pathogens are transmitted, as are species of *Plasmodium* (causing malarias) and some microfilaria (Nematoda).

(b) Family Simuliidae (Black Flies; Turkey Gnats; Buffalo Gnats)

These flies are found worldwide, and most females need a blood meal for egg develop-ment. Most of the 1450 species feed on birds but mammals are also used. A dozen or more species of *Simulium* are regular pests of live-stock in both tropical and temperate regions, and turkeys, chickens and ducks are par-ticularly vulnerable. Livestock actually suffer from blood loss because of the vast numbers of flies that occur, and the harassment can disturb cattle to the extent that they are put off feeding, yields are reduced and the animals can die. Cattle filariasis, caused by *Onchocerca gutturosa*, is spread by *S. ornatum* and others in Europe and North America, and *S. venustum* transmits *Leucocytozoon anctis* in

North America. Five other species also carry *Leucocytozoon* in poultry. Historically there have been very serious losses of livestock in the southern parts of the USA (lower Mississippi River) and in Eastern Europe, attributable to species of *Simulium*. River drainage control has, however, reduced such risks.

(c) Family Ceratopogonidae (Biting Midges)

The midges themselves are not quite as impor-tant regarding livestock as they are regarding humans, but a couple of species of *Culicoides* transmit parasites in sheep in Africa (Table 5.4). Others are vectors of arboviruses in livestock and haematozoa in poultry.

(d) Family Tabanidae (Horse Flies; Deer Flies; Clegs)

This worldwide group of large blood-sucking flies plagues cattle and other livestock, espe-cially in warmer regions. Some of the African species are as large as Bumble Bees. Harass-ment of livestock is a major factor, and in addition these flies act as mechanical vectors of various blood-dwelling parasites, espe-cially Trypanosomes, some filarial worms (Nematoda) and some bacteria. Tularaemia occurs in cattle as well as in humans, and is carried by *Chrysops* and other tabanids. *Trypanosoma evansi* causes 'surra' in camels, but can also attack cattle, buffalo, horses and dogs; transmission is by biting Tabanidae and the disease is often fatal. There is evidence for the mechanical transmission of the rinderpest virus in Africa. The harassment factor is quite considerable and when flies are present in large numbers the attacked livestock are very distressed.

(e) Family Braulidae (Bee 'Lice')

These small, wingless flies live in beehives where the larvae tunnel in the wax combs.

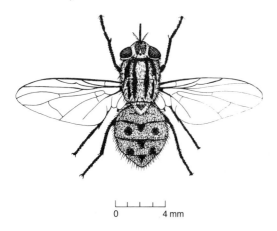

```
0          4 mm
```

Fig. 5.27 Stable Fly (*Stomoxys calcitrans* – Dipt.; Muscidae); Hong Kong.

Little actual damage is done, although they are of nuisance to apiarists.

(f) Family Muscidae (Sweat Flies; Stable Flies)

The subfamily Stomoxinae are characterized by an elongate hard proboscis projecting anteriorly; they are blood-sucking flies and serious pests of cattle and other livestock worldwide. The two main genera are *Stomoxys* (*S. calcitrans* is the Stable Fly or Biting House Fly (Fig. 5.27)) and *Haematobia* (*H. irritans* is the Horn Fly).

Both are cosmopolitan species and both sexes are blood-sucking. They feed mainly on cattle and buffalo, but also on other animals and sometimes humans. Heavy infestations reduce milk yields of dairy cattle, sometimes by up to 50%; beef cattle gain weight more slowly. Fly densities of 1000 or more per beast are not uncommon.

Other muscoid flies that attack livestock include *Hydrotaea irritans* (the Headfly) that attacks the eyes, mouth, nose and ears to feed on secretions (including blood). *Hydrotaea* is a Holarctic genus, often referred to as Sweat Flies, that seek body fluids from cattle and other domestic livestock (and also humans). They have a rasping proboscis which can break the skin to cause a blood flow. *Morellia*

are also called Sweat Flies – these are cosmopolitan in distribution. *Musca sorbens* (Eyefly; Bush Fly) has only small prestomal teeth but the females require protein for ovary development and they take body secretions and fluid from faeces and carrion. *Musca crassirostris* (Blood-sucking House Fly) in Eurasia and Africa can be a serious pest of cattle. Several of these muscoid flies are important as mechanical transmitters of the bacterium *Corynebacterium pyogenes*, which is responsible for mastitis in cattle. Many muscoid flies will feed at wound sites and also at sites where tabanids and ticks have pierced the skin, and passive transmission of pathogens can occur. Figure 5.28 shows a modest level of attack of muscoid flies on cattle.

A recent note in *Spore* (No. 53, October 1994) states that diseases of livestock spread by flies (mostly House Flies) in the USA cost farmers an estimated US$60 million a year. Research in Canada found that using Muscovy ducks in calf and pig pens reduced fly numbers by 80–98%.

Fly harassment of humans is fairly easily assessed, as almost everyone has experience of this nowadays, but with animals it is not so straightforward. An interesting example quoted by Kettle (1990, page 234) refers to the Ngorongono Crater in Tanzania in 1962. Here, a severe outbreak of *S. calcitrans* caused a change in lion behaviour, when they started to climb trees to avoid being bitten while resting.

(g) Family Gasterophilidae (Horse Bot Flies, etc.)

This is a tiny group of only 18 species confined mostly to the Equidae as hosts. The Horse Bots are all species of *Gasterophilus* and the larvae develop in the host alimentary canal. Eggs may be laid on the host body and are licked up by the horse into its mouth; others are laid on grasses and are eaten by the host. Heavily infested animals may die, partly as a

Fig. 5.28 Muscoid flies feeding on the flank of a cow, at Gibraltar Point NNR, UK.

result of intestinal irritation or by obstruction of the alimentary canal.

(h) Family Oestridae (Warble and Bot Flies) (34 species)

Another group of stout furry flies, resembling Bumble Bees, whose larvae are endoparasites of mammals. They appear to have originated as parasites of wild Bovidae (cattle) and Cervidae (deer) and have now adapted to live in domestic cattle, sheep and goats. Most of the common species have occasionally been recorded infesting humans, causing myiasis of one type or another. The burrowing larvae feed on serous exudates from the host tissues.

The Ox Warble Fly is typical of the group. Eggs are laid on hairs on the legs of the cow and the hatching larvae bore into the skin. They remain under the skin for several months and come to rest for a while near the oesophagous. Boring continues until the larvae finally come to rest under the skin along the back. The skin is pierced so that the posterior spiracles are in contact with the air. The fully developed larvae are large and inhabit large swellings (warbles) along each side of the spine; they finally push their way out through the skin to pupate in the soil. The injuries to the cattle include perforation of the hide, lowered rate of weight gain, reduced milk yield and a general deterioration of meat quality; economic losses can be high. The sound of the beating wings (buzzing) of the female flies can drive cattle to panic and stampedes are often recorded.

The main pest species include:

- *Crivellia silenus* (Goat Warble Fly) – on sheep and goats in North Africa and warm temperate Eurasia.
- *Gedoelstia* spp. (Antelope Nasal Bot Flies) – in Africa these attack antelope, the larvae

penetrating the eye orbit through a vein and lodging in the frontal sinuses. The wild hosts tolerate the infestation but in sheep – and occasionally in humans – first instar larvae (which develop no further) may cause blindness.

- *Hypoderma* spp. (Ox Warble Flies) – Holarctic in a series of species, these flies attack cattle and deer. They can cause human myiasis which gives great discomfort and can lead to paralysis.

(i) Family Calliphoridae (Blow Flies; Bluebottles; Greenbottles, etc.)

These stout hairy-bodied flies, often metallic blue or green in colour, are important pests of livestock worldwide. The adults only take liquid food, and some are very important as pollinators of open-flowered crops. The maggots are also reared as fishing bait. The larvae have evolved from being saprophagous on corpses and dung to infesting wounds and eventually to the invasion of living flesh. Sheep suffer particularly from myiasis, for the flies are attracted to the dung-soiled wool around the anus and the maggots often end up boring the soft skin around this area. Such attacks are called 'fly-strike'. The culmination of evolution is seen in the obligatory flesh parasites of the Screw-worms where the larvae bore in the flesh of the host, causing extensive damage and sometimes death.

The total range of flies causing 'sheep-strike' and both primary and secondary myiasis in livestock worldwide is considerable. In this family, half a dozen different genera and a dozen or more species are involved.

The more important Blow Flies and Screw-worms are as follows:

- *Calliphora* spp. (Blow Flies; Bluebottles) – a Holarctic genus found also in temperate Australia; several species are strongly synanthropic and several are associated with 'sheep-strike' in Australia (Fig. 5.15).

- *Chrysomya* spp. (Blow Flies; Screw-worms, etc.) – found in the Old World tropics; widespread and abundant. *C. bezziana* is the Old World Screw-worm which is common in Africa, tropical Asia and the Pacific, but not yet in Australia. It attacks all types of livestock and also humans. Several other species are involved in secondary myiasis in Africa and Asia.

- *Cochliomyia* spp. (=*Callitroga*) (Screw-worms) – New World obligate parasites in livestock and sometimes humans; these are bluish metallic flies with orange eyes. *C. hominivorax* is the New World Screw-worm, which causes primary myiasis in livestock and sometimes humans. This was the pest used to demonstrate the 'sterile male' population control technique on the island of Curaçao in 1954. *C. macellaria* is the New World secondary Screw-worm – a very common species, said to outnumber the primary Screw-worm by almost 600 to 1.

C. hominivorax caused great concern quite recently, for in 1988 it was reported in Libya, the first time outside the New World. Apparently it arrived in Africa from South America as larvae in a consignment of live goats. During its first year in Libya it infested more than 200 people and 2000 domestic animals (Wall and Stevens, 1990). In the New World in the 1950s and 1960s the Screw-worm caused damage costing US$100 million a year, but control measures generally have proved so successful that in the USA the species is now quite scarce. There was great concern that this pest might invade the whole of North Africa with its 70 million head of livestock, and then eventually it would spread south of the Sahara. International aid was co-ordinated through the Food and Agriculture Organization (FAO) and a US$80 million integrated control project was launched. The crucial factor was the release of thousands of sterile male flies over a period of a year or more. After a couple of years the project was announced a success

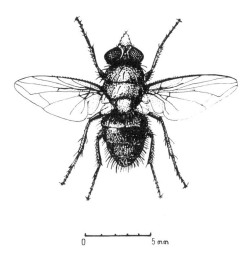

Fig. 5.29 Greenbottle Fly (*Lucilia* sp. – Dipt.; Calliphoridae); Hong Kong.

– the immigrant New World Screw-worm was eradicated in North Africa.

- *Lucilia* spp. (Greenbottles; Fig. 5.29) – cosmopolitan flies, many of which are metallic green. Several species have larvae that will infect wounds and invade the body cavities of both humans and livestock. *L. cuprina* is the Green Sheep Blowfly responsible for 80% of sheep-strike in Australia; parasitized sheep may refuse to feed and will then die. In 1969/70 it was estimated that sheep-strike losses amounted to Australian $28 million.

(j) Family Sarcophagidae (Flesh Flies, etc.)

Two genera are recorded as occasional pests of livestock. Both are larviparous and this can make them more effective as parasites:

- *Sarcophaga* spp. (Flesh Flies) – a worldwide genus with many species (Fig. 5.16), a few of which are occasionally recorded as causing myiasis in both humans and livestock.

- *Wohlfartia* spp. (Screw-worms) – several species are obligate parasites of warm-blooded animals (Vertebrata) in Africa, Asia and North America; the larvae cause tissue destruction but only occasionally the death of the host.

(k) Family Glossinidae (Tsetse)

These flies basically attack wild animals and show different degrees of host-specificity – some prefer reptiles, others birds, some pigs (Suidae) and still others antelope (Bovidae). Both sexes are obligate ectoparasites and feed only on vertebrate blood. Several species (out of the total of 22) have adapted to feeding on livestock and on humans. The genus *Glossina* is divided into a series of sub-genera. Their main importance as pests is that they transmit (and are intermediate hosts to) various species of *Trypanosoma*, causing serious disease. They are seldom numerous enough to be a problem as biting flies for livestock. The main disease of livestock is called 'Nagana', and is caused by a series of species of *Trypanosoma* (*brucei* and others). This occurs in cattle, horses, camels, buffalo, pigs, sheep, goats, donkeys, cats and dogs. The disease 'surra' is caused by *T. evansi* in camels and horses. Some of the trypanosomes are less pathogenic than others and some infections are benign.

(l) Family Hippoboscidae (Keds; Louse Flies, etc.)

An interesting group of ectoparasitic, blood-sucking flies, which produce single, fully grown larvae one at a time (similar to the Glossinidae). These immediately pupate (and are deposited as pupae, hence the former name of 'Pupipara'). Most adults are fully winged and fly strongly, but the more specialized forms are wingless. Several species will be regularly encountered on livestock and other animals, but only the Sheep Ked is commonly given pest status. The flies most

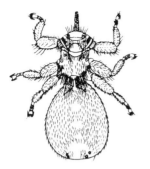

Fig. 5.30 Sheep Ked (*Melophagus ovinus* – Dipt.; Hippoboscidae) (Hutson, 1984).

likely to be encountered on domesticated animals include:

- *Hippobosca* spp. – a cosmopolitan genus of fully winged flies; several species occur on different hosts including horses, cattle, camels, ostrich and dogs.
- *Lipoptena* spp. (Deer Keds) – these are Holarctic on deer and goats; they are winged initially but these are shed after finding a host.
- *Molophagus ovinus* (Sheep Ked) – cosmopolitan; apterous pests found on sheep and sometimes goats (Fig. 5.30).
- *Pseudolynchia canarienis* (Pigeon Louse Fly) – now cosmopolitan, found on domestic pigeons; a pest of pigeons reared for food in the Orient.

5.2.6 Class Arachnida, Order Acarina (Ticks and Mites)

These eight-legged arthropods are quite serious pests of livestock, partly as direct pests sucking blood but also as vectors of a range of pathogenic organisms. Many have a special piercing hypostome covered with curved hooks with which they attach themselves to the host and suck the blood. Cattle suffer especially from attack by ticks, and in parts of East Africa, for example, tick pests are probably the main limiting factor in cattle

production. Poultry also suffer quite severely from attacks by Acarina. Some of the more important pests are mentioned below.

(a) Family Dermanyssidae (Red Mites, etc.)

Most live on wild birds and rodents, but some can also attack humans. Two important pest species are:

- *Dermanyssus gallinae* (Poultry Red Mite) – a cosmopolitan species on domestic fowl; vector of various virus diseases; heavy infestations can be lethal.
- *Ornithonyssus bursa* (Tropical Fowl Mite) – attacks all types of poultry in the tropical regions.

(b) Family Argasidae (Soft Ticks)

These are soft-bodied, distendable ticks which can swell to a large size; they are cosmopolitan and feed on a wide range of Vertebrata, especially birds. *Argas persicus* is the Fowl Tick which causes anaemia and transmits several disease organisms. About 90 species of *Ornithodorus* are known, some of which are pests of livestock and vectors of pathogens.

(c) Family Ixodidae (Hard Ticks)

A large group, quite cosmopolitan, and very important pests of livestock. Infestations on cattle can be very large so that blood loss is appreciable – some cows lose up to 100 kg of blood in a single season (Harwood and James, 1979). Many ticks are the natural vectors of parasitic protozoa, viruses, spirochaetes and Rickettsias, which cause serious diseases in cattle (and humans) which may even be fatal. There are some grassland areas on all the continents that are rendered unsuitable for cattle-rearing owing to the large numbers of disease-carrying ticks that occur there. These large populations are maintained by the local wildlife (deer, etc.)

which have evolved tolerance to the ticks. Some of the most important genera include:

- *Amblyomma* (100 spp.) – large tropical ticks.
- *Boöphilus* (five spp.) – small in size (when unfed) and often missed in quarantine inspections.
- *Dermacentor* (30+ species) – mostly occur in North America.
- *Haemaphysalis* (150 spp.) – worldwide distribution, found on many different hosts.
- *Ixodes* (250 spp.) – worldwide on many hosts; vectors of several serious diseases of livestock.
- *Rhipicephalus* (60+ spp.) – Old World, mostly found in Africa and tropical Asia (Fig. 5.31).

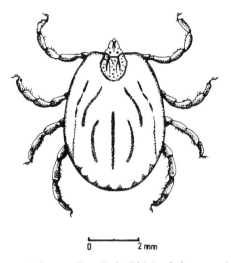

Fig. 5.31 Brown Dog Tick (*Rhipicephalus sanguineus* – Acarina; Ixodidae), engorged female, ex dog; Hong Kong.

(d) Family Demodicidae (Hair Follicle Mites)

These tiny worm-like mites live in hair follicles on the skin of various mammals (including humans). Different species of *Demodex* infest the skin of cattle (*D. bovis*), of dogs (*D. canis*), deer, horses, goats, pigs, cats and rodents. Some dogs are known to have died because of such infestation.

(e) Family Trombiculidae (Chiggers, etc.)

The larvae of *Trombicula* pierce the skin of birds and mammals and cause severe itching. Economic losses are recorded – chickens, turkeys, quail, horses and other livestock are attacked.

(f) Family Sarcoptidae (Mange Mites)

These are mites which burrow in the skin of various birds and mammals, and the major pest species are now quite cosmopolitan. The burrowing of *Cnemidocoptes* spp. (the Bird Skin Mites) causes 'scalyleg' on domestic fowl; other species can cause feather-shedding. *Sarcoptes scabei* (the Mange Mite) occurs as a series of different races (varieties): on cats and dogs; var. *bovis* on cattle; var. *equi* on horses, etc. The mites burrow in soft skin, making breeding galleries and cause intense itching.

(g) Family Psoroptidae (Scab Mites, etc.)

Hair bases are the favoured infestation sites, where the mites pierce the skin and cause inflammation (scab mange). Several species of *Chorioptes* and *Psoroptes* attack cattle and other livestock, and *Otodectes* (Fig. 5.32), the Carnivore Ear Mite, is commonly found in cats and dogs.

(h) Class Crustacea, Order Branchiura

A recent arthropod pest of interest is the Fish Louse (*Argulus*) which has become of great economic importance as a pest of commercially reared salmon in both the UK and Norway. Its control is currently posing many problems.

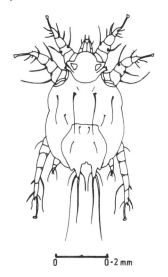

(a)   (b)

Fig. 5.32 (a) Carnivore Ear Mite (*Otodectes* sp. – Acarina; Psoroptidae); and (b) dog with mange.

5.3 HOUSEHOLD AND STORED PRODUCTS PESTS

Household and stored products pests can be categorized in three separate but related ways. Stored products refers to post-harvest grain, pulses and root crops, which are basically annual crops that have to be kept dried in storage for continued use throughout the remaining year. In the Western world post-harvest storage is now very sophisticated and grain, pulses and oil seeds (such as rape) are stored in bulk in regional co-operatives under very carefully monitored conditions (Fig. 5.33) (Hill, 1990). More perishable crops are kept in refrigerated stores. In the Third World, however, much produce is kept in on-farm stores (Fig. 5.34), in small quantities under quite primitive conditions. In a tropical climate deterioration of stored material can be rapid, aided by insect damage and moulds. Even regional and commercial stores suffer problems because of the climate and mechanical maintenance, and pest infestations are often far too high, causing extensive losses.

The term 'household' refers to dwellings and associated buildings/structures. In a primitive village there is probably only a series of huts and perhaps a single 'dukka' that serves as a café as well as a shop. In the UK, on the other hand, a whole village, town or city is included in the urban domestic habitat, and food pests include those insects that feed on meats, vegetables, fruits, confectionery, etc., as well as grain and grain products (such as flour and bread).

In the domestic situation there are also insect pests that attack the actual building itself – not too many attack bricks and concrete, but wood and wattle are very susceptible. The furnishings of a house or dwelling-place also have their share of insect pests.

These insect and mite pests are mostly synanthropic to some extent – in a number of cases the association is intimate and long-standing, but at the other extreme it can be very casual. Human foodstuffs fall roughly into two categories: the fresh and the preserved. Fresh foods are consumed immediately or else within a few days. With a domestic refrigerator, short-term food storage is possible, but this is usually only for a matter of a few days. The simplest type of

Fig. 5.33 Bulk grain store (Union Grain Co-operative, Orby, Lincolnshire, UK); a six-bay storage facility of 20 000 ton capacity.

Fig. 5.34 Small on-farm grain stores; Alemaya, Ethiopia.

food preservation is drying, either air-drying or with a kiln or oven. Smoking, pickling and salting are alternative methods of preservation, often associated with canning and, of course, nowadays freezing (to −20°C) has become a most important method.

The common and important insect and mite groups in domestic premises which cause damage to foodstuffs, furnishings and building structures are listed in Table 5.5.

The recent trend in temperate Europe and North America towards building new houses and converting the old, with double-glazed windows and efficient central heating, has virtually created a whole series of new permanent habitats for urban insects and mites. The favourable microclimates thus created allow the insects to feed and breed all the year round instead of just during the summer months. Thus several major pest species (such as the Hide Beetle, Carpet Beetle, Clothes Moth, Book Lice and Flour Moth) that had become relatively scarce in the UK during the last few decades (owing to increased levels of hygiene) are now becoming more common and causing serious damage. One of the Royal Palaces recently had to contract for pest control to kill Carpet Beetles that were destroying antique tapestries.

Another domestic problem is the increase in the animal flea population; both Cat and Dog Fleas now occur in larger numbers in most houses in the UK, and it is quite common to find fleas on pets in December and January – pet flea infestations used to be only a summer problem. Another major domestic pest is the House Dust Mite, which has really only become known since the advent of warmer homes, and is now a pest of serious consequence. The end result is that several domestic insect pests that were generally declining in importance are now becoming ever-increasingly abundant and causing more and more damage. This trend is likely to continue.

These domestic pests are presented in more detail below.

Fig. 5.35 Silverfish (*Lepisma saccharina* – Thysan.; Lepismatidae); Hong Kong.

5.3.1 Order Thysanura (Silverfish etc.)

These are primitive wingless insects with long tail filaments and cerci, and biting mouthparts. A few species are common domestic pests, eating foodstuffs and organic debris, and are now more or less worldwide in distribution. They include the Silverfish (*Lepisma saccharina*) (Fig. 5.35) which can also eat paper and several species of *Ctenolepisma*. The Firebrat (*Thermobia domestica*) is found in heated premises such as bakeries.

5.3.2 Order Collembola (Springtails)

In most parts of the world there are a few local species of springtail to be found in

Table 5.5 Major insects and mites as domestic pests

Group	*Common name*
Order Thysanura	Silverfish, etc.
Family Lepismatidae	Silverfish
Order Collembola	springtails
Order Orthoptera	
Family Gryllidae	Crickets
Order Dictyoptera	Cockroaches
Family Blattidae	
Epilampridae	
Order Isoptera	Termites
Family Termitidae	Subterranean Termites
Rhinotermitidae	Wet-wood Termites
Kalotermitidae	Dry-wood Termites
Order Pscoptera	Book- and Bark lice
Several families are involved	
Order Hemiptera	Bugs
Several families can be involved	
Order Coleoptera	Beetles
Family Dermestidae	Hide/Larder Beetles
Anobiidae	Wood Beetles
Ptinidae	Spider Beetles
Bostrychidae	Black Borers
Lyctidae	Powder-post Beetles
Cleridae	Checkered Beetles
Nitidulidae	Sap Beetles
Cucujidae	Flat Bark Beetles
Silvanidae	Flat Grain Beetles
Terebrionidae	Flour Beetles
Cerambycidae	Longhorn Beetles
Bruchidae	Seed Beetles
Anthribidae	Fungus Weevils
Apionidae	Seed Weevils, etc.
Curculionidae	Weevils
Order Diptera	Flies
Family Drosophilidae	Small Fruit Flies
Anthomyidae	Root Flies, etc.
Muscidae	House Flies
Calliphoridae	Blow Flies; Bluebottles, etc.
Piophilidae	Cheese Skipper
Tephritidae	Fruit Flies
Order Lepidoptera	Moths
Family Tineidae	Clothes Moths
Oecophoridae	House Moths
Gelechiidae	
Pyralidae	
Order Hymenoptera	
Family Formicidae	House Ants
Vespidae	Wasps
Colletidae	Plasterer Bees
Xylocopidae	Carpenter Bees
Order Acarina	Mites
Family Acaridae (*sensu lato*)	House Mites

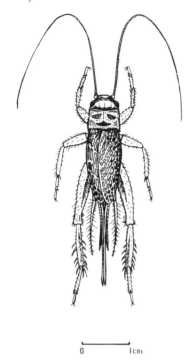

Fig. 5.36 House Cricket (*Acheta domesticus* – Orth.; Gryllidae), adult female; Hong Kong.

houses and other buildings, where they feed on food fragments, organic debris and fungi. They are seen occasionally in sinks and on floors but usually are of no importance.

5.3.3 Order Orthoptera (Crickets)

This group contains all the grasshoppers that are agricultural pests, but only a few crickets (family Gryllidae) are domestic pests. In the warmer parts of the world it is very common to have a House Cricket (*Acheta domesticus*) on the premises (Fig. 5.36); the nocturnal chirruping produced by the males is sometimes described as 'annoying' and sometimes as 'comforting'. In the UK they can be found in bakeries and some hotel kitchens where the perpetual warmth makes a suitable microclimate for their survival.

5.3.4 Order Dictyoptera (Cockroaches) (4000 species)

Formerly all the cockroaches were placed in the single family, Blattidae, but they are now divided into four families. These are scavengers of the forest floor, basically living in leaf litter – although some occur in caves and a number have invaded human dwellings and are established domestic pests. They are sometimes found in enormous numbers.

Direct damage by cockroaches is not often serious but their mere presence is disturbing and food contamination is very important on commercial premises. When present in large numbers they have an unpleasant characteristic smell. Direct damage includes the eating of book bindings, and some fabrics are also chewed. Adults can be very long-lived (*Periplaneta* lives for 12–18 months) and eggs are laid inside protective oöthecae. They are active at night and avoid daylight, and have large biting mouthparts. The three most important pest species are *Periplaneta americana* (the American Cockroach) (Fig. 5.37), now cosmopolitan and abundant in all warmer regions and in heated buildings in temperate countries. *Blatta orientalis* is the so-called Oriental Cockroach (native to North Africa!) (Fig. 5.38) with the flightless black female, now also found in Europe and North America. The smaller *Blatella germanica* (Fig. 5.39) is now placed in the family Epilampridae; it is completely cosmopolitan, very small in relation to the others, agile and active. It is also very adaptable, and is found as far north as Alaska in heated buildings. It is a common species on ships, even the steel warships of the British Royal Navy. The Blatellinae contains some 1300 species, so there are many other similar species worldwide.

5.3.5 Order Isoptera (Termites)

This is a tropical group that extends into Southern Europe and the southern USA. It

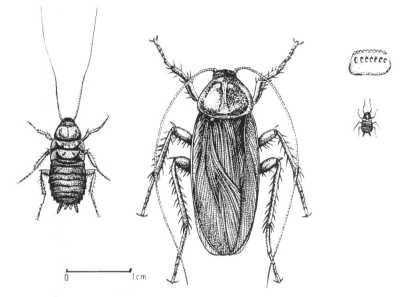

Fig. 5.37 American Cockroach (*Periplaneta americana* – Dict.; Blattidae), adult, nymphs and oötheca; Hong Kong.

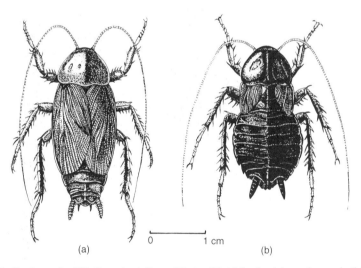

Fig. 5.38 Oriental Cockroach (*Blatta orientalis* – Dict.; Blattidae), (a) male and (b) female; Hong Kong.

is renowned for a complex social life and polymorphism and, in the hot tropics, for very spectacular nests. Termites are remarkable in that they can digest cellulose; they feed on growing crops, trees and wooden structures (buildings, bridges, etc.) according to the species concerned. Fletcher (1974) estimated that worldwide termite damage to growing crops, trees and wooden structures amounted to some US$500 million.

0 ————————— 1cm

Fig. 5.39 German Cockroach (*Blatella germanica* – Dict.; Epilampridae); Hong Kong.

The young adults are winged and undergo mating/dispersal flights on warm, wet evenings in the tropics. After mating, the wings are shed (at special basal fracture lines) and the young adults start to build a new nest and colony, usually underground. The female becomes the queen and her abdomen swells with eggs and fat body. She lives deep underground (in the Termitidae) together with the male (king) inside a special royal chamber (Fig. 5.40). The other castes are workers and soldiers, and are wingless and sterile. Workers are usually small and pale, while the soldiers are larger with a dark, sclerotized head capsule and long jaws. Soldiers defend the nest and also watch over the foraging workers. Daylight is avoided, so activity is either nocturnal or subterranean; the 'bark termites' build an enveloping sheath of mud and frass particles over the lower part of tree trunks and they remove the bark underneath this shielding cover.

In the family Termitidae the nest is underground, usually with a large surface mound (*Macrotermes*, etc.). Plant material is collected by the foraging workers and taken back to the

Fig. 5.40 Opened termite mound (*Macrotermes bellicosus* – Isoptera; Termitidae) in Kampala, Uganda; position of royal chamber (see Fig. 2.19) just below ground level in centre of mound.

Fig. 5.41 Fence post eaten away by underground termites (probably *Macrotermes* sp. (Isoptera; Termitidae); Alemaya, Ethiopia.

nest, where it is chewed and mixed with saliva to make a honeycomb-like structure called a 'fungus garden'. This cellulose matrix is inoculated with a special fungus (*Termitomyces*, etc.). The fungus breaks down the cellulose into its constituent starches and sugars and, as the mycelium develops, some hyphal ends become swollen with this food material. These swellings (called brometia) are eaten by the termites.

The domestic pests are those that destroy timbers and wooden structures, although some subterranean species of Termitidae can enter buildings and food stores through underground cracks. They will remove any stored grains and other foodstuffs to which they have access. External structural timbers underground are often attacked by *Macrotermes* and others (family Termitidae); the structures most susceptible are probably fence posts (Fig. 5.41), and wooden piles and supports for buildings and food stores. Many huts and houses in the tropics are built on stilts with concrete bases (Fig. 5.42) to make them safe from termite attack.

The two main domestic groups are:

(a) Kalotermitidae (Dry-wood Termites) (250 species)

These small species make a small colony inside completely dry wood – they have an intestinal microfauna of symbiotic protozoa, etc., that can break down cellulose. The termites can also fully utilize metabolic water, so they are effectively 'eating' dry wood. They can be very damaging to timbers and wood inside buildings, including furniture. The two main genera are *Cryptotermes* (Fig. 5.43) and *Kalotermes*.

(b) Rhinotermitidae (Wet-wood Termites)

These are also basically forest species, found on the forest floor where they tunnel into wet tree stumps, dead roots and fallen trunks. From this origin they have adapted to live in moist structural timbers such as wooden bridges and building foundations. Any house in the tropics with a leak in the roof is likely to have a Wet-wood Termite infestation in the roof timbers. They also occur in modern concrete blocks of flats (such as in Hong Kong) in airing cupboard timbers wetted by

Fig. 5.42 House in Malaya built on stilts with concrete bases to make the building safe from termite attack.

leaking water pipes and parquet floor blocks splashed by water from a kitchen sink. They are energetic and very opportunistic in seeking infestation sites. The two main pest genera are *Coptotermes* (45 species) (Fig. 5.44) in South East Asia and Australasia, and the

northern temperate *Reticulitermes* in Europe and North America. Both can digest cellulose directly via the micro-organisms in the gut. In 1994 an infestation of *R. lucifugus* was found in a house in rural Devon – the first record of a colony in the UK.

5.3.6 Order Pscocoptera (Bark- and Book Lice)

These tiny, soft-bodied insects may be either winged or apterous. Their presence in domestic premises is often overlooked because they are so small and inconspicuous, but they are becoming increasingly encountered and their damage to museum collections has long been known. They have weak biting mouthparts, and usually feed on fungal mycelium, lichens and algae, although some feed on general organic debris. In buildings some species are to be found on damp walls feeding on the microflora, and

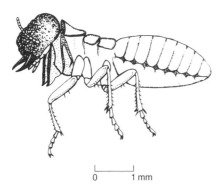

Fig. 5.43 Dry-wood Termite (*Cryptotermes brevis* – Isoptera; Kalotermitidae); soldier; Hong Kong.

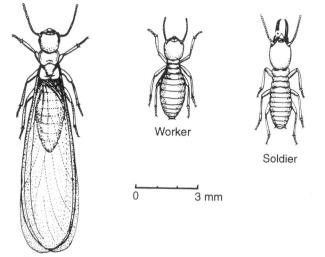

(a)

Worker

Soldier

Alate adult

0   3 mm

(b)

Fig. 5.44 Wet-wood Termite (*Coptotermes formosanus* – Isoptera; Rhinotermitidae); (a) alate adult, worker and soldier; Hong Kong; and (b) dead infested tree trunk.

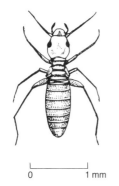

Fig. 5.45 Book Louse – typical wingless domestic species (Psocoptera); Hong Kong.

others invade foodstuffs in storage (Fig. 5.45). The name 'Book Lice' refers to their tendency to feed on damp bookbindings.

5.3.7 Order Hemiptera (Bugs)

A few Heteroptera are regularly found on domestic premises, and they are blood-sucking pests of humans, rats and pets. Some Homoptera are found on foodstuffs where they sometimes cause concern, but damage is usually slight and more aesthetic than actual. Vegetables and fresh fruits are often picked, collected or bought contaminated with aphids, mealybugs, soft or hard scales, and their presence is often overlooked because of their small size. Commercially infested produce could be downgraded or rejected and, in prepared foods, could contravene hygiene regulations. The Cabbage Aphid is a regular contaminant on Brassica plants. Potatoes in storage and chitting potatoes are often attacked by aphids; several species are regularly involved and because they are also virus vectors their infestation can be serious. In the UK, leaves of the bay tree, used as a spice, are often attacked by the Bay Psyllid. The leaf is deformed and usually rejected for culinary purposes, but the *Coccus hesperidum* scales can lie along the midrib on the ventral surface (Fig. 5.46) and escape notice.

5.3.8 Order Coleoptera (Beetles)

In terms of pests of foodstuffs and wood/timber this order contains most of the

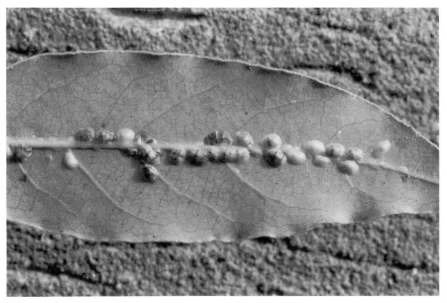

Fig. 5.46 Bay leaf with Soft Brown Scales (*Coccus hesperidum* – Hem.; Coccidae) along the midrib underside; Skegness, UK.

most important pest species. A small number of species in a large number of families (15 plus) are involved, but overall this is the largest group of insects to be classed as domestic pests.

(a) Family Dermestidae (Hide Beetles, etc.)

These are scavengers of corpses in the wild, adapted to feeding on dried skin, hair, fur, horns, hooves and feathers, but also dried fish, dried meat, bacon and cheese. In households wool, carpets and furnishings, as well as edible animal materials, are eaten. Both adults and larvae feed and cause damage, but the latter are the more serious.

One of the few herbivorous species is the Khapra Beetle (*Trogoderma granarium*) (Fig. 5.47), now cosmopolitan throughout warmer parts of the world, and a very serious pest of stored grains. Other major pests include *Dermestes lardarius* (the Larder Beetle; Fig. 5.48) which feeds on a wide range of animal and plant materials in warmer regions, and *D. maculatus* (the Hide Beetle; Fig. 5.49) which is more typical of the genus with its diet restricted to dried animal protein. It is a major pest in the tropics on sun-dried fish.

Anthrenus and *Attagenus* have species called Carpet Beetles, widespread in distribution, which are damaging to carpets, furs and some types of clothing.

(b) Family Anobiidae (Wood Beetles)

Most of the 1100 species are timber beetles and are found in dead wood. The larvae are rather scarabaeiform, with well-developed thoracic legs, but are sluggish. The adults are

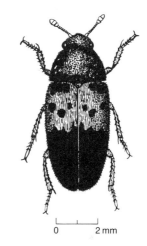

Fig. 5.48 Larder Beetle (*Dermestes lardarius* – Col.; Dermestidae).

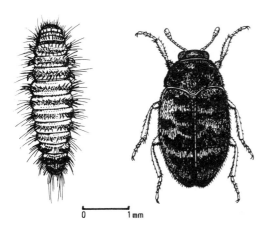

Fig. 5.47 Khapra Beetle (*Trogoderma granarium* – Col.; Dermestidae), larva and adult; Mombassa, Kenya.

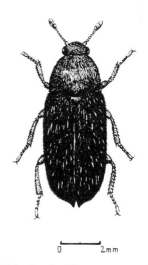

Fig. 5.49 Hide Beetle (*Dermestes maculatus* – Col.; Dermestidae).

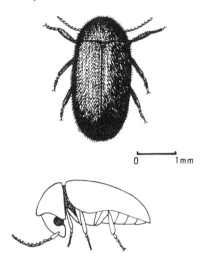

Fig. 5.50 Tobacco Beetle (*Lasioderma serricorne* – Col.; Anobiidae); Hong Kong.

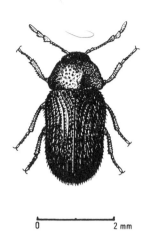

Fig. 5.51 Drugstore Beetle (*Stegobium paniceum* – Col.; Anobiidae), ex coriander seeds; Calcutta, India.

hard-bodied, with a large hooded pronotum shielding a deflexed head – this is typical of many timber-boring insects, so that the more delicate head is protected. Four species are important domestic pests – two in timber and wood, and two polyphagous foodstuff-feeders:

- *Anobium punctatum* (Common Furniture Beetle) – a temperate species also found in subtropical regions; very damaging to house timbers, floorboards, furniture and anything made of old softwood. It is also to be found in dead trees and branches. The larvae bore deep tunnels in wood (usually softwood), typically taking a year for development. The emerging adults leave a small, round exit hole together with a small pile of fine frass.
- *Lasioderma serricorne* (Tobacco Beetle) (Fig. 5.50) – a tropical species with larvae which are polyphagous on a wide range of stored foodstuffs. They are particularly damaging to tobacco.
- *Stegobium paniceum* (Drugstore Beetle) (Fig. 5.51) – more temperate and quite cosmopolitan, but feeding only on material of

plant origin. It is damaging to dried herbs, spices and some drugs.
- *Xestobium rufovillosum* (Deathwatch Beetle) – a larger, temperate species that attacks oak and other hardwoods where fungal decay is present. It is found in Europe and the USA, typically in churches and old buildings with oak rafters and beams.

(c) Family Ptinidae (Spider Beetles)

Out of 700 species some 24 are pests of stored foodstuffs worldwide. The body is globular, and with long legs and long antennae the general appearance is somewhat spider-like (Fig. 5.52). They feed on dried material of both plant and animal origin; in the wild they inhabit nests and lairs of birds and mammals in warmer regions.

(d) Family Bostrychidae (Black Borers; Auger Beetles)

These dark-bodied (black or brown), cylindrically shaped beetles, with a large hooded and toothed prothorax, are mostly borers of trees and woody shrubs. However, two species

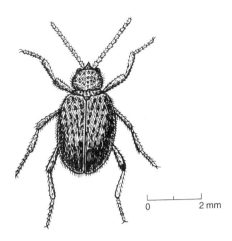

Fig. 5.52 Australian Spider Beetle (*Ptinus tectus* – Col.; Ptinidae); Hong Kong.

are important pests of stored grain in the tropics:

- *Rhizopertha dominica* (Lesser Grain-borer) – this is South American in origin but is now pantropical, as well as being found in heated stores in Europe and the USA (Fig. 5.53). It is a primary pest of grain (it can penetrate intact grains), both as adult and as larva. Post-harvest cereal crop losses can be very high.
- *Prostephanus truncatus* (Greater/Larger-Grain-borer) (Fig. 5.54) – this is only

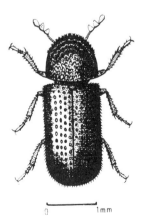

Fig. 5.53 Lesser Grain-borer (*Rhizopertha dominica* – Col.; Bostrychidae); Kenya.

minutely larger than the previous species, but is of great importance as an invader of tropical Africa from Central America. In its new location, in the absence of its usual predators and parasites, it is proving to be a devastating pest of stored maize (and to a lesser extent of stored cassava tubers). Its accidental introduction into Tanzania took place in 1975; it is now well established and spreading – post-harvest losses of maize of 70–80% in four months have been recorded. However, there is now a concerted international effort to find suitable natural enemies in Central America to bring to Africa, and other aspects of control are being studied in the hope of stopping the spread of this pest.

Some other species which are basically wood-borers can be found in food stores and domestic premises; these will have emerged from timbers and bamboos etc. used in the building's construction or from wooden containers (cf. Cerambycidae).

(e) Family Lyctidae (Powder-post Beetles)

These beetles have larvae that tunnel in the sapwood of various hardwood timbers from broad-leafed trees. The wood is often so tunnelled that all that remains is a fine, powdery frass. They are widespread but not common, and can cause considerable damage occasionally.

(f) Family Trogossitidae

A diverse group of beetles that contains one species of domestic importance. *Tenebriodes mauritanicus* is the Cadelle (Fig. 5.55), to be found in stored foodstuffs. The larvae and adults are somewhat omnivorous and feed on grains, processed foodstuffs and insects. Larvae can survive for months without food, and adults live for one to two years, which accounts to some extent for their pest status.

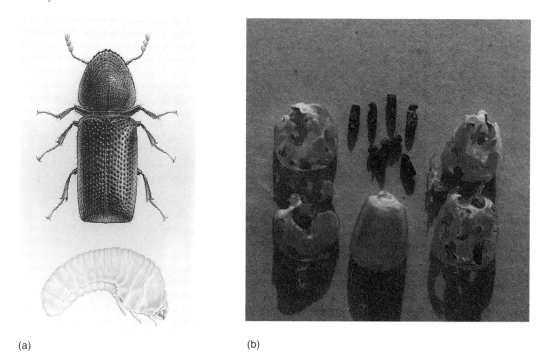

(a)   (b)

Fig. 5.54 Larger Grain-borer (*Prostephanus truncatus* – Col.; Bostrychidae); (a) adult and larva (courtesy of ICI); and (b) maize grains after attack.

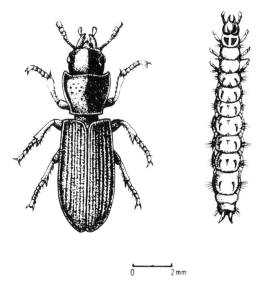

0   2 mm

Fig. 5.55 Cadelle, adult and larva (*Tenebriodes mauritanicus* – Col.; Trogossitidae).

(g) Family Cleridae (Checkered Beetles)

A large tropical group mostly with both adults and larvae preying on wood-boring beetles (especially Scolytidae). However, *Necrobia* occurs on dead animals and will also eat fly maggots in the carcass. Two species are known as 'Ham Beetles' (Fig. 5.56) and are serious pests on stored hams and bacon, and copra in the tropics.

(h) Family Nitidulidae (Sap Beetles)

A large family of variable form and habits. A few are pests, but they are not very important. Several species of *Carpophilus* are known as Dried Fruit Beetles (Fig. 5.57) – these attack ripe fruits in the field and dried fruits in storage, and often carry fungi on their bodies. Damaged pineapples are often

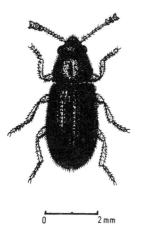

Fig. 5.56 Copra/Ham Beetle (*Necrobia rufipes* – Col.; Cleridae); Hong Kong.

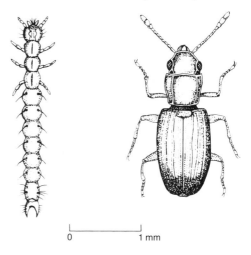

Fig. 5.58 Flat Grain Beetle (*Cryptolestes ferrugineus* – Col.; Cucujidae).

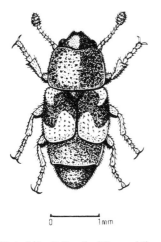

Fig. 5.57 Dried Fruit Beetle (*Carpophilus hemipterus* – Col.; Nitidulidae); Makerere, Uganda.

infested with these small beetles and their presence causes problems in canning factories.

(i) Family Cucujidae (Flat Bark Beetles)

These small, flattened beetles mostly live under the bark of dead trees; some feed on dead insects in spiders' webs while others are phytophagous. The genus *Cryptolestes* contains Flat Grain Beetles (Fig. 5.58) which are

of some importance as secondary grain pests worldwide.

(j) Family Silvanidae (Flat Grain Beetles)

Very closely related to the former group, this family is also of varied habits. A couple of important pest species attack stored grains and foodstuffs. *Oryzaephilus* occurs as two species; these are important secondary pests of grain worldwide and on all types of cereal products. *Ahasverus* is also found worldwide but is a fungus-feeder, usually on damp produce.

(k) Family Tenebrionidae (Flour Beetles; Darkling Beetles, etc.)

A large group comprising 15 000 species, which are found worldwide but are most abundant in the tropics, both in forests and hot deserts (Sahara, Namibia, etc.). More than 100 species have been recorded on stored foodstuffs worldwide, but the most important domestic pests are quite few in number:

- *Blaps* (Churchyard Beetles) – found in cellars of buildings in Europe and Asia.

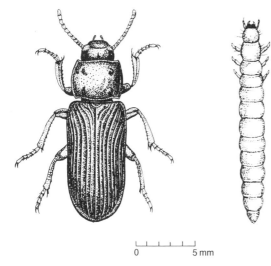

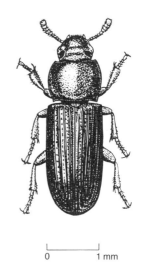

Fig. 5.59 Yellow Mealworm Beetle and larva (*Tenebrio molitor* – Col.; Tenebrionidae); Hong Kong.

Fig. 5.60 Flour Beetle (*Tribolium* sp. – Col.; Tenebrionidae); Hong Kong.

- *Gnathocerus* (Horned Flour Beetles) – two species, both pantropical and temperate, are found on a wide range of foodstuffs, but as with others of this family they are only secondary pests in stored grains.
- *Latheticus oryzae* (Long-headed Flour Beetle) – found in South East Asia.
- *Palorus* (Depressed Flour Beetle, etc.) – pantropical, and often omnivorous, feeding both on broken grains and other insects.
- *Tenebrio* (Mealworms) – two large species; temperate in Europe and the USA (Fig. 5.59); mostly on cereal products. The larvae (mealworms) are reared as food for birds and other animals.
- *Tribolium* (Flour Beetles) – three main species, two pantropical and one temperate (Fig. 5.60). These are very important secondary pests in grains, and are widespread in a range of cereal products and flours.

(l) Family Cerambycidae (Longhorn Beetles)

A large group of forestry pests which are occasionally found in domestic situations.

The larvae are long-lived (one to three years) and can be present in untreated timber and bamboos. They can thus emerge from domestic structures, such as doors, rafters and beams. *Hylotrupes bajulus* (the House Longhorn Beetle) bores in coniferous timbers in Europe and South Africa and is found in houses often enough to warrant its common name.

(m) Family Bruchidae (Seed Beetles) (1300 species)

A worldwide group with larvae that develop inside seeds – mostly in the Leguminosae. They are very serious pests of stored dried legumes (pulses), especially in warmer countries. Some start as pests of the ripening legume crop in the field and then carry on with continuous breeding in the store. However, others are only to be found in the field and will not breed on dried pulses. The more important pests of stored pulses are as follows:

- *Acanthoscelides* (Bean Bruchid, etc.) (Fig. 5.61) – in the tropics this can destroy

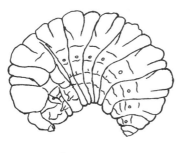

Larva

Egg

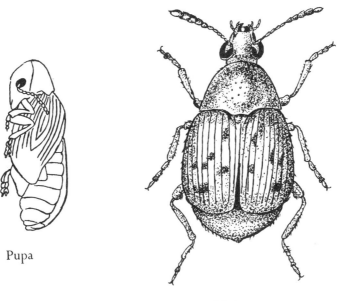

Pupa

Adult

Windowed beans

Holed beans

Fig. 5.61 Bean Bruchid (*Acanthoscelides obtectus* – Col.; Bruchidae); adult and immature stages; Kenya.

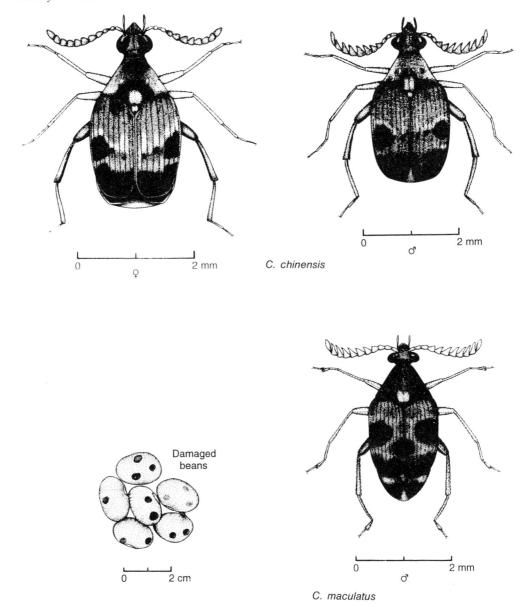

C. chinensis

Damaged beans

C. maculatus

Fig. 5.62 Cowpea Bruchids (*Callosobruchus* spp. – Col.; Bruchidae).

(infest) up to 80% of stored beans in six to eight months.

- *Callosobruchus* (Cowpea Bruchids, etc.) (Fig. 5.62) – these attack most grain legumes; most are pantropical but some are cosmopolitan.

- *Caryedon serratus* (Groundnut-borer) – pantropical; mostly on stored groundnuts, but also on other produce.

- *Zabrotes* (Mexican Bean Beetle, etc.) – mostly found on beans in Central and South America, and parts of Africa.

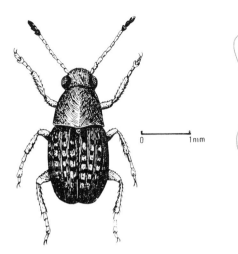

Fig. 5.63 Nutmeg/Coffee Bean Weevil (*Araecerus fasciculatus* – Col.; Anthribidae); Hong Kong.

(n) Family Anthribidae (Fungus Weevils)

The genus *Araecerus* has several species that attack various dried seeds and fruits. *A. fasciculatus*, the Nutmeg or Coffee Bean Weevil (Fig. 5.63), is now found throughout the tropics.

(o) Family Apionidae (Seed/Pod Weevils, etc.)

A common, worldwide group with larvae that develop in the seeds of Leguminosae etc. Species of *Apion* are Seed or Pod Weevils, and most are found in the smaller-seeded legumes, such as clovers. Two species of *Cylas* are Sweet Potato Weevils and are very damaging to this crop (Fig. 5.64), both in the field and to some extent in storage (most infested tubers rot in storage).

(p) Family Curculionidae (Weevils proper)

An enormous group of beetles, with many serious agricultural and forestry pests, and a few stored foodstuffs pests. The damage is done by the feeding larvae, but most adult females use the long 'snout' to make a feeding hole which is then used as the egg-site. Thus the egg is protected and larval survival ensured.

The few stored products pests are species of *Sitophilus* – three species are important (Fig. 5.65), two pantropical and one more temperate. They are very damaging primary pests of stored grains; the tropical species starting as field infestations (adults can fly for about half a mile) on the ripening grain.

5.3.9 Order Diptera (Flies)

A few flies are dominating in the human domestic situation, but really the group is most important as medical and veterinary pests. The domestic (synanthropic) species are mostly Muscoidea where the adults have a sucking/absorbing proboscis and feed on liquids, and the reduced legless larvae (maggots) are saprophagous. Adult flies transmit a number of pathogens passively, particularly by food contamination. Some species regurgitate a mixture of saliva and gut contents in order to digest solid foodstuffs externally, and some bacteria pass through the fly alimentary canal and are voided with the faeces. The larvae can be responsible for spoilage of foods, and they sometimes cause myiasis of the alimentary tract. Sometimes they are used to break down tough animal muscle. In the UK the practice of 'hanging' game animals arose primarily because people in cities became used to eating 'high' (starting to decay) game meat. In the days of horse-drawn transport, game birds such as grouse and animals such as Red Deer were commonly shot in Scotland and transported to London, a journey that took one to two weeks, so that the game was starting to decay by the time it reached its destination. On the other hand, some slight decay was probably desirable, for the flesh of wild creatures is typically tough – both muscle and connective tissue are difficult to tear and chew. In the modern meat industry it is the practice to hang beef carcasses for a few days before sale in order to break down some connective

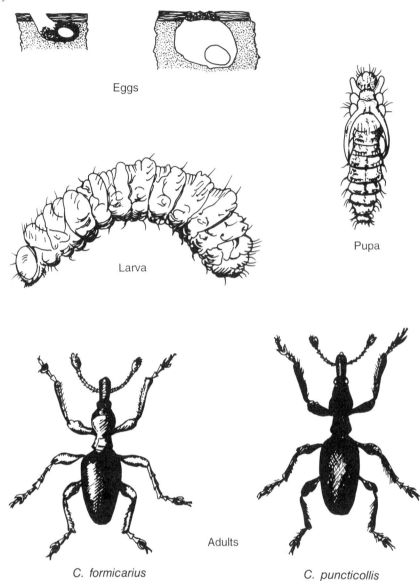

Eggs

Pupa

Larva

Adults

C. formicarius   *C. puncticollis*

Fig. 5.64 Sweet Potato Weevils (*Cylas* spp. – Col.; Apionidae); adult and immature stages; Kenya.

tissues so as to make the meat more edible. There are probably a few wealthy elderly gentlemen who still hang their shot pheasants and grouse until the 'maggots fall out'. Similarly, a few probably keep their Stilton (and other) cheeses until the maggots (Cheese Skipper) are evident (see section (f)).

The more important domestic flies belong to the following groups:

(a) Family Drosophilidae (Small Fruit Flies; Vinegar Flies)

These are the tiny flies (Fig. 5.66) so important in genetic research; they are attracted by

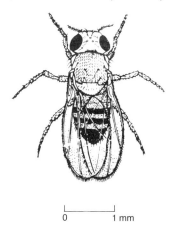

Fig. 5.66 Vinegar Fly (*Drosophila* sp. – Dipt.; Drosophilidae); Hong Kong.

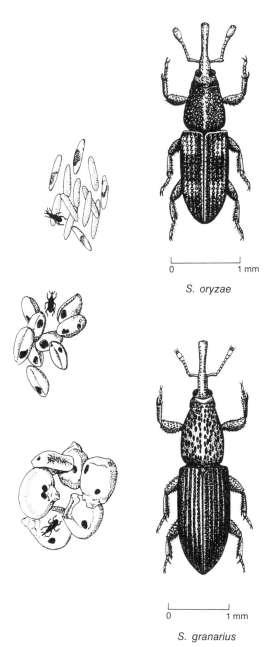

S. oryzae

S. granarius

Fig. 5.65 Grain Weevils (*Sitophilus* spp. – Col.; Curculionidae); Hong Kong.

various byproducts of fermentation and fly to over-ripe fruits where they often feed on the fungus rather than the fruit itself. An entomological colleague once 'lost' a

ripe banana inside his car and for a month its interior was filled with *Drosophila* flies. In domestic premises they are found occasionally as small, white maggots inside over-ripe fruits.

(b) Family Anthomyiidae (Root Flies, etc.)

Most are crop pests but a few species are encountered in stored produce, mostly vegetables, although they do not breed in storage. The occasional Onion Fly (*Delia antiqua*) is found in storage, but most infested bulbs are removed at inspection prior to packaging. Similarly, Cabbage Root Fly (*D. radicum*) maggots can be found in Brussels sprouts (either fresh or frozen) and radishes.

(c) Family Muscidae (House Flies, etc.)

A large group of flies (comprising nearly 4000 species), many of which are synanthropic; they appear to have evolved from a compost-feeding ancestor and have diversified into plant-feeders, dung- and carrion-feeders and (less common) blood-feeders. Domestically both adults and larvae can be pests; the adults mostly by spreading fungi and bacteria

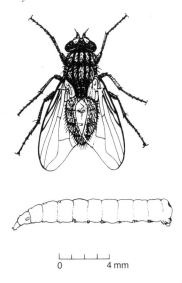

Fig. 5.67 House Fly and maggot (*Musca domestica* – Dipt.; Muscidae); Hong Kong.

on to foodstuffs which are then rapidly spoilt. If eggs are laid on the foodstuffs then the feeding larvae will start putrefaction and encourage development of saprophagous micro-organisms. In commercial food and catering premises, the presence of fly larvae and adult flies is sufficient to cause alarm and could lead to legal prosecution.

The species most concerned with domestic nuisance is clearly *Musca domestica* (the House Fly) (Fig. 5.67) which is totally synanthropic and now cosmopolitan. However, in the hot tropics other species may be more abundant. *Ophyra* are known as Tropical House Flies and several species are pantropical.

(d) Family Calliphoridae (Blow Flies; Bluebottles; Greenbottles, etc.)

These larger flies are not so abundant in domestic habitats, but many species are synanthropic. They are saprophagous carrion- and dung-feeders, and the adults also take nectar from flowers. Domestically *Calliphora* are very opportunistic and uncovered meat quickly attracts flying adults. An egg

mass of a dozen yellow eggs can be laid in a matter of minutes. Such meat is then known as 'fly-blown'. One of the common tropical genera is *Chrysomya*, found mostly in Asia and Africa. *Lucilia* are the distinctive Greenbottles but these are not common in domestic premises.

(e) Family Sarcophagidae (Flesh Flies)

These large and distinctively striped grey/black flies (Fig. 5.68) are quite important domestically for they are larviparous. Thus an infestation by *Sarcophaga* develops very rapidly as first instar larvae are deposited by the female and they can quickly burrow into the food material to hide. Each female *Sarcophaga* can deposit 40–80 larvae in her lifetime.

(f) Family Piophilidae (Skippers)

Only six species exist; they are saprophagous and usually live in carrion. However, *Piophila casei*, the Cheese Skipper (Fig. 5.69), is now cosmopolitan, found infesting cheeses and fatty bacon and hams. A few epicures claim to enjoy their cheese infested with maggots but the majority of consumers find this

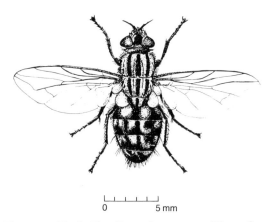

Fig. 5.68 Flesh Fly (*Sarcophaga* sp. – Dipt.; Sarcophagidae); Hong Kong.

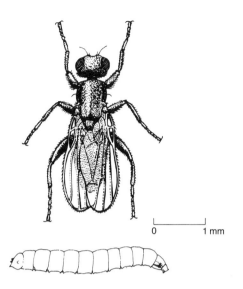

Fig. 5.69 Cheese Skipper Fly and larva (*Piophila casei* – Dipt.; Piophilidae); Hong Kong.

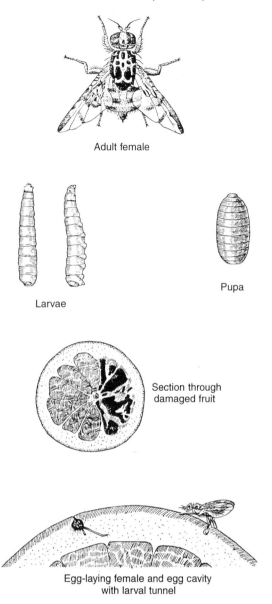

Adult female

Larvae

Pupa

Section through damaged fruit

Egg-laying female and egg cavity with larval tunnel

Fig. 5.70 Medfly (*Ceratitis capitata* – Dipt.; Tephritidae) and immature stages; Kenya.

objectionable; there have been many re-corded cases of human intestinal myiasis due to these maggots.

(g) Family Tephritidae (Fruit Flies)

These are brightly coloured flies that are serious pests of fruit crops and some vegetables. The maggots develop inside the fruits, usually without any obvious external signs. Some fruits may fall prematurely but many stay on the tree, and the maggots will not be seen until the fruit is opened; often this takes place in the kitchen. In Hong Kong (New Territories) guava is planted in places partly as a fruit tree but also to stabilize the soil on some steep and delicate hillsides. However, as a fruit crop it is not a success – many a luscious-looking ripe fruit has been bitten into deeply and resulted in a mouthful of large white maggots! Figure 5.70 shows a female Medfly (Mediterranean Fruit Fly) with the characteristic striped body and marbled wings. The other common genus, *Dacus*, is a drabber insect with hyaline wings (often with a terminal spot).

5.3.10 Order Lepidoptera (Moths)

There are two groups of moths that are serious domestic pests; only the larval stage (caterpillar) is damaging and the damage is done by the larvae feeding with their

(a)

(b)

Fig. 5.71 (a) Buffalo horns bored and eaten by larvae of *Ceratophaga* sp. (Lep. Tineidae); Queen Elizabeth Park, Uganda; (b) close-up of the caterpillars under the tunnel-tube mass; Alemaya, Ethiopia.

biting and chewing mouthparts. They are widespread in the warmer parts of the world, but also common in heated premises in the colder temperate regions. With the advent of central heating and double-glazing in the UK, the following species are becoming ever more common in houses.

- Grain/flour, etc. eaters – the genera *Ephestia*, *Cadra*, *Pyralis* and *Plodia*, etc. in the family Pyralidae.
- *Sitotroga* (Gelechiidae) and some Oecophoridae (House Moths) – these are phytophagous and polyphagous in the main.
- Hair-/wool-/carpet-eaters – the main culprits are the Clothes Moths (family Tineidae), especially *Tinea* and *Tineola* whose larvae have enzymes in the gut capable of breaking down keratin. In the wild they feed on cadavers and dried corpses, eating mainly the dried skin, horns (Fig. 5.71), fur and feathers of Bovidae, antelopes and birds. Several species of Pyralidae also feed on these materials in the wild and on owl pellets.

These are sometimes found in domestic situations.

The main groups of importance are reviewed below:

(a) Family Tineidae (Clothes Moths, etc.) (2400 species)

A worldwide group found on dried animal corpses in the wild and on domestic premises (also in fungal bodies, a wide range of dried plant materials and birds' nests). The more important pests are in the genera *Ceratophaga* in tropical Africa (Fig. 5.71), and the Common Clothes Moth (*Tineola*) (Fig. 5.72). The Corn Moth (*Nemapogon granella*) is a cosmopolitan polyphagous pest on dried plant materials including grains, nuts, fruits and dried fungi.

(b) Family Oecophoridae (House Moths, etc.) (3000 species)

Quite a large group, of varied habits, with a few eating seeds and grains in storage and

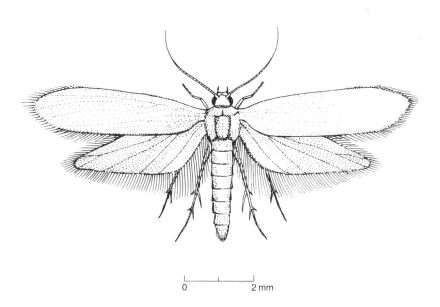

```
0                    2 mm
```

Fig. 5.72 Common Clothes Moth (*Tineola bisselliella* – Lep.; Tineidae); Hong Kong.

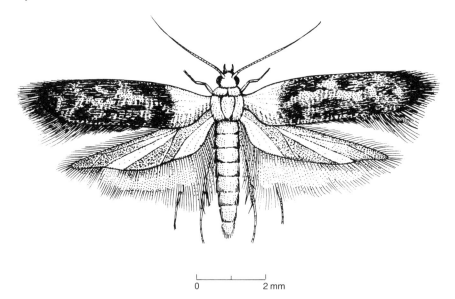

Fig. 5.73 White-shouldered House Moth (*Endrosis sarcitrella* – Lep.; Oecophoridae); Skegness, UK.

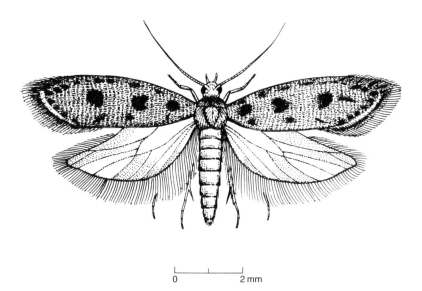

Fig. 5.74 Brown House Moth (*Hofmannophila pseudospretella* – Lep.; Oecophoridae); Skegness, UK.

two common household pests: *Endrosis sarcitrella* (the White-shouldered House Moth; Fig. 5.73) and *Hofmannophila pseudospretella* (the Brown House Moth; Fig. 5.74). The larvae feed on vegetable materials and organic debris, and sometimes some animal materials such as leather but usually not on clothing. The two are very common throughout the UK, and are both regarded as cosmopolitan.

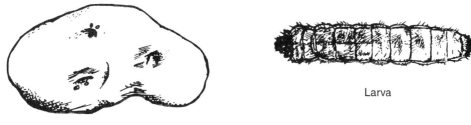

Eggs

Larva

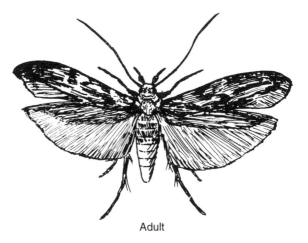

Pupa

Adult

(a)

(b)

Fig. 5.75 (a) Potato Tuber Moth (*Phthorimaea operculella* – Lep.; Gelechiidae) and immature stages; Kenya; (b) damaged potato tubers from Alemaya, Ethiopia.

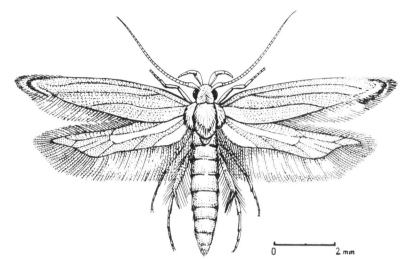

(a)

(b)

Fig. 5.76 (a) Angoumois Grain Moth (*Sitotroga cerealella* – Lep.; Gelechiidae); and (b) infested sorghum panicle; Alemaya, Ethiopia.

(c) Family Gelechiidae (4000 species)

This is a large, diverse group with 400 genera found worldwide. Many larvae bore in plant stems, fruits or seeds and leaves. Two species are of particular importance so far as stored produce is concerned:

- *Phthorimaea operculella* (Potato Tuber Moth) (Fig. 5.75) – this starts as a field infestation and infested tubers may be taken into stores where development continues. The

life cycle can continue in storage indefinitely, which makes this a very serious potato pest throughout much of the warmer part of the world. A similar pest, confined to South and Central America, is *Scrobipalpopsis solanivora*.

- *Sitotroga cerealella* (Angoumois Grain Moth) (Fig. 5.76) – a major pest of stored grains throughout the world in warmer regions and in heated stores in temperate countries. Infestation usually starts un-

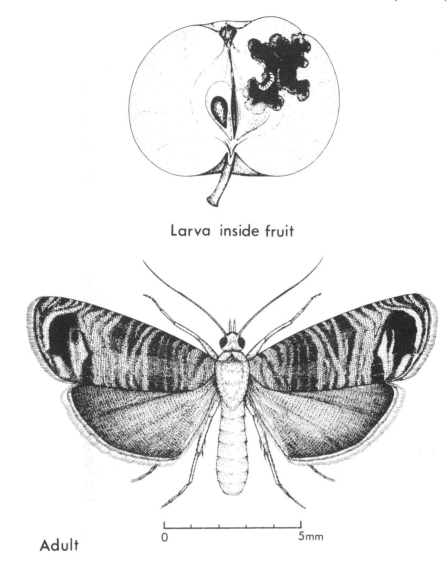

Larva inside fruit

Adult

0   5mm

Fig. 5.77 Codling Moth (*Cydia pomonella* – Lep.; Tortricidae); Cambridge, UK.

noticed in the field, and the life cycle continues in storage. It is equally damaging to sorghum, maize and wheat in different regions – in eastern Ethiopia (Alemaya University farm) sorghum was apparently preferred.

(d) Family Tortricidae

A large group of small moths renowned for their damage to growing crops. A few are encountered in the kitchen. In the UK the commonest is probably the Pea Moth (*Cydia nigricana*); infested pods show no outward sign but when opened they reveal one or two fat little caterpillars, several eaten peas and a pile of faecal frass (Fig. 5.182). The related Codling Moth (*Cydia pomonella*) (Fig. 5.77) is a very common pest of apples and pears worldwide, but a discoloured hole in the skin surface gives away the fact that the fruit is

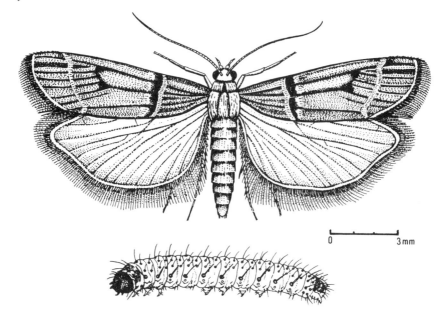

Fig. 5.78 Dried Currant/Tropical Warehouse Moth and larva (*Cadra cautella* – Lep.; Pyralidae); Hong Kong.

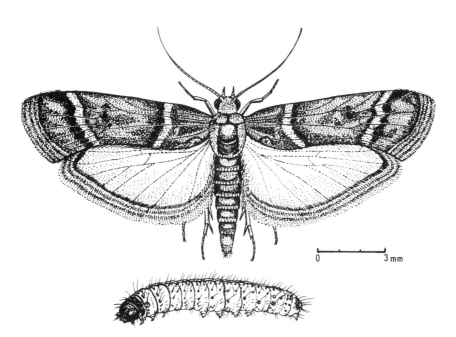

Fig. 5.79 Warehouse Moth and larva (*Ephestia eleutella* – Lep.; Pyralidae); Hong Kong.

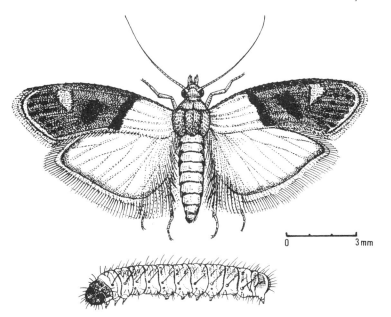

Fig. 5.80 Indian Meal Moth and larva (*Plodia interpunctella* – Lep.; Pyralidae); Hong Kong.

infested (Fig. 4.2), and such fruit (which may fall prematurely) is not kept for storage. A few other species of Tortricidae occur in fruits and nuts but are only rarely encountered domestically.

(e) Family Pyralidae (Snout Moths)

A very large group with a wide range of crop pests causing extensive damage worldwide. Most are agricultural pests but a few are important pests of stored foodstuffs. The species most likely to be encountered domestically, worldwide, include:

- *Herculia nigrivitta* – larvae destroy dried palm leaves woven into the coarse, woven sheets known as 'atap' and used as a roof covering on buildings in South East Asia.
- *Hypsopygia* – several species feed on dried grasses (hay) and damage thatched roofs and haystacks in Europe.
- *Pyralis* – several species are widespread pests of stored foodstuffs, although some also feed on meats.

- *Cadra* – several species feed on dried fruits, beans, nuts and grains (Fig. 5.78).
- *Ephestia* – two main species attack a range of stored foodstuffs (Fig. 5.79).
- *Plodia interpunctella* (Indian Meal Moth) (Fig. 5.80) – larvae prefer food of vegetable origin; cosmopolitan in warmer regions.

In the Western world, some of the popular farinaceous products, such as breakfast cereals and biscuits, some confections and dried nuts and fruits are sometimes packaged complete with moth eggs. If the produce is eaten promptly no one is any the wiser, but if a package is left for a long while a small cloud of moths may be released upon opening or a seething mass of caterpillars may be revealed inside. Despite ever-increasing levels of commercial hygiene in food preparation, such infestations are surprisingly common; it is only cool storage and prompt consumption that prevents many more instances being observed.

A few other caterpillars can be found on foodstuffs in the kitchen, but only occasionally, because of produce inspection. These include:

- Cabbage Caterpillars (*Pieris* spp.; *Plutella xylostella*; *Mamestra brassicae*) – these occur on all cultivated Cruciferae; usually as caterpillars, although the Diamond-back Moth pupates in a silken cocoon securely fastened to the plant and has been recorded ruining an otherwise prime crop of broccoli heads.
- Pyralidae and Noctuidae – several species have caterpillars that bore into capsicum fruits (such as sweet peppers) and feed inside.

A few species are known as 'Eye Moths', and throughout South East Asia visit mammals at night to suck lachrymal fluids from their eyes. They sometimes enter houses to seek sleeping people.

5.3.11 Order Hymenoptera (Ants, Wasps, Bees)

A very large group of parasitic or stinging insects, some of which are social and live in large communities (nests). They are not serious household pests as a group but a few species will be encountered.

(a) Family Formicidae (Ants) (15 000 species)

These are social insects where the reproductive female (queen) stays in the nest. Polymorphism is usual, with small wingless workers and large aggressive soldiers. The foraging workers make and follow pheromone trails. Damage and nuisance are caused by the workers and soldiers, but control measures have to be directed against the nest colony and in particular the queen. Most species are omnivorous.

A few species are regarded as 'Garden Ants', such as *Lasius niger* (Common Black Ant in the UK); these often invade the ground floor of buildings causing nuisance, but the nest is invariably subterranean outside. Other synanthropic species may have the nest inside the building or outside in close proximity, such as the tropical *Iridomyrmex* (200 spp.) (Fig. 5.81) known as House or Meat Ants, and the cosmopolitan *Tapinoma* (60 spp.) called House or Sugar Ants. The tiny Pharaoh Ant, *Monomorium pharaonis* (Fig. 5.82), is one of 300 species; it is a cosmopolitan pest of warm buildings. In the UK it can be a serious pest in hospitals; it usually

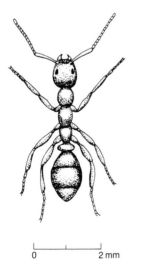

Fig. 5.81 The Tropical House (Sugar) Ant (*Iridomyrmex* sp. – Hym.; Formicidae); Hong Kong.

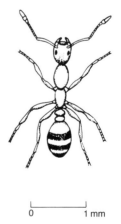

Fig. 5.82 Pharaoh Ant (*Monomorium pharaonis* – Hym.; Formicidae); Hong Kong.

nests in some inaccessible basement or foundation corner and foraging workers have trails throughout the establishment. They have been shown to transmit *Salmonella* and other bacteria and so·are especially undesirable in hospitals.

(b) Family Vespidae (Wasps)

These are large, yellow and black (sometimes reddish) winged insects that live in large colonies within a papery nest. They are carnivorous, but are also attracted to sugar, jam and ripe fruits, and the sting is used to immobilize prey – usually other insects. They are very aggressive in defence of the nest and accidental intruders may be savagely attacked. Both wasps and bees secrete an 'attack' pheromone when they sting, which accounts in part for their ferocity.

In the tropics several species, such as *Balanogaster*, build tiny nests of about a dozen cells, under the eaves of buildings or attached to ceilings, and if accidentally disturbed the wasps readily sting. The large tropical wasps (*Vespa* spp.) usually nest in trees, often in gardens, but high up and out of reach cause no nuisance. The temperate wasps (*Vespula* spp., etc.), however, often nest inside buildings. Favourite sites include lofts and attics, and also garden sheds, but some species prefer to nest underground and choose garden rockeries and similar sites. The presence of a wasps' nest in a house or garden almost invariably causes great alarm and regrettably usually ends with the killing of the colony.

(c) Family Sphecidae (Mud-dauber Wasps, etc.)

These solitary wasps, with a distinctive long, narrow petiole (waist), build a small nest of mud in the corners or crevices in buildings. Some smaller species use keyholes, bookcase shelf holes and other tiny spaces in which to nest, and they then just seal off the entrance. *Sceliphron* is a common larger wasp found throughout the tropics and is a real mud-

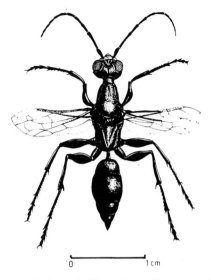

Fig. 5.83 Cockroach Wasp (*Ampullex* sp. – Hym.; Sphecidae); Hong Kong.

dauber, making the whole nest cell of mud. The prey are invariably spiders. *Trypoxylon* are smaller and use natural cavities and crevices; the prey are also small spiders. If accidentally disturbed, these solitary wasps will occasionally sting people. The Cockroach Wasps (*Ampullex*) (Fig. 5.83) have not been recorded stinging man.

(d) Family Eumenidae (Potter Wasps; Mason Wasps)

These are solitary species found worldwide, but are more common in warmer climates. There are 22 species in the UK. *Eumenes* are the Potter Wasps, which build beautiful small urn-shaped mud nests and provision them with paralysed small caterpillars.

Mason Wasps include species of *Odynerus* that use natural cavities and will nest in holes in walls, holes in concrete posts, etc. *Ancistrocerus* are less urban but may sometimes occur in gardens.

These small wasps are attractive in appearance and pose no threat to the house occupants, but are often viewed with alarm because they look like 'wasps'.

Fig. 5.84 Nest tunnel of *Colletes daviesanus* in house brickwork mortar; Skegness, UK, 1972.

(e) Family Colletidae (Membrane/Plasterer Bees)

Colletes is a Holarctic genus of solitary bees that sometimes nest gregariously. *C. daviesanus* is recorded in the UK, making horizontal tunnels in the mortar between bricks. Several houses are recorded as suffering from the attack of several hundred bees (Fig. 5.84) which caused structural damage; their buzzing also frightened the house occupants. The nest tunnel is usually 5–10 cm deep and contains between two and eight separate cells, each containing a mixture of honey and pollen and a single egg.

(f) Family Xylocopidae (Carpenter Bees)

These tropical bees are large, hairy, stout-bodied and often either black or rufous in colour. The large mandibles are used to excavate breeding tunnels in solid wood. Favoured sites include structural timbers (beams) projecting in the eaves of huts and houses (Fig. 5.85), and also telephone and electricity poles. Occasionally, beams are

seen with enough of the 1 cm diameter holes (10–20 cm deep) actually to weaken them physically. The flying bees are very large, conspicuous and noisy and may cause alarm by their presence.

(g) Family Apidae (Bees; Social Bees)

The only species encountered in houses are the tiny *Trigona* (Mosquito/Stingless Bees). In Kampala, bungalow water-tank overflow pipes can be colonized by nests of *Trigona* – but they are really harmless.

Honey Bees are domesticated and usually regarded as beneficial insects, but they can be aggressive and a nuisance if a neighbour has hives in the garden or under the eaves of their home.

5.3.12 Class Arachnida, Order Acarina (Mites)

Because of the small size of these arthropods and a dearth of more obvious anatomical

(a)

(b)

Fig. 5.85 (a) Carpenter Bee (*Xylocopa* sp. – Hym.; Xylocopidae) (body length 23 mm); south China; (b) nest tunnels in dead (fallen) tree trunk.

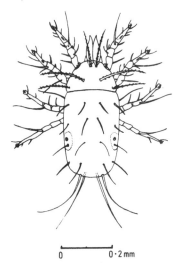

0   0·2mm

Fig. 5.86 Flour Mite (*Acarus siro* – Acarina; Acaridae); Hong Kong.

features (such as wings and mandibles, etc.) the classification of this order is subject to a great deal of dispute. With regard to domestic pests two different families are mentioned, but in other books and other regions other group names might be used.

(a) Family Acaridae (Astigmata partim)

A diverse group of free-living mites found in foodstuffs and in houses and other human habitations. Most are tiny species and can be inhaled into the lungs and swallowed into the intestine. On the skin or scalp they can cause dermatitis. Several species are responsible for what is called 'Grocers' Itch'. Despite their tiny size, they can occur in such vast populations that food materials can be completely destroyed during a couple of months' storage. And of course their tiny size means that most infestations escape notice, at least initially.

The main domestic pests include:

- *Acarus siro* (Flour Mite) (Fig. 5.86) – cosmopolitan, feeds on a wide range of foodstuffs.
- *Glycyphagus domesticus* (House Mite) – widespread.

- *Rhizoglyphus* spp. (Cheese Mites) – several species; widespread.
- *Tyrophagus* spp. – widespread and polyphagous.

(b) Family Pyroglyphidae

This contains the now very serious domestic pest *Dermatophagoides pteronyssinus* (House Dust Mite), which causes widespread asthma and respiratory allergic disorders in humans. They apparently feed on tiny flakes of skin (keratin) and general organic debris – they are not generally pests of stored foodstuffs.

5.4 AGRICULTURAL PESTS

In the context of this book, agriculture refers to the cultivation of plants for food and products (such as cotton, sisal, rubber and coffee) as well as for ornamental purposes – this is agriculture *sensu stricta*. In other words, it covers the cultivation of large-scale field crops, horticulture and ornamentals. Tree production is dealt with later under 'Forestry'.

Most of the damage done to growing plants by insects is regarded ecologically as grazing, and in point of fact this grazing represents the basic consumer level of most of the main terrestrial food webs. When considering the world biota it is clear that by far the greatest proportion of members of the animal kingdom are phytophagous grazers. Thus out of the total of 29 insect orders, most have some members that are plant-eaters, and to the agriculturalist most of these phytophagous insects are pests when they are eating cultivated plants.

The main groups of insects and mites that are regarded as agricultural pests (*sensu stricta*) are listed in Table 5.6.

Some insect species are particularly important in that they are vectors of disease-causing organisms. Table 5.7 lists a few of the insect pests that transmit disease-causing fungi, bacteria and viruses.

Table 5.6   Major insects and mites as agricultural pests

Group	Common name
Order Collembola	Springtails
Several families are involved	
Order Orthoptera	
Family Tettigoniidae	Long-horned Grasshoppers
Gryllidae	Crickets
Gryllotalpidae	Mole Crickets
Pyrgomorphidae	Stink Grasshoppers
Acrididae	Locusts; Grasshoppers
Order Phasmida	
Family Phasmidae	Stick Insects
Order Isoptera	Termites
Family Hodotermitidae	Harvester Termites
Termitidae	Mound-building Termites
Order Hemiptera, Suborder Homoptera	Plant Bugs
Family Cicadellidae	Leafhoppers
Delphacidae	Planthoppers
Aleyrodidae	Whiteflies
Aphididae	Aphids
Pseudococcidae	Mealybugs
Coccidae	Soft Scales
Diaspididae	Hard/Armoured Scales
Order Hemiptera, Suborder Heteroptera	Animal/Plant Bugs
Family Miridae	Capsid/Mosquito Bugs
Coreidae	Brown Bugs
Pentatomidae	Stink Bugs
Scutelariidae	Shield Bugs
Order Thysanoptera	Thrips
Family Thripidae	Thrips
Phlaeothripidae	Leaf-rolling Thrips
Order Coleoptera	Beetles
Family Scarabaeidae	Cockchafers; White Grubs, etc.
Coccinellidae	Epilachna Beetles; Ladybirds
Meloidae	Flower/Pollen Beetles
Cerambycidae	Longhorn Beetles
Bruchidae	Seed Beetles
Chrysomelidae	Leaf Beetles
Apionidae	Pod/Seed Weevils, etc.
Cuculionidae	Weevils proper
Order Diptera	Flies
Family Cecidomyiidae	Gall Midges
Anthomyiidae	Root Flies, etc.
Muscidae	Shoot/Stem Flies
Agromyzidae	Leaf-miners, etc.
Chloropidae	Gout/Frit Flies
Tephritidae	Fruit Flies
Order Lepidoptera	Moths and Butterflies
Family Gracillariidae	Leaf-miners
Gelechiidae	Stem-borers, etc.
Limacodidae	Stinging/Slug Caterpillars

Table 5.6   *(cont.)*

Group	Common name
Tortricidae	Tortrix Moths, etc.
Pyralidae	Snout Moths
Lycaenidae	Blue Butterflies
Pieridae	White Butterflies
Papilionidae	Swallowtails
Hesperiidae	Skippers
Geometridae	Loopers, etc.
Lasiocampidae	Tent Caterpillars
Sphingidae	Hawk Moths; Hornworms
Arctiidae	Tiger Moths, etc.
Noctuidae	Owlet Moths
Lymantriidae	Tussock Moths
Order Hymenoptera	
Family Tenthredinidae	Sawflies
Formicidae	Ants
Vespidae	Wasps
Class Arachida, Order Acarina	Mites
Family Eriophyidae	Gall/Rust Mites
Tetranychidae	Spider Mites

5.4.1 Order Collembola (Springtails)

These small, primitive, wingless insects have weak biting mouthparts and can only damage soft plant tissues. Some seedling stems can, however, be weakened, and a few species are pests in temperate greenhouses. *Sminthurus* is called the 'Lucerne Flea' in Europe, but is in fact worldwide.

5.4.2 Order Orthoptera

These primitive insects are mostly large, with long, simple antennae and large terminal cerci. They have large biting mouthparts with which they eat plant foliage.

(a) Family Tettigoniidae (Long-horned Grasshoppers; Katydids)

A large group (5000 species) of cryptic insects, which are mostly nocturnal, tropical and winged. The males stridulate, making a chirping noise. They are basically forest/bush insects but a few live in Gramineae and can be damaging to cereal crops. A dozen or more genera are recorded as minor crop pests – most are polyphagous, but individually are seldom serious pests. *Anabrus* includes the Mormon 'Cricket' of the USA; *Conocephalus* is widespread in Asia; and *Elimaea* can be damaging in South East Asia (Fig. 5.87). *Homorocoryphus* are the Edible Grasshoppers of Eastern Africa; they are unusual in that they feed on Gramineae, swarm seasonally and fly towards light at night.

(b) Family Gryllidae (Crickets)

These nocturnal insects either live underground in a nest or in leaf litter, under stones, etc. They emerge at night to feed on plants, being especially damaging to soft-stemmed seedlings. In covered seed-beds they can be very destructive to the seedlings. Males stridulate loudly at night as they sit at the entrance to their tunnels. Major pest species include:

- *Acheta* spp. (Two-spotted Cricket, etc.) – worldwide.

Table 5.7 Some plant diseases transmitted by insects and mites

Vector	Disease	Parasite	Transmission	Distribution
Antestiopsis spp.	coffee berry rot	*Nematospora* spp.	Feeding punctures;	Africa
Dysdercus spp.	cotton staining	*Nematospora* spp.	biological and mechanical	Pantropical
Calidea spp.	cotton staining	*Nematospora* spp.	transmission	Africa
Nezara viridula	cotton staining	*Nematospora* spp.		Pantropical
Scolytus spp.	Dutch elm disease	*Ceratostomella ulmi*	Mechanical; on insect body	Europe, N. America
Platygaster sp.	coffee leaf rust	*Hemileia vastratrix*	Spores on insect body	Africa
Frankliniella occidentalis	ear rot of corn	*Fusarium moniliforme*	Feeding	USA Europe, N.
Aphids and Leafhoppers	fire blight	*Erwinia amylovora* (bacteria)	Contamination	America
Aphis craccivora	groundnut rosette	Virus	Feeding	Pantropical
Myzus persicae (100 diseases)	cauliflower mosaic	Virus	Feeding	Europe, USA
	potato leaf roll	Virus	Feeding	Europe, USA
	sugar beet mosaic	Virus	Feeding	Europe, USA
	sugar beet yellows	Virus	Feeding	Europe, USA
Pentalonia nigronervosa	banana bunchy top	Virus	Feeding	Old World tropics
Toxoptera citricidus	tristeza disease	Virus	Feeding	Africa
Bemisia tabaci	cassava mosaic	Virus	Feeding	Pantropical
	cotton leaf curl	Virus	Feeding	Pantropical
	sweet potato B	Virus B	Feeding	Pantropical
Nephotettix spp.	rice dwarf	Virus	Feeding	India, South East
	rice yellow dwarf	Virus	Feeding	Asia to China
	transitory yellowing	Virus	Feeding	and Japan
Sogatella furcifera	rice yellows	Virus	Feeding	India through
	stunt disease	Virus	Feeding	to China
Laodelphax striatella	northern cereal mosaic	Virus	Feeding	Europe, Asia
	oat rosette	Virus	Feeding	Europe, Asia
Pseudococcus njalensis	rice stripe	Virus	Feeding	Asia
	swollen shoot of cocoa	Virus complex	Feeding	W. Africa
Thrips tabaci	tomato spotted wilt	Virus	Feeding (nymph)	Cosmopolitan
Cecidophyopsis ribis	currant reversion	Virus	Feeding	Europe
Eriophyes tulipae	wheat streak mosaic	Virus	Feeding	Cosmopolitan

Taken from Hill, 1994.

Fig. 5.87 Long-horned Grasshopper (*Elimaea* sp. – Orth.; Tettigoniidae) on hibiscus leaf; Hong Kong.

Fig. 5.88 Big-headed Cricket (*Brachytrypes* sp. – Orth.; Gryllidae); body length 25 mm; Hong Kong.

- *Brachytrypes* spp. (Big-headed Cricket, etc.) – Africa and Asia (Fig. 5.88).
- *Teleogryllus* spp. (Field Crickets) – Asia and Australia.

(c) Family Gryllotalpidae (Mole Crickets)

A small group especially adapted for burrowing with large shovel-like front legs. Their lives are subterranean, except for dispersal flights, and they feed mostly on plant roots, tubers and underground stems. Two genera are of particular importance as pests; these are *Gryllotalpa* (Fig. 5.89), found as several species worldwide and *Scapteriscus*, a New World genus, common in the USA.

(d) Family Pyrgomorphidae (Stink Grasshoppers)

These are brightly coloured (aposematic), large, diurnal species that flaunt their presence and rely on their stink glands to repel predators. Several species are

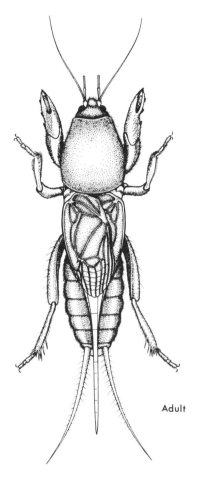

Adult

Fig. 5.89 African Mole Cricket (*Gryllotalpa africana* – Orth.; Gryllotalpidae); length 20 mm; Makerere, Uganda.

major pests and defoliate fruit trees, cotton and cereal crops; generally most are polyphagous. Some common pests include:

- *Chrotogonus* spp. (Surface Grasshoppers) – Africa and India.
- *Phymateus* spp. (Bush Locusts) – Africa; polyphagous.
- *Pyrgomorpha* spp. – Africa and India, found on many crops.
- *Valanga* spp. – found on Palmae and other crops in South East Asia.

- *Zonocerus* spp. (Variegated Grasshoppers) – on many different crops throughout Africa (Fig. 5.90).

(e) Family Acrididae (Grasshoppers and Locusts) (9000 spp.)

A large group, worldwide, but most abundant in warmer regions. As many as 500 species can be regarded as agricultural pests. As a group they are diurnal grassland insects, with large hind wings and legs for jumping. In each of the great grassland habitats of the world there is a complex of Acrididae that graze the grasses, and when these areas are ploughed for cereal production the insects move on to the cereal crops.

Some species show great fecundity, and population explosions are frequent. This is often associated with a migratory dispersal of swarms of insects. Some species also show special social behaviour and congregate in their social phase to disperse as a swarm of locusts. Locust swarms have devastated warmer parts of the world since time immemorial, with almost one-third of the land surface of the world under threat from these voracious pests. There is a vast literature on locusts and grasshoppers, such as the book by the Centre for Overseas Pest Research (COPR, 1982), now available from the Natural Resources Institute at Chatham, Kent, UK.

The main locust species breed in temporary habitats, often at the edge of deserts, where after the rains there is a lush growth of grass as food for the nymphs. As the habitat shrinks, the nymphs collect in a swarm and disperse, looking for new food sources. As the nymph swarm 'marches' it feeds voraciously and destroys vegetation. Most locusts show a feeding preference for Gramineae and thus are most damaging to cereal crops, but some are more polyphagous. After the nymphs metamorphose the young adults are winged and will fly. The largest locust swarms have been estimated to contain up

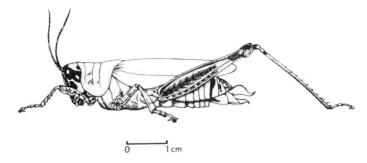

0   1 cm

(a)   Adult ♀

(b)

Fig. 5.90 Variegated Grasshopper (*Zonocerus variegatus* – Orth.; Pyrgomorphidae); (a) female from Kenya; (b) mating pair of brachypterous forms on Citrus tree; Malawi.

to 1000 million insects; large swarms have spread for up to ten years across Africa and Asia, eventually covering 1000 miles or more. The larger swarms are estimated to eat 1000 tonnes of fresh green vegetation each day.

The importance of these pests, and especially their international nature, has necessitated their monitoring and control through an international approach. In 1960 the Desert Locust Control Organization (DLCO) was

Fig. 5.91 Desert Locust (*Schistocerea gregaria* – Orth.; Acrididae), body length 50 mm; Ethiopia.

established; the main centre for co-ordination of locust study is now FAO headquarters in Rome.

The most important locust species are as follows:

- *Locusta migratoria* (Migratory Locust) – four subspecies are known, and together they cover Africa, Asia and Australasia. They prefer Gramineae but will eat other plants when hungry.
- *Nomadacris septemfasciata* (Red Locust) – found in tropical Africa.
- *Schistocerea gregaria* (Desert Locust) (Fig. 5.91) – the area inhabited includes Africa and the Middle East to India, and there are usually several generations per year. The DLCO is hampered somewhat by political differences and internal discord in some countries. In the late 1980s the FAO satellite surveillance project gave warning of the likelihood of a Desert Locust outbreak in Africa, but on the ground some countries failed to carry out the required monitoring and control duties and swarms developed. The FAO's US$400 million locust programme thus failed. Luckily, however, there was a wind change and strong winds blew the swarm into the ocean where the locusts perished. This incident clearly shows that locusts remain serious potential pests. As many

parts of Africa and Asia are at risk, continuous vigilance and international co-operation are required.

- *Locustana pardalina* – the South African Brown Locust.
- *Chortoicetes terminifera* – the Australian Plague Locust.
- *Patanga succincta* (Bombay Locust) – occurs from India to China.
- *Dociostaurus maroccanus* – the Mediterranean Locust.
- *Schistocerca* – are found in the USA and Canada.
- *Melanoplus* – particularly important in North America; some dozen or more species are regular pests on the prairies, both on pasture grasses and crops.

5.4.3 Order Phasmida (Stick and Leaf Insects)

Both groups are leaf-eaters, but damage to crops is not common. *Graeffea* stick insects, however, occasionally defoliate coconut palms in parts of the Pacific region.

5.4.4 Order Isoptera (Termites)

These are crop pests in all parts of the tropics, but are also equally important as pests of trees and wooden structures. Wet-wood Termites (Rhinotermitidae) such as *Coptotermes*

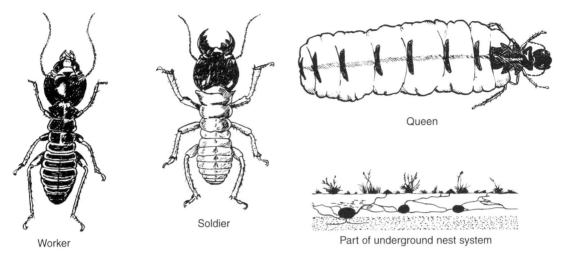

Worker   Soldier   Queen

Part of underground nest system

Fig. 5.92 Harvester Termite (*Hodotermes mossambicus* – Isoptera; Hodotermitidae) from East Africa.

Fig. 5.93 Termite mound about 1.5 m tall (*Macrotermes* sp. – Isoptera; Termitidae); central Ethiopia.

have been recorded damaging many crop plants throughout South East Asia, but other groups are more important as crop pests, including:

(a) Family Hodotermitidae (Harvester Termites)

Hodotermes makes an underground nest in arid areas of Africa and between the Middle East and India. They forage for grasses which are cut and taken back to the nest as food (Fig. 5.92).

(b) Family Termitidae (Mound/Subterranean Termites)

This is the largest group of termites (1300 species), found in the tropics and subtropics.

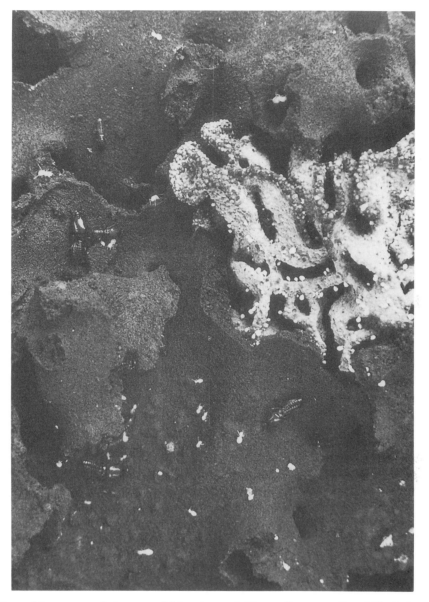

Fig. 5.94 Part of fungus comb in nest of *Macrotermes bellicosus* (Isoptera; Termitidae) showing solider termites, workers and the fungal bromatia; Kampala, Uganda.

The nest is mostly underground but has a central mound which can be as tall as 2–4 m (Fig. 5.93). In cooler areas (Hong Kong, Alemaya in Ethiopia) there is no visible nest mound, the entire nest being underground. These termites usually forage either underground or on tree trunks covered with a shield of earth/mud, frass and saliva – they avoid light when foraging. The bark, roots and foliage are taken into the nest and chewed finely, mixed with saliva and made into fungus combs (Fig. 5.94) on which the edible fungus is cultivated. The precise mode of feeding in some species is not clear, for

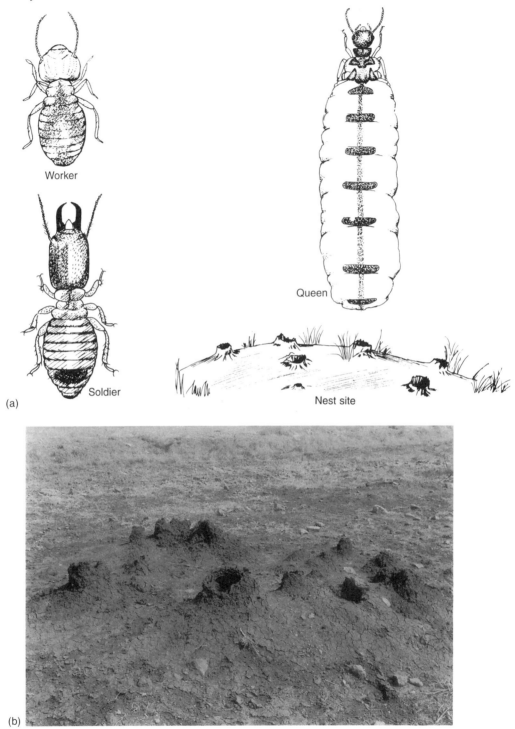

Worker

Soldier

Queen

Nest site

(a)

(b)

Fig. 5.95 (a) Crater Termite (*Odontotermes badius* – Isoptera; Termitidae), Kenya; and (b) nest entrance on an earthen road, Rift Valley, Ethiopia.

some appear to have intestinal micro-organisms while others possibly produce cellulases.

The more important genera include:

- *Macrotermes* (Mound Termites; Bark-eating Termites) – some ten species occur in the Old World tropics (except Australasia).
- *Microcerotermes* (Live Wood-eating Termites) – found in Africa, India, South East Asia and, less commonly, in South America. They can apparently invade living wood and plant tissues; nesting habits are varied.
- *Microtermes* (Seedling Termites) – small termites inhabiting small nests underground, but can be very abundant and crop seedlings may be destroyed. Found in Africa and South East Asia.
- *Nasutitermes* – a large group of arboreal species found in tropical Asia, Africa and South America. They make carton nests on the trunks of palms (coconut, oil palm, etc.) and eat into the trunk tissues. Other nests are made in the canopy of the palm and other trees.
- *Odontotermes* – subterranean species, usually without a nest mound; 23 species have been recorded as pests of crops in the Old World (Harris, 1971). *O. badius* is the Crater Termite of Africa with its distinctive nest (Fig. 5.95).
- *Pseudacanthotermes* (Sugarcane Termites) (Fig. 5.96) – two closely related species are found in tropical Africa, on sugarcane and tea bushes. Harris (1971) recorded 59 species of termite attacking sugarcane (especially setts) worldwide.

5.4.5 Order Hemiptera, Suborder Homoptera (Plant Bugs)

A very large group – the order contains 38 families and 56 000 or more species – and of great importance agriculturally. The bugs are often responsible for more damage than their numbers would warrant, for some species are vectors of disease-causing micro-organisms;

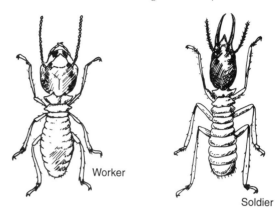

Worker

Soldier

Section of damaged cane

1 m

Nest mound

Fig. 5.96 Sugarcane Termite (*Pseudacanthotermes militaris* – Isoptera; Termitidae); Kenya.

for example, all the plant viruses are transmitted by feeding bugs (see Table 5.7). The mouthparts are modified into a piercing and sucking proboscis (rostrum) and the insect feeds on plant fluids, mostly sap. Most species are winged and some engage in quite spectacular annual migrations, but the scale insects (Coccoidea) are wingless and anatomically reduced.

There are several small families, each containing a few pest species, but to save space

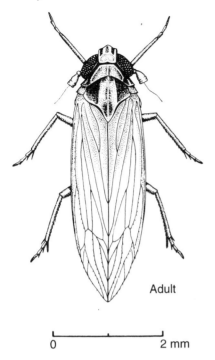

Adult

0   2 mm

Fig. 5.97 Small Brown Planthopper (*Laodelphax striatella* – Hem.; Delphacidae).

these will be omitted from the present account (for more details see Hill, 1994). The major pest groups include the following:

(a) Family Delphacidae (Planthoppers) (1300 spp.)

As a group these prefer Gramineae as hosts, and are more abundant in warmer regions. Several species are major pests of cereals (especially rice) and sugarcane. Some pests' saliva is somewhat toxic to the host plant, and heavy infestations can cause 'hopperburn' when foliage turns brown and dry. A number of species transmit viruses causing diseases in the crop; some are migratory, and regularly invade southern Japan in the summer from South East Asia and China. Major pests include:

- *Laodelphax striatella* (Small Brown Planthopper) – a Palaearctic pest of cereal crops and sugarcane (Fig. 5.97).

- *Nilaparvata lugens* (Brown Rice Planthopper) – a serious pest of rice throughout South East Asia, from India to China.
- *Peregrinus maidis* (Corn Planthopper) – pest of maize, sorghum and sugarcane, found throughout the tropics.
- *Perkinsiella saccharicida* (Sugarcane Planthopper) – Australia and Hawaii on sugarcane; 22 other species occur in the tropics.
- *Sogatella furcifera* (White-backed Planthopper) – on rice in South East Asia.

(b) Family Cercopidae (Froghoppers and Root Spittlebugs)

This and the Aphrophoridae are common insects and are frequently seen on a wide range of plants. However, they are not often serious pests, although in the West Indies, and Central and South America, the complex of *Aeneolamia* and *Tomaspis* (Fig. 5.98) is very damaging to sugarcane. Adult Froghoppers have toxic elements in their saliva that cause foliage 'burning' ('Froghopper blight') and the nymphs live in a large spittle mass on the surface of the root stool. In the UK the distinctive black and red *Cercopis vulnerata* is very common on the foliage of many different plants, but the nymphs are seldom seen.

(c) Family Aphrophoridae (Froghoppers; Spittlebugs)

These are the very common 'spittlebugs' or 'cuckoo-spit insects' whose nymphs live inside a mass of white froth on aerial parts of plant foliage (Fig. 5.99). The common species in the UK is *Philaenus spumatius* (Common Meadow Spittlebug) – the small grey/brown adult is not very conspicuous but the spittle masses are very evident on a wide range of plants.

(d) Family Cicadellidae (=Jassidae) (Leafhoppers) (8500 spp.)

A large, widespread group of ecological importance in most parts of the world, which

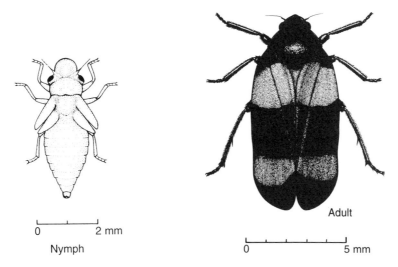

Fig. 5.98 Sugarcane Spittlebug (*Tomaspis* sp. – Hem.; Cercopidae); Trinidad.

Fig. 5.99 Common Meadow Spittlebug (*Philaenus spumatius* – Hem.; Aphrophoridae), adult and nymph on lavender; Skegness, UK.

includes many crop pests. Some species are found only on Gramineae, but in the UK, fruit trees, roses and brambles are important hosts. Many of these species hibernate over winter as adults on the underside of leaves.

On warm winter days they may be active for a few hours, which makes them very conspicuous at times when insects are scarce.

The grass-infesting species often have population irruptions when they disperse in

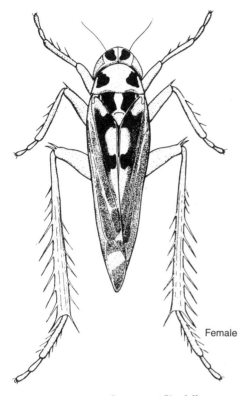

Fig. 5.100 Potato Leafhopper (*Cicadella aurata* – Hem.; Cicadellidae); length 6 mm; Cambridge, UK.

large numbers. In South East Asia several rice pests migrate annually up the east coast of China to Japan; similar movement takes place in North America.

Most of these bugs cause little damage other than leaf-curling and slight discoloration (as on roses in the UK), but some cause foliage-browning ('hopperburn') and several are important virus vectors. A few common pests include:

- *Cicadella* spp. (Green/White Leafhoppers) – found on rice, potato, etc. in both Europe and the USA (Fig. 5.100).
- *Cicadulina* spp. (Maize Leafhoppers) – common in tropical Africa, on maize, other cereals and sugarcane; vectors of the Maize Streak Virus.

- *Edwardsiana* spp. (Rose/Fruit Tree Leafhoppers) – in Europe, Asia, the USA, South America and Australasia.
- *Empoasca* spp. (Green Leafhoppers) – cosmopolitan on a wide range of host plants.
- *Nephotettix* spp. (Green Rice Leafhoppers) – on Gramineae in South East Asia.
- *Zygina* spp. – common in Europe and Asia on a wide range of hosts, as well as in temperate greenhouses.

(e) Family Psyllidae (Jumping Plant Lice) (1300 spp.)

An especially interesting group, showing great ecological diversity with respect to host plants and their effects in the plants. Usually they are quite host-specific, producing a wide range of galls and leaf distortions on the plants. Many different genera are known, many of which include pest species. World-wide one large group of species is associated with *Ficus* (Moraceae) and another with the family Malvaceae. Generally, damage to the host plant is not serious, although it is often spectacular, with distorted foliage covered in white, waxy filaments. Some important pests include:

- *Diaphorina citri* (Citrus Psylla) – found in South East Asia and South America.
- *Heteropsylla cubana* (Leucaena Psyllid) – indigenous to Central America, this is now a serious pest in India and parts of Africa on *Leucaena leucocephala*.
- *Psylla* spp. – found on apple, pear and other woody hosts in Europe, Asia and North America.
- *Trioza* spp. – found on a wide range of hosts throughout the world; the most common pest is probably *T. eritreae* (Citrus Psyllid) in tropical Africa (Fig. 5.101).

(f) Family Aleyrodidae (Whiteflies)

These tiny winged bugs are usually white, but some have dark wings, and the nymphs

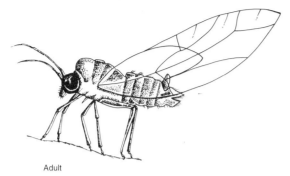

Adult

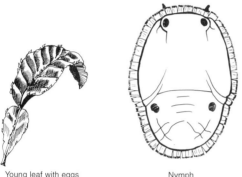

Young leaf with eggs   Nymph

Damaged leaf

Fig. 5.101 Citrus Psyllid (*Trioza eritreae* – Hem.; Psyllidae); Kenya.

are flat and scale-like with a fringe of white waxy filaments. Most of the 1200 species are tropical, but a few are temperate and some are major pests in greenhouses. Sap losses can be heavy, and most infestations are associated with sooty moulds which grow on the excreted honey-dew. Many of the pests are virus vectors. Hosts are equally often

woody plants and herbaceous, but grasses are not often used (except for *Bambusa*). Infestation levels are often very high and the underside of leaves may be covered with the tiny white insects.

Some important pests include:

- *Aleurocanthus* spp. – polyphagous on citrus, coffee, mango, etc.; pantropical.
- *Aleyrodes* spp. – found on a wide range of hosts throughout the world.
- *Bemisia tabaci* (Tobacco/Cotton Whitefly) – polyphagous on many different crops; pantropical (Fig. 5.102).
- *Dialeurodes* spp. (Citrus Whiteflies) – also polyphagous and pantropical.
- *Trialeurodes* spp. – polyphagous and cosmopolitan, important in greenhouses in Europe and North America (Fig. 5.103).

(g) Family Aphididae (Aphids; Greenfly; Plant 'Lice', etc.) (3500 spp.)

One of the largest and most important groups of pest insects; totally worldwide but most abundant in north temperate regions. They are pests on all types of plant – woody trees and shrubs, herbs and grasses. Although small in size, they occur in vast numbers. Only rarely do they have toxic saliva, but many are virus vectors and very damaging to crops such as potato and sugar beet. Many produce copious honey-dew, and sooty moulds are commonly associated with their infestation. In temperate regions they occasionally have spectacular annual dispersal flights in the summer when the air is filled with millions of flying insects. Many species practise facultative parthenogenetic vivipary by which enormous wingless populations are built up – as the aphid population grows, the increasing population density triggers the female to produce winged individuals, and when a particular population density is reached, the next couple of generations are mostly winged in readiness for the summer dispersal flight. Many species also practise

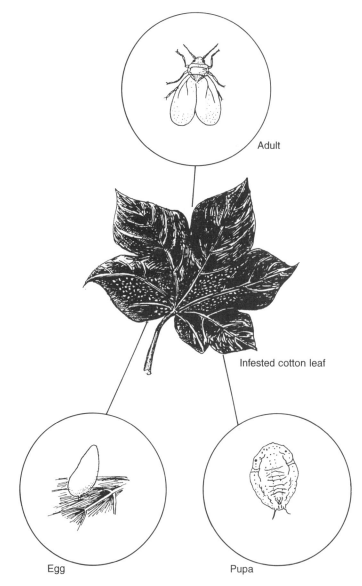

Adult

Infested cotton leaf

Egg

Pupa

Fig. 5.102 Tobacco Whitefly (*Bemisia tabaci* – Hem.; Aleyrodidae); Kenya.

alternation of hosts, so that they overwinter as eggs on woody twigs (trees and bushes) and then in late spring as winged, adult females they move on to herbaceous plants where food is more plentiful and more easily obtained. A typical European example is the Peach-Potato Aphid, *Myzus persicae* (Fig. 5.104), although in the tropics this polyphagous species is called the Green Peach Aphid and it does not practise host alternation, because eggs (and males) are not produced.

In the wild, most aphids are quite host-specific, such as the Elder Aphid and Sea Aster Aphid, although most of the important crop pest species are polyphagous, found on

Fig. 5.103 Greenhouse Whitefly (*Trialeurodes vaporariorum* – Hem.; Aleyrodidae) on tobacco leaf; Alemaya, Ethiopia.

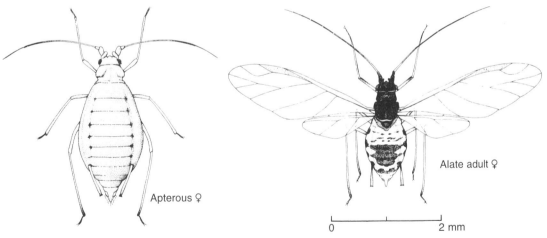

Fig. 5.104 Peach-Potato Aphid (*Myzus persicae* – Hem.; Aphididae); Cambridge, UK.

young shoots and underneath leaves (see Fig. 5.106). A major publication covering this topic is Blackman and Eastop (1984). Out of dozens of important pest species of Aphididae a few of the most serious and widespread are listed below:

- *Acyrthosiphon pisum* (Pea Aphid) – on pea and other legumes; temperate, cosmopolitan.
- *Aphis craccivora* (Groundnut Aphid) – polyphagous and cosmopolitan (Fig. 5.105).

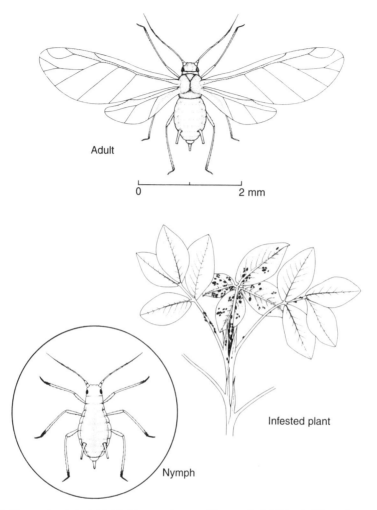

Adult

0   2 mm

Infested plant

Nymph

Fig. 5.105 Groundnut Aphid (*Aphis craccivora* – Hem.; Aphididae); Uganda.

- *Aphis fabae* (Black Bean Aphid) – polyphagous and cosmopolitan.
- *Aphis pomi* (Green Apple Aphid) – found on apple, pear, etc. in Europe and North America.
- *Aulacorthum solani* (Potato Aphid) – polyphagous and widespread on potatoes.
- *Brevicaryne brassicae* (Cabbage Aphid) – cosmopolitan on Cruciferae.
- *Diuraphis noxia* (Russian Wheat Aphid) – on barley and wheat; Palaearctic but has now spread to Africa and Argentina where it is becoming a very serious pest.
- *Macrosiphum rosae* (Rose Aphid) – on *Rosa* and *Dipsacus*; worldwide.
- *Myzus persicae* (Peach-Potato Aphid) – polyphagous and worldwide (Fig. 5.104).
- *Rhopalosiphum maidis* (Maize/Corn Leaf Aphid) – on Gramineae in warmer regions; other species also on Gramineae (Fig. 5.106).
- *Schizaphis graminum* (Wheat Aphid) – on Gramineae; widespread.

Fig. 5.106 Maize/Corn Leaf Aphid (*Rhopalosiphum maidis* – Hem.; Aphididae) on maize leaf; Alemaya, Ethiopia.

(h) Family Pemphigidae (Woolly Aphids)

A small group without long evident siphunculi. The nymphs produce large quantities of wax as long filaments. Several species are important pests on palms and on sugarcane. Two species are conspicuous in the UK:

- *Eriosoma lanigerum* (Apple Woolly Aphid) – on apple and closely related Rosaceae; almost cosmopolitan (Fig. 5.107).
- *Pemphigus* spp. (Root Aphids) – Palaearctic; overwinter as eggs on trees, and in the spring make leaf-petiole galls (mostly on poplar – Fig. 5.108). In the summer they live on the roots of some vegetables such as lettuce.

(i) Family Phylloxeridae (Phylloxeras)

A very small group, but one which includes the dreaded Vine/Grape Phylloxera (*Viteus vitifoliae*), the scourge of the wine industry.

On grapevine this insect makes leaf-galls; it also attacks the roots which can be destroyed. This pest almost destroyed the European wine industry when in about 1850 it entered Europe from the New World. Within the next decades it spread and destroyed 2 million hectares of vineyards in France, and was similarly damaging in Italy, Spain, Germany, South Africa and New Zealand. Other species of grape native to North America are tolerant of Phylloxera through vigorous compensatory root growth, and there developed the modern practice of grafting European grapevines on to American vine rootstocks. This has kept the Phylloxera at bay for many years. Such was the level of complacency that in recent developments in California, the growers and breeders grew careless in their breeding programme and included varieties of rootstock that were not properly tested for resistance. Growers wanted to use rootstocks from French/American hybrids in the belief

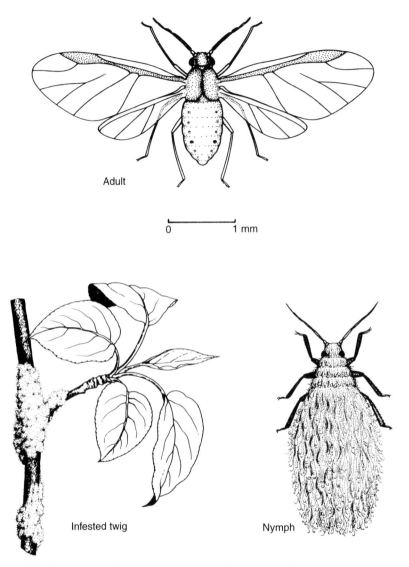

Adult

0 ⊢————————⊣ 1 mm

Infested twig   Nymph

Fig. 5.107 Apple Woolly Aphid (*Eriosoma lanigerum* – Hem.; Pemphigidae); Kenya.

that the wine would be of superior quality, but some of these hybrids were not sufficiently resistant to Phylloxera. The end result was that a couple of years ago vines in California stated dying as the Phylloxera struck (Lewin, 1993) and destroyed the roots. It now appears that a large part of the best vineyards in California (at least 20 000 ha) will have to be cleared, fumigated and replanted, at a cost of US$1–2000 million!

During the last decade there have also been occasional outbreaks of Phylloxera in parts of Europe when growers grew careless in their choice of new rootstocks.

(j) Family Margarodidae (Fluted Scales)

A small group but with one genus very important in the tropics, it being polyphagous on a wide range of trees,

Fig. 5.108 Lettuce Root Aphid (*Pemphigus bursarius* – Hem.; Pemphigidae) in poplar leaf petiole galls; Cambridge, UK.

bushes and palms on which it can be debilitating, killing shoots. The genus is *Icerya* and the most important species is *I. purchasi* (Cottony Cushion Scale) (Fig. 5.109).

(k) Family Pseudococcidae (Mealybugs)

An extensive tropical group with a few species adapted for life in temperate greenhouses. They are wingless (except for adult males) and anatomically reduced, with the reddish body surface covered with wax (usually white) and sometimes a few long waxy filaments (Fig. 5.111). Mostly they are found on aerial parts, especially in leaf axils, by leaf veins and other sites where a little physical protection is offered, although some make underground root infestations. Honeydew is produced and ants are usually in attendance; sooty moulds are also common. Some of the more important pest species include:

- *Dysmicoccus* spp. (Pineapple/Sugarcane Mealybugs) – pantropical.
- *Ferrisia virgata* (Striped Mealybug) – polyphagous and pantropical.
- *Phenacoccus* spp. – found on apple in Europe and cassava in South America; now occurs in Africa.
- *Planococcus* spp. (Citrus/Root Mealybugs, etc.) – polyphagous and pantropical (Fig. 5.110).
- *Pseudococcus* spp. – found on a wide range of crops; cosmopolitan (Fig. 5.111).

Fig. 5.109 Cottony Cushion Scale (*Icerya purchasi* – Hem.; Margarodidae) on Cassia in Hong Kong; inset on *Rosa*, Alemaya, Ethiopia.

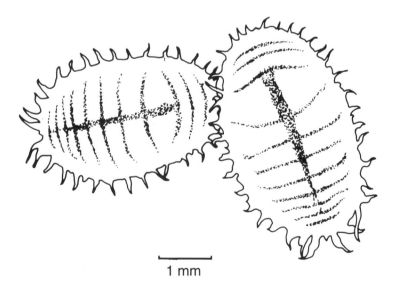

1 mm

Fig. 5.110 Citrus/Root Mealybug (*Planococcus citri* – Hem.; Pseudococcidae); Kenya.

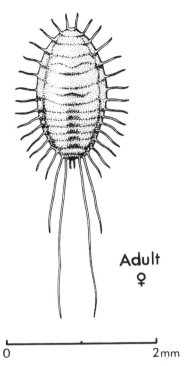

Adult ♀

0   2mm

Fig. 5.111 Long-tailed Mealybug (*Pseudococcus longispinus* – Hem.; Pseudococcidae); Hong Kong.

- *Saccharicoccus sacchari* (Pink Sugarcane Mealybug) – pantropical.

(l) Family Coccidae (Soft Scales)

A very important group of crop pests, abundant throughout the tropics and subtropics with a few species adapted for temperate regions; also found in greenhouses. Males are often rare and usually winged (see Fig. 5.115), but the female is reduced and scale-like with a dorsal skeletal shield (scale). Some scales are naked and others are covered with a layer of wax. As with other Coccoidea the first instar nymphs do not feed but actively disperse – they are called 'crawlers'. Honey-dew production is variable and so ants and sooty moulds may or may not be present. However, some species are always found with ants in attendance, and some are ac-

tually kept inside small ant sub-nests. Some common pest species include:

- *Ceroplastes* spp. (Waxy Scales) – found in the tropics and subtropics; the body is hidden under a thick layer of wax; often pink in colour.
- *Coccus* spp. (Soft Green Scales, etc.) – about a dozen pest species found on an enormous range of trees, bushes and palms; the genus is distributed world-wide – most species are tropical but *C. hesperidum* occurs in the UK (Fig. 5.46).
- *Gascardia* spp. (White Waxy Scales) – pantropical and polyphagous.
- *Parthenolecanium* spp. (Brown Scales, etc.) – warmer regions and temperate; polyphagous.
- *Pulvinaria* spp. (Woolly Scales) – cosmopolitan and polyphagous.
- *Saissetia* spp. (Helmet Scales, etc.) – pantropical on woody hosts (Fig. 5.112).

(m) Family Diaspididae (Hard or Armoured Scales)

A large and very important group of pests worldwide, mostly on fruit and other trees. Some pests are very damaging and can kill small fruit trees. The scale on the female body is separate and can be lifted off (see Fig. 5.115). The small or tiny scales can escape notice and many pests have been widely distributed throughout the world on planting stock. The list of pest species is long, so only a few can be mentioned here:

- *Aonidiella* spp. (Red Scale, etc.) – pantropical, on fruit trees, palms and others (Fig. 5.113).
- *Aspidiotus* spp. (Ivy/Coconut Scales, etc.) – cosmopolitan and polyphagous.
- *Aulacaspis tegalensis* (Sugarcane Scale) – Old World tropics on sugarcane.
- *Chrysomphalus* spp. (Purple Scale, etc.) – on trees and shrubs worldwide.
- *Lepidosaphes* spp. (Mussel/Oystershell Scales) – cosmopolitan on a wide range of

Fig. 5.112 Helmet Scale (*Saissetia coffeae* – Hem.; Coccidae) on leaf of sago palm; Hong Kong.

Fig. 5.113 Red Scale (*Aonidiella aurantii* – Hem.; Diaspididae) on oranges; Dire Dawa, Ethiopia.

Fig. 5.114 Mussel Scale (*Lepidosaphes ulmi* – Hem.; Diaspididae) encrusting apple branch; Skegness, UK.

trees and shrubs; serious pests (Fig. 5.114).

- *Parlatoria* spp. (Black Scale, etc.) – some species host-specific on *Citrus*, Date Palm, etc., others polyphagous; pantropical.
- *Quadraspidiotus perniciosus* (San José Scale) – polyphagous and very damaging on a wide range of deciduous fruit trees and shrubs, and ornamentals (700 hosts recorded) (Fig. 5.115).

5.4.6 Order Hemiptera, Suborder Heteroptera (Animal/Plant Bugs)

These are usually larger insects, fully winged and agile, and all have toxic saliva that kills tissues during feeding. Some feed on leaves and shoots, but more show a preference for sucking sap from fruits and seeds. One group of bugs is predatory on terrestrial animals and humans, and another includes fiercely predacious aquatic insects, whose toxic saliva is also used to kill prey. The more important agricultural pests include:

(a) Family Miridae (= Capsidae) (Capsid Bugs; Plant Bugs) (6000 spp.)

A large group of smallish, flattened bugs, some delicate in appearance and called Mosquito Bugs. Some feed almost continuously and can make 20–150 feeding punctures in a day, each resulting in a dark necrotic spot. Necrotic lesions in young leaves lead to a characteristic 'tattering' as the leaves enlarge. Damage to the host plant can be serious; eggs are laid embedded in the plant tissues so they are protected. Pest species include:

- *Calocoris* spp. (Potato Capsid, etc.) – polyphagous in Europe and North America.
- *Helopeltis* spp. (Mosquito Bugs) – polyphagous on cotton, tea, cocoa, etc. in the tropics of the Old World.

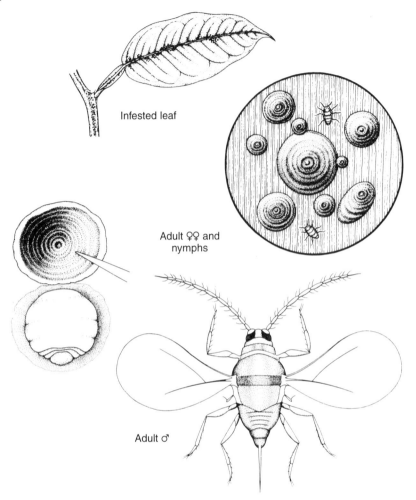

Infested leaf

Adult ♀♀ and nymphs

Adult ♂

Fig. 5.115 San José Scale (*Quadraspidiotus perniciosus* – Hem.; Diaspididae); Kenya.

- *Lygocoris* spp. (Common Green Capsid, etc.) – polyphagous in Europe, North America and Africa (Fig. 5.116).
- *Lygus* spp. (Tarnished Plant Bugs) – polyphagous and cosmopolitan (Fig. 5.117)

(b) Family Lygaeidae (Seed Bugs) (2000 spp.)

A smallish group of tiny bugs of varied habits – some are predatory, but most are phytophagous and usually feed on fruits and seeds. A few species (genera) are common and widespread and occasionally are serious pests on crops. These include:

- *Blissus* spp. (Chinch Bugs, etc.) – on wheat and other cereals in parts of Africa and Asia, but mostly in North America (Fig. 5.118).
- *Nysius* spp. (Seed Bugs) – on cotton and many other crops; cosmopolitan.
- *Oxycarenus* spp. (Cotton Seed Bugs) – on cotton and other Malvaceae; cosmopolitan in warmer regions.

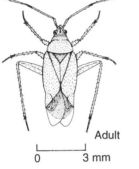

Fig. 5.116 Common Green Capsid (*Lygocoris pabulinus* – Hem.; Miridae) on marigold flower; Skegness, UK.

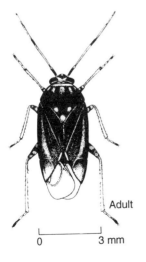

Fig. 5.117 Tarnished Plant Bug (*Lygus rugulipennis* – Hem.; Miridae).

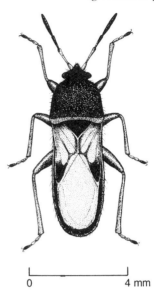

Fig. 5.118 Chinch Bug (*Blissus leucopterus* – Hem.; Lygaeidae).

(c) Family Pyrrhocoridae (Cotton Stainers; Red Bugs)

A small pantropical group of bugs that suck sap from fruits and seeds. They are well known agriculturally because of the many species of *Dysdercus* known as Cotton Stainers (Fig. 5.119); these are more important as pests in the Old World but are abundant in the New.

(d) Family Coreidae (Brown Bugs, Twig-wilters, etc.) (2000 spp.)

Large brown bugs mostly; strong fliers; often with a pungent smell; all phytophagous and with toxic saliva which kills plant tissues. A few feed on soft, young woody shoots which usually die distally to the feeding point, and these are often called 'Twig-wilters'. A few pest species include:

- *Acanthocoris* spp. (Squash bugs) – on Cucurbitaceae in South East Asia.
- *Anoplocnemis* spp. (Giant Twig-wilters) – polyphagous in Africa and tropical Asia.

Fig. 5.119 Cotton Stainer (*Dysdercus* sp. – Hem.; Pyrrhocoridae) on leaf of *Althaea*; Alemaya, Ethiopia.

- *Clavigralla* spp. (Spiny Brown Bugs) – on league pods in Africa.
- *Leptoglossus* spp. (Leaf-footed Plant Bugs) – polyphagous and pantropical (Fig. 5.120).
- *Mictis* spp. (Giant Twig-wilters) – on fruit trees in tropical Asia and Australasia.

(e) Family Alydidae

These are slender-bodied Coreids, of importance on rice and other cereals and legume pods in tropical Asia and Africa. They include:

- *Cletus* spp. (Cletus Bugs) – polyphagous in Africa and Asia.
- *Leptocorisa* spp. (Asian Rice Bugs) – found on Gramineae in tropical Asia.
- *Stenocoris southwoodi* (African Rice Bug) – found on rice, etc., in tropical Africa.

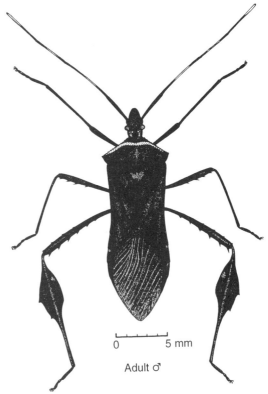

Fig. 5.120 Leaf-footed Plant Bug (*Leptoglossus australis* – Hem.; Coreidae).

(f) Family Scutelleridae (Shield Bugs)

Quite large bugs, often brightly coloured, with scutellum long and covering the entire abdomen. Most abundant in the tropics but present in temperate regions, where the adults overwinter in hibernation (as do the Stink Bugs, section (g)). Both groups have stink glands from which a fine stream of stinking liquid can be ejected. Feeding sites show tissue necrosis where the toxic saliva has killed cells, and they are also used as invasion sites for pathogenic fungi. Several pest species transmit the fungus *Nematospora* on important crop plants, such as cotton and coffee. A few important pest species include:

- *Calidea* spp. (Blue Bugs) – on cotton, etc. in tropical Africa (Fig. 5.121).

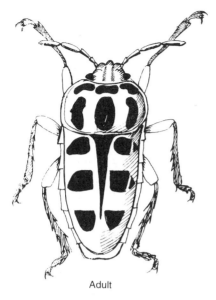

Adult

Fig. 5.121 Blue Bug (*Calidea* sp. – Hem.; Scutelleridae); Kenya.

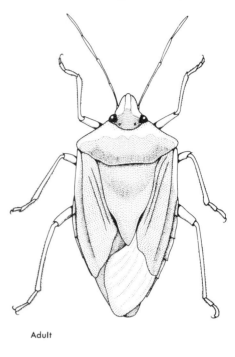

Adult

Fig. 5.122 Green Stink Bug (*Nezara viridula* – Hem.; Pentatomidae); Hong Kong.

- *Euryaster* spp. (Wheat Shield Bugs) – on cereals; Europe and Asia.
- *Poecilocoris* spp. (Tea Shield Bugs) – on tea, Camellia; South East Asia.
- *Scotinophara* spp. (Black Rice Bugs) – on rice; Asia.

(g) Family Pentatomidae (Stink Bugs) (2500 spp.)

An interesting and diverse group, totally worldwide, but again most abundant in the tropics. Several distinctive subfamilies are established. The scutellum only covers the first half of the abdomen. The number of pest species is relatively low, and includes:

- *Antestiopsis* spp. (Antestia Bugs of coffee) – on arabica coffee; tropical Africa and South East Asia.
- *Bagrada* spp. (Harlequin Bugs) – on Cruciferae; Africa and South Asia.
- *Nezara* spp. (Green Stink Bugs) – polyphagous; cosmopolitan (Fig. 5.122).

- *Oebalus* spp. (Rice Stink Bugs) – on rice and Gramineae; North, Central and South America.
- *Rhynchocoris* (Citrus Stink Bugs) – on Citrus; South East Asia.
- *Tessaratoma* spp. (Litchi Stink Bugs) – on litchi and longan; South East Asia.

5.4.7 Order Thysanoptera (Thrips) (5000 spp.)

Very small slender insects with narrow fringed wings and mouthparts that rasp the surface of plant bodies. Most are plant-feeders, on the foliage of higher plants, but some are fungivorous and some are predacious. They are worldwide in distribution and at times very common. In the UK they

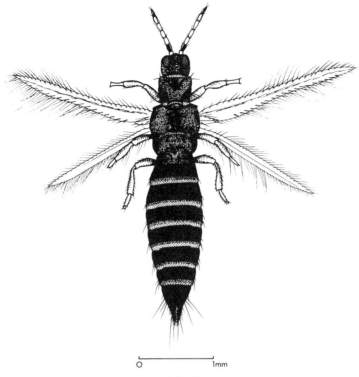

Adult ♀

Fig. 5.123 Flower Thrips (*Frankliniella* sp. – Thy.; Thripidae).

have extensive dispersal flights annually in the summer, and for a few days the air is filled with them – they get the name of 'Thunder-flies' from this habit. They are found on a wide range of plants and sometimes their damage to crop plants can be serious. These species include the following:

(a) Family Thripidae (Thrips) (2000 spp.)

- *Aptinothrips* spp. (Cereal/Grass Thrips) – on Gramineae in Europe, the USA and Canada.
- *Caliothrips* spp. – on a wide range of crops; pantropical.
- *Frankliniella* spp. (Flower Thrips) – on many cultivated plants; worldwide; some

act as vectors for viruses and fungal spores (Fig. 5.123).

- *Heliothrips haemorrhoidalis* (Black Tea Thrips, etc.) – polyphagous on many tropical crops and in greenhouses in temperate regions; cosmopolitan (Fig. 5.124).
- *Limothrips* spp. (Cereal and Grass Thrips) – a cosmopolitan group mostly found on Gramineae.
- *Scirtothrips* spp. (Citrus/Tea Thrips, etc.) – several species are serious pests on a wide range of crops; pantropical.
- *Taeniothrips* spp. (Pear Thrips, etc.) – in flowers of fruit trees and other cultivated plants; in both tropics and temperate regions.
- *Thrips* spp. – a large genus, many species

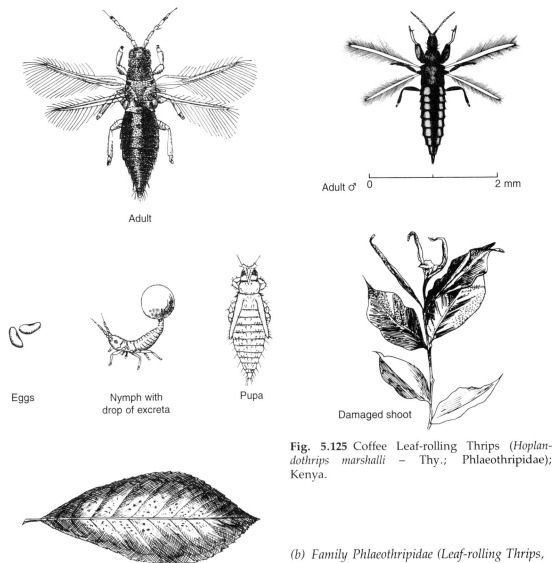

Adult

Adult ♂   0   2 mm

Eggs   Nymph with drop of excreta   Pupa

Damaged shoot

Fig. 5.125 Coffee Leaf-rolling Thrips (*Hoplandothrips marshalli* – Thy.; Phlaeothripidae); Kenya.

Damaged leaf

Fig. 5.124 Black Tea Thrips (*Heliothrips haemorrhoidalis* – Thy.; Thripidae); Kenya.

to be found in flowers of a wide range of cultivated plants and sometimes a pest in greenhouses; several hundred species are known; they are worldwide but most abundant in temperate regions.

(b) Family Phlaeothripidae (Leaf-rolling Thrips, etc.) (2000 spp.)

Some 200 genera show a great diversity of habits, some causing leaf distortion on cultivated plants. Relatively few are pests, but these include:

- *Haplothrips* spp. (300 spp.) – some in flowers, others on leaves, of a wide range of cultivated plants; found worldwide.
- *Hoplandothrips marshalli* (Coffee Leaf-rolling Thrips) – this species is found on coffee in East Africa (Fig. 5.125).

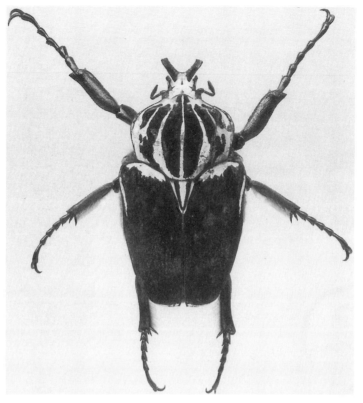

Fig. 5.126 Goliath Beetle (*Goliathus goliathus* – Col.; Scarabaeidae), ex Budongo Forest, Uganda (body length 100 mm).

5.4.8 Order Coleoptera (Beetles) (330 000 spp.)

This is the largest order of insects, with many new species being described each year. Beetles range in size from the minute to the very large (6 inches long), and they show a vast range of habits. Often there is a great difference between the habits of larvae and adults. Some are carnivorous, others saprophagous, but many are phytophagous and feed on different parts of the plant body in different ways. The more important agricultural pest groups are reviewed below:

(a) Family Scarabaeidae (Cockchafers, etc.; White Grubs) (17 000 spp.)

A distinctive group with many crop pests. The adults have strong biting mandibles and can damage both foliage and fruits; sometimes flowers are destroyed and some destroy seedlings by gnawing through the stem below ground. Larvae are soft-bodied, white and C-shaped with a slightly distended abdomen. They live in the soil where they eat plant roots and attack underground stems. There are six large and well-defined sub-families, whose members have different habits and some basic anatomical differences. Some of these beetles are very spectacular and beautiful insects; as an example, the Goliath Beetle (*Goliathus goliathus*) from Uganda (Fig. 5.126) is almost four inches in body length, and is to be found only in the canopy of the tropical rain forest. Some of the more important and widespread crop pests are listed below.

- *Amphimallon* spp. (Chafers) – in Europe,

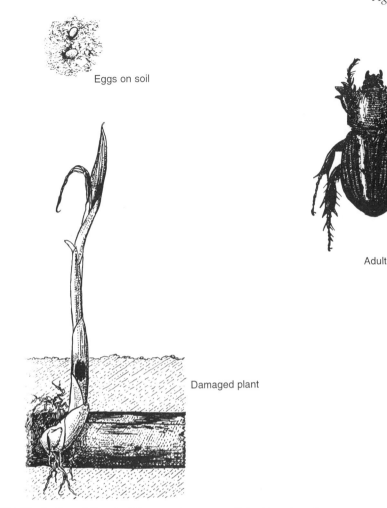

Eggs on soil

Adult

Damaged plant

Fig. 5.127 Black Cereal Beetle (*Heteronychus* sp. – Col.; Scarabaeidae); Kenya.

Asia and North America, these are pests of pastures and some crops (larvae mostly).

- *Anomala* spp. (Chafers, etc.) (100 + spp.) – a widespread genus, found worldwide attacking many crops both as adults and larvae.
- *Heteronychus* spp. (Black/Sugarcane Beetles, etc.) – adults bite plant stems below ground; Africa and Australasia (Fig. 5.127).
- *Holotrichia* spp. (Chafers; Chafer Grubs) – larvae polyphagous; Europe, Asia.
- *Melolontha* spp. (Cockchafers) – larvae

polyphagous root-eaters; adults bite leaves and young apples; Palaearctic.
- *Oryctes* spp. (Rhinoceros Beetles) – on Palmae and sugarcane; tropical Asia and Africa (Fig. 5.128).
- *Phyllophaga* spp. (20 spp.) (June Beetles, etc.) – pastures and crops damaged by larvae in North, Central and South America.
- *Popillia* spp. – adults eat flowers and foliage of many plants; *P. japonica* (Japanese Beetle) is a serious pest in Asia and North America (Fig. 5.129).

Fig. 5.128 Rhinoceros Beetle (*Oryctes monoceros* – Col.; Scarabaeidae); Mahé, Seychelles. Top left and right, leaf damage; bottom left, larva; bottom right, adults (left and right).

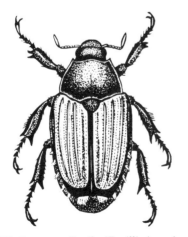

Fig. 5.129 Japanese Beetle (*Popillia japonica* – Col.; Scarabaeidae); Washington, USA.

(b) Family Buprestidae (Jewel Beetles;
Flat-headed-borers) (11 500 spp.)

A large group of predominantly tropical timber pests, often brightly coloured (some are brilliant and iridescent and can be made into ornamental brooches; Fig. 5.130). However, a number bore into stems and branches of shrubs (cotton, cowpea, etc.), fruit trees and also sugarcane. The legless larvae bore in the wood of the host plant for one to two years, causing extensive tunnelling which weakens the stem, trunk or branch. A striking feature of this family is that as presently interpreted, a few genera contain a large

Fig. 5.130 Jewel Beetles made into brooches (*Sternocera aequisignata* – Col.; Buprestidae) from Cheng Mai, Thailand.

number of species – examples include *Agrilus* with 700 species; the Australian *Stigmodera* with 400 species, and both *Chrysobothris* and *Sphenoptera* with about 300 species each.

The main agricultural pests include:

- *Agrilus* spp. (Citrus-borers, etc.) – on many different hosts; worldwide (Fig. 5.131); some in temperate regions in Europe, Asia and the USA.
- *Capnodis* spp. (Fruit Tree-borers) – bore in stems of rosaceous fruit trees; Europe, Africa and Asia.
- *Chrysobothris* spp. (Flat-headed-borers) – many deciduous trees and shrubs are attacked worldwide.
- *Sphenoptera* spp. (Cotton-borers, etc.) – found on cotton, cowpea, various trees and other shrubs; Africa and India.
- *Stigmodera* spp. (Jewel Beetles) – on a wide range of hosts in Australia.
- *Trachys* spp. (Buprestid Leaf-miners) – tiny larvae mine leaves of fruit trees, shrubs, legumes, etc.; Europe and Asia.

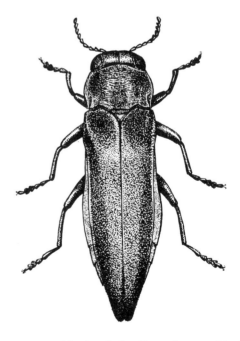

Fig. 5.131 Flat-headed Citrus-borer (*Agrilus auriventris* – Col.; Buprestidae); Canton, south China (body length 18 mm).

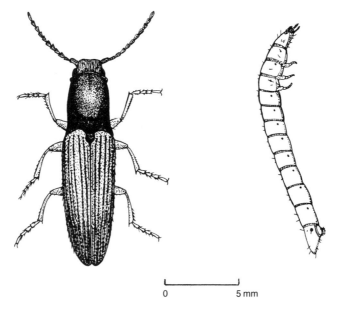

Fig. 5.132 Common Click Beetle and Wireworm (*Agriotes* sp. – Col.; Elateridae); Cambridge, UK.

(c) Family Elateridae (Click Beetles; Wireworms) (7000 spp.)

The larvae are the notorious 'Wireworms' of temperate agriculture – they were very serious pests during and after World War II in the UK when extensive pasture land was ploughed to grow cereals and potato crops. The soil-dwelling larvae eat the roots and stems of cereal seedlings and bore into potato tubers making them unsaleable. The adults are mostly smallish and drab brown in temperate regions, but some tropical species are brightly metallic and quite large (up to 4 cm). Development is slow and in the UK species of *Agriotes* may take up to five years to mature. Adults overwinter in hibernation. In the wild these are mostly grassland pests and occasionally pastures are damaged by the feeding larvae.

Some of the more important agricultural pests include:

- *Agriotes* spp. (Click Beetles; Wireworms) – polyphagous in Europe, Asia and North America (Fig. 5.132).

- *Athous* spp. (Garden Wireworms) – polyphagous in Europe.
- *Conoderus* spp. (Sugarcane Wireworms) – on sugarcane roots; China, Australia, the USA and West Indies.
- *Ctenicera* spp. (Grassland Wireworms) – in pastures; Europe and the USA.
- *Lacon* spp. (Tropical Wireworms) – polyphagous; pantropical.
- *Limonius* spp. (Sugar Beet Wireworms) – on sugar beet; USA.
- *Melanotus* spp. On surgarcane, sweet potato, etc.; Asia and the USA.

(d) Family Bostrychidae (Black Borers; Auger Beetles) (430 spp.)

The adults are usually black or brown, with hard cylindrical bodies and a large, hooded prothorax sheltering the deflexed head. The female beetle makes a breeding tunnel in wood (either branches or trunks of trees or bushes) and when the wood is dying, eggs are laid. In Africa coffee and cocoa are attacked by *Apate* beetles that bore the stems

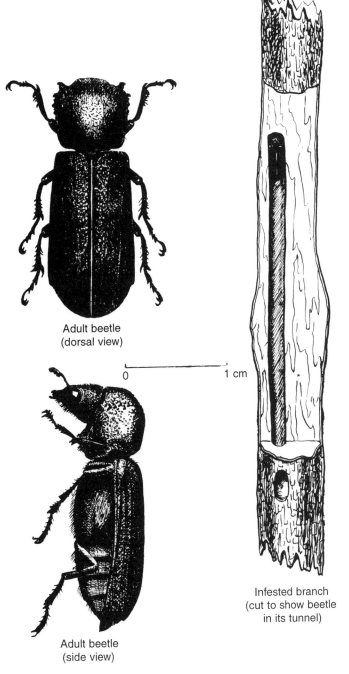

Adult beetle
(dorsal view)

0   1 cm

Adult beetle
(side view)

Infested branch
(cut to show beetle
in its tunnel)

Fig. 5.133 Black Borer (*Apate monacha* – Col.; Bostrychidae); Kenya.

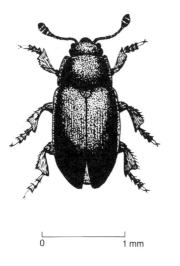

0 _____ 1 mm

Fig. 5.134 Blossom Beetle (*Meligethes* sp. – Col.; Nitidulidae); Cambridge, UK.

but do not breed in these plants; they breed in dead wood on other trees and bushes. Host trees can be killed by the boring.

Some of the species regarded as agricultural pests include:

- *Apate* spp. (Black Borers) – adults bore in a very wide range of woody hosts, and are pantropical in distribution (Fig. 5.133).
- *Bostrychopsis* spp. – on orchard trees in Australia and coffee in South East Asia and Africa.
- *Sinoxylon* spp. – bore in a wide range of woody hosts in Africa and Asia.

(e) Family Nitidulidae (Sap Beetles) (2200 spp.)

As already mentioned, the genus *Carpophilus* includes the Dried Fruit Beetles; many of these are to be found on either flowers or rape, or damaged fruits throughout the tropics. In the UK the genus *Meligethes* (Blossom Beetles) (Fig. 5.134) has recently become very abundant with the extensive cultivation of rape.

(f) Family Coccinellidae (Ladybird Beetles; Epilachna Beetles)

These beetles are renowned as predators of aphids and other crop pests, both as larvae and adults. The subfamily Epilach-ninae, however, is phytophagous, and the several species of *Epilachna* feed on the leaves of various Cucurbitaceae, Solanaceae and some legumes in tropical Africa and Asia (Fig. 5.135). *E. varievestis* is the Mexican Bean Beetle and can be quite damaging on *Phaseolus* in the New World.

(g) Family Meloidae (Blister/Oil Beetles) (2000 spp.)

These are soft-bodied beetles with a thin neck region and often a distinctive coloration – many are red/orange and black. They are diurnal and conspicuous, as befits their poisonous nature. The group is also un-usual in that the soil-dwelling larvae have a predacious and very active first instar called a 'triungulin', which seeks out the under-ground nest pods of Short-horned Grasshop-pers and Locusts (Acrididae), or in some cases the nests of solitary bees and wasps. The developing larvae stay with the eggs of the prey and become quite inactive as they grow. The adult beetles are often found on the flowers of legume crops, sweet potato, or ornamentals such as *Hibiscus* in large num-bers and all the flowers may be eaten away. However, the genus *Epicauta* is different in that they feed on leaves. Some adults of other genera prefer the tassel of maize and the developing panicle of other cereal plants.

The main pest genera are as follows:

- *Coryna* spp. (Pollen Beetles) – polyphagous on anthers and flowers of many crop plants in tropical Africa.
- *Cylindrothorax* spp. – found on the tassel of maize and panicle of rice and millets, and the flowers of some legumes; Africa and India.

Fig. 5.135 Epilachna Beetle (*Epilachna sparsa* – Col.; Coccinellidae) on *Solanum* leaves; Hong Kong. Top left, eggs; top right, larvae; bottom, adult.

- *Epicauta* spp. (Striped Blister Beetles, etc.) – adults eat leaves of many different plants; both Old World and New World tropics (Fig. 5.136).
- *Mylabris* spp. (Banded Blister Beetles) – adults eat flowers of legumes, cotton, *Hibiscus*, sweet potato, maize, etc.; Africa, India and warmer parts of Asia (Fig. 5.137).

(h) Family Cerambycidae (Longhorn Beetles) (20 000 spp.)

Most important as timber pests, but some attack tree crops and are serious agricultural crop pests – for example, cocoa bushes are recorded being attacked by 104 species of Cerambycidae (Entwhistle, 1972) and coffee 54 species (Le Pelley, 1968). Eggs are laid on

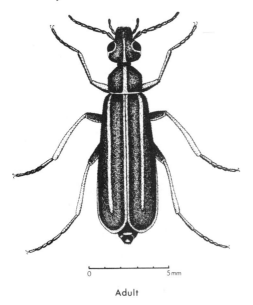

Adult

0 ———————— 5 mm

Fig. 5.136 Striped Blister Beetle (*Epicauta albovittata* – Col.; Meloidae); Kenya.

Fig. 5.137. Large Yellow-banded Blister Beetle (*Mylabris phalerata* – Col.; Meloidae); south China (body length 28 mm).

the tree bark and the larvae bore into the wood. A few species – these are usually small in size – attack herbaceous plants. Pupation occurs in the host tissues just beneath the bark and the emerging adult typically leaves a round hole in the bark. Some adults gnaw

the bark or the sapwood of the tree after emergence, and in some species they often actually girdle the tree/stem causing its death. The family is completely worldwide but more abundant in the tropical forests.

Some of the more important crop pests include:

- *Aeolesthes* spp. (Apple Stem-borer, etc.) – bore trunks and branches of fruit and nut trees; India and South East Asia.
- *Anoplophora* spp. (Citrus Longhorn, etc.) – bore *Citrus*; South East Asia (Fig. 5.138).
- *Anthores* spp. (White Coffee-borer, etc.) – on coffee; Africa.
- *Apriona* spp. (Jackfruit Longhorn) – bore Moraceae, apple, etc.; South East Asia.
- *Batocera* spp. (Spotted Longhorns, etc.) – fruit and nut trees bored; India to China and Australia.
- *Dirphya* spp. (Yellow-headed-borers) – bore coffee; East Africa (Fig. 5.139).
- *Glenea* spp. (Cocoa Stem-borers) – on cocoa; Africa and South East Asia.
- *Paranaleptes reticulata* (Cashew Stem Girdler) – on cashew, citrus, etc.; East Africa (Fig. 5.140).
- *Saperda candida* (Round-headed Apple Tree-borer) – in apple, etc.; USA and Canada.
- *Stenodontes* spp. – larvae bore a wide range of hosts; Africa.
- *Sthenias* spp. – bore many fruit and nut trees; tropical Africa and India.

(i) Family Bruchidae (Seed Beetles) (1300 spp.)

These are most damaging as post-harvest pests of pulses, but some species are only found in the field and some of the storage pests start their infestation in the field. The growing plant is not attacked – eggs are laid on the developing pod (or, as with *Callosobruchus*, the adults go to ripe pods that have opened and oviposit on the exposed seeds). Thus the damage to field crops consists of seed destruction – one seed per beetle

(a)

(b)

Fig. 5.138 (a) Citrus Longhorn Beetle (*Anoplophora chinensis* – Col.; Cerambycidae) (body length 30 mm); and (b) damaged *Citrus* stem; Hong Kong.

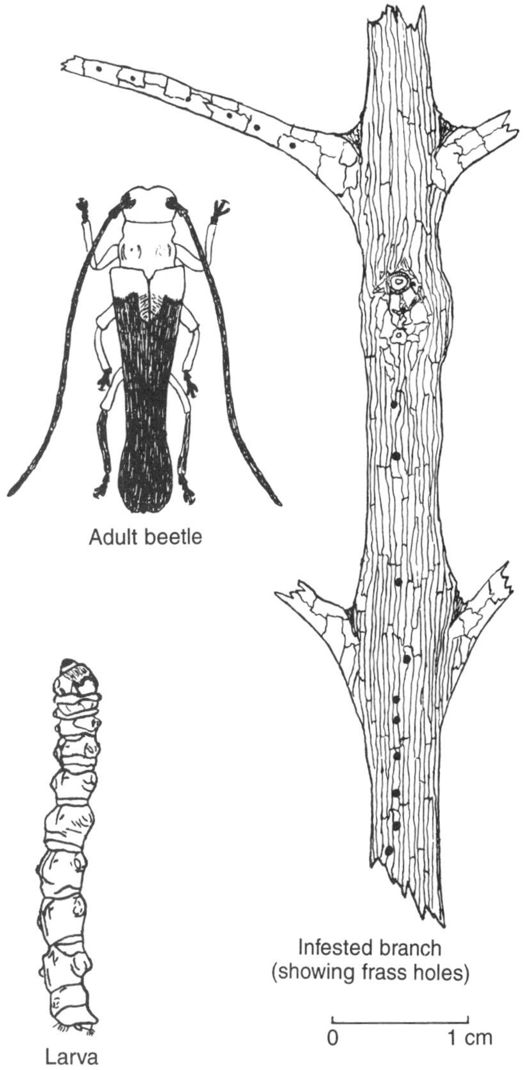

Adult beetle

Larva

Infested branch
(showing frass holes)

0   1 cm

Fig. 5.139 Yellow-headed-borer of coffee (*Dirphya nigricornis* – Col.; Cerambycidae); Tanzania.

larva – and as such is not often serious. The main pest species concerned are as follows:

- *Acanthoscelides obtectus* (Bean Bruchid) – young *Phaseolus* pods are bored by larvae in the field, but this species can live and breed in stores indefinitely; now cosmopolitan (Fig. 5.61).
- *Bruchus* spp. (Pea and Bean Beetles) – on pea, beans, lentil, vetch, etc.; in the field only; as a group worldwide (Fig. 5.141).
- *Bruchidius* spp. – on different legumes, both pulses and forage; mostly in the Mediterranean region.

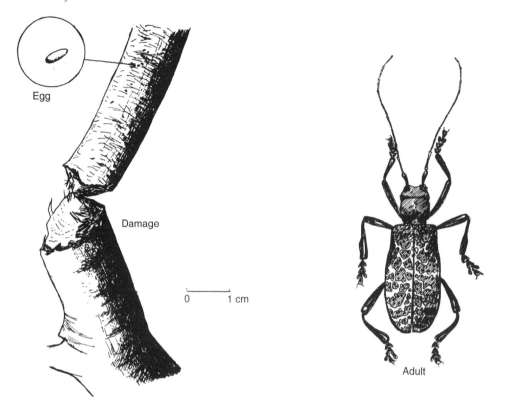

Fig. 5.140 Cashew Stem Girdler (*Paranaleptes reticulata* – Col.; Cerambycidae); Mombassa, Kenya.

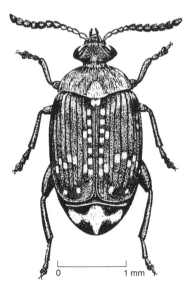

Fig. 5.141 Pea Bruchid (*Bruchus pisorum* – Col.; Bruchidae); Boston, UK.

- *Callosobruchus* spp. (Cowpea Bruchids, etc.) – eggs laid on ripe pods or exposed seeds in the field and development continues in storage; pantropical (Fig. 5.62).
- *Caryedon serratus* (Groundnut-borer) – attacks groundnut in the field in West Africa and tree legumes in India; in food stores worldwide.

(j) Family Chrysomelidae (Leaf Beetles) (20 000 spp.)

A very large group showing great diversity, and with many pests of cultivated plants. It is divided into a number (10–15) of different subfamilies, and it is difficult to generalize about the crop pests. They do, however, all feed on the leaves of the host plant, and both adults and larvae typically feed together. A

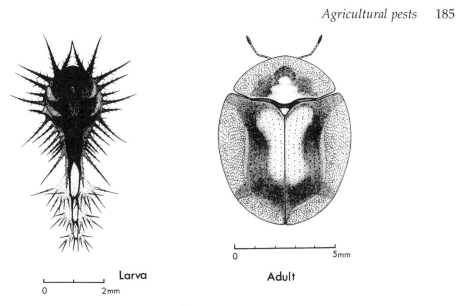

Larva

0   2mm

Adult

0   5mm

A. cincta

Fig. 5.142 Sweet Potato Tortoise Beetle (*Aspidomorpha cincta* – Col.; Chrysomelidae); Makerere, Uganda.

few of the more damaging and widespread species include:

- *Aspidomorpha* spp. (Sweet Potato Tortoise Beetles) – on *Ipomoea* and other Convolvulaceae and some *Solanum*; Africa, Asia and the West Indies (Fig. 5.142).
- *Cassida* spp. (Beet and Sweet Potato Tortoise Beetles) – Asia and Europe.
- *Crioceris* spp. (Asparagus Beetles) – Europe, Africa, Asia, North America.
- *Diabrotica* spp. (Cucumber Beetles; Corn Rootworms) – many species are serious pests on cucurbits, potato, legumes and maize; North, Central and South America (Fig. 5.143).
- *Epitrix* spp. (Potato/Tobacco Flea Beetles) – on solanaceae; Africa, Australia and North America.
- *Leptinotarsa decemlineata* (Colorado Beetle) – the main pest of potato worldwide; USA, Canada and now Europe (Fig. 5.144).
- *Monolepta* spp. (Spotted Leaf Beetles) – polyphagous minor pests, widespread in tropical Asia and Africa.

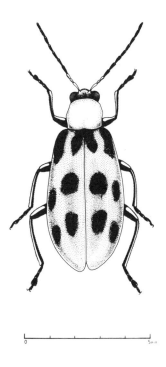

0   5mm

Fig. 5.143 Spotted Cucumber Beetle (*Diabrotica undecimpunctata* – Col.; Chrysomelidae).

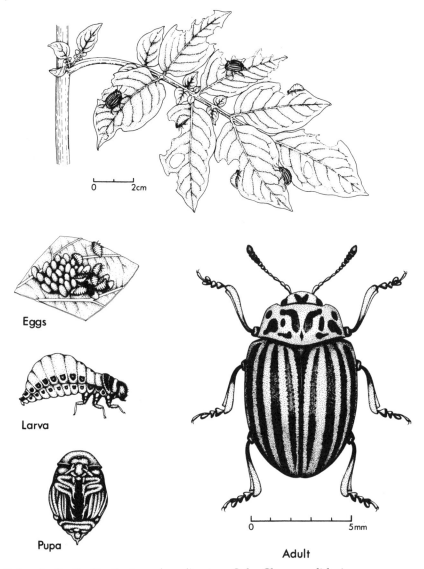

Eggs

Larva

Pupa

Adult

Fig. 5.144 Colorado Beetle (*Leptinotarsa decemlineata* – Col.; Chrysomelidae).

- *Oulema* spp. (Cereal Leaf Beetles) – on cereals and some other crops; Europe, Asia, Africa and the USA (Fig. 5.145).
- *Phyllotreta* spp. (Cabbage Flea Beetles, etc.) – larvae in soil but adults hole foliage and can destroy seedlings of Cruciferae; Europe, Asia, Africa, Australia and North America (Fig. 5.146).

- *Trichispa* spp. (Rice Hispid) – on rice in Africa (Fig. 5.147).

(k) Family Attelabidae (= Rhynchitinae) (300 spp.)

A small group of weevils recently separated from the bulk of the weevils; the antennae

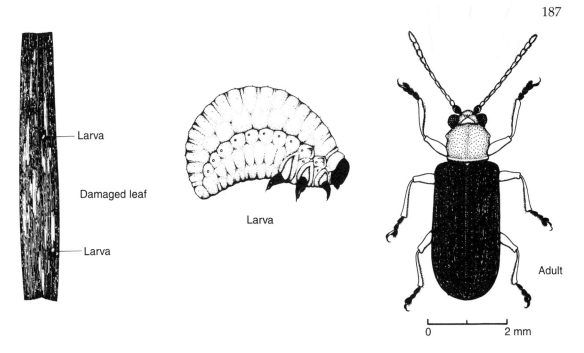

Larva

Damaged leaf

Larva

Larva

Adult

0   2 mm

Fig. 5.145 Cereal Leaf Beetle (*Oulema melanopus* – Col.; Chrysomelidae); Cambridge, UK.

are clubbed but not geniculate. The few horticultural pests belong to the genus *Rhynchites* – buds, shoots and young fruits can be destroyed (Fig. 5.148).

(l) Family Apionidae (Seed/Pod Weevils, etc.) (1000 spp.)

A small group, worldwide, characterized by black globular body, long pointed rostrum and antennae clubbed but not geniculate. *Cylas* and some others do, however, have a more elongate body. The two main pest genera are:

- *Apion* (Seed/Pod Weevils) – many species, most of which develop inside the seeds of legumes and other plants. Clover seeds are regular hosts in the UK (Fig. 5.149).
- *Cylas* (Sweet Potato Weevils) – pantropical and very damaging to sweet potato; larval tunnels in the tubers are usually infected by fungi (Fig. 5.150).

(m) Family Curculionidae (Weevils proper) (60 000 spp.)

A very large group, all phytophagous, with many damaging pests of agriculture. Both adults and larvae feed and cause damage, but seldom together; the larvae live either inside the plant tissues or in soil where they eat roots of many plants. The female often uses her long rostrum to make a deep feeding-hole in the plant tissues into which an egg is laid. Thus the protected larvae can be especially damaging to crop plants. All parts of the plant body are eaten by weevils. Adults typically eat leaves and buds and most are long-lived. Larval feeding habits are very varied but two types are most common: those that live in the soil eat plant roots, tubers, etc. and many others bore into the plant tissues. There are, however, very few leaf-miners.

There are dozens of important agricultural pests in this group – too many to list; but a

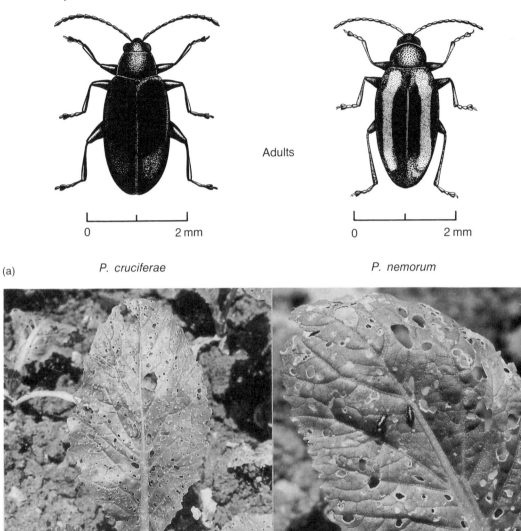

Adults

0   2 mm   0   2 mm

(a)   *P. cruciferae*   *P. nemorum*

(b)

Fig. 5.146 (a) Cabbage Flea Beetles (*Phyllotreta* spp. – Col.; Chrysomelidae); south China; and (b) damage caused.

small selection, chosen to represent the range of habits, is given below.

- *Alcidodes* spp. (Stem-girdling Weevils, etc.) – larvae bore stems and adults gnaw stems of cotton, beans, sweet potato, etc.; Africa, India.
- *Anthonomus grandis* (Cotton Boll Weevil) –

major pest of cotton; USA and Central and South America (Fig. 5.151). Other species attack fruit buds in Europe.
- *Ceutorhynchus* spp. (Turnip/Cabbage Weevils) – larvae found in seed pods of rape or stem galls of Cruciferae; Europe, the USA and Canada.
- *Cosmopolites sordidus* (Banana Weevil) –

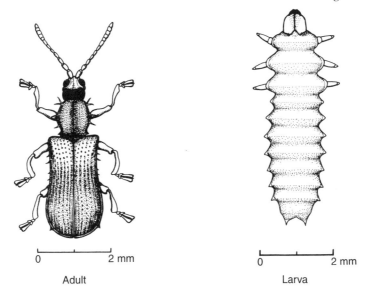

Fig. 5.147 Rice Hispid (*Trichispa serica* – Col.; Chrysomelidae), adult and larva; Kenya.

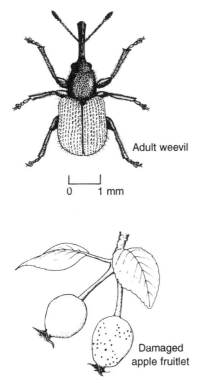

Fig. 5.148 Apple Fruit Rhynchites (*Rhynchites aequatus* – Col.; Attelabidae); Cambridge, UK.

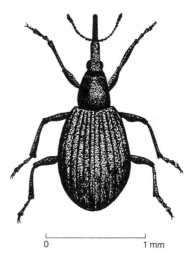

Fig. 5.149 Seed Weevil (*Apion* sp. – Col.; Apionidae); Cambridge, UK.

larvae bore rhizome of bananas; pantropical (Fig. 5.152).
- *Curculio* spp. (Nut Weevils) – larvae inside hazelnuts, chestnuts, etc.; Europe and North America.

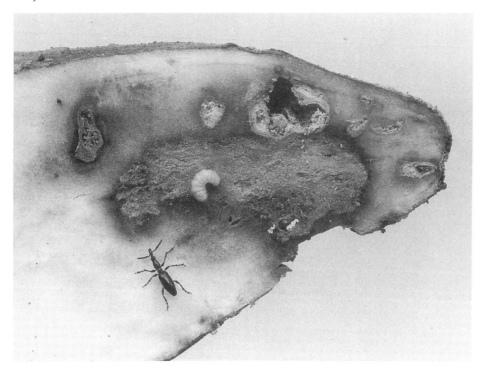

Fig. 5.150 Sweet Potato Weevil (*Cylas formicarius* – Col.; Apionidae) and larva on damaged tuber; south China.

Larva in boll   Adult on flower   Adult

Fig. 5.151 Cotton Boll Weevil (*Anthonomus grandis* – Col.; Curculionidae).

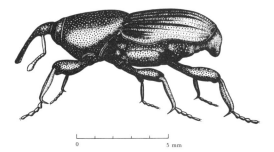

Fig. 5.152 Banana Weevil (*Cosmopolites sordidus* – Col.; Curculionidae); Uganda.

- *Graphognathus* spp. (White-fringed Weevils) – polyphagous; Australasia, the USA and South America.
- *Lixus* spp. (Beet/Cabbage Weevils) – larvae bore in roots and stems of vegetables; Africa, Asia, Mediterranean and the USA.
- *Myllocerus* spp. (Grey Weevils) – adults eat foliage, larvae eat roots of many crops; India to Japan.
- *Otiorhynchus* spp. (Root Weevils, etc.) – both adults and larvae eat many crops; cosmopolitan (Fig. 5.153).
- *Phyllobius* spp. (Common Leaf Weevils) – adults eat leaves of many fruit and nut trees, etc.; larvae in soil eat grass roots; Europe, Asia and the USA.
- *Rhynchophorus* spp. (Palm Weevils) – larvae bore crown of palms; pantropical (Fig. 5.154).
- *Sitophilus* spp. (Maize/Rice Weevils) – infestations of cereal panicles/cobs start in the field; cosmopolitan (see Figs 5.65 and 5.155).

(n) Family Scolytidae (Bark Beetles; Ambrosia Beetles)

Mainly forestry pests but a few attack crops, mostly as bark-borers on fruit and nut trees. A few of the crop pests include:

- *Hypothenemus hampei* (Coffee Berry-borer) – adults bore ripe berries of coffee; a

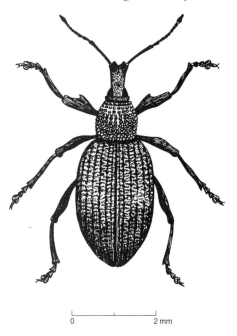

Fig. 5.153 Vine Weevil (*Otiorhynchus sulcatus* – Col.; Curculionidae); Cambridge, UK.

serious pest in Africa, South East Asia and now South America (Fig. 5.156).
- *Scotylus* spp. (Fruit Tree Bark Beetles) – many fruit trees, etc. attacked worldwide.
- *Xyleborus* spp. (Shot-hole-borers, etc.) – polyphagous and pantropical (Fig. 5.157).

5.4.9 Order Diptera (Flies) (85 000 spp.)

The group is of most importance as medical and veterinary pests, but some are pests of cultivated plants and the larvae of some can cause damage of economic proportions. Adults very rarely cause damage themselves.

The pest families include:

(a) Family Tipulidae (Crane Flies; Leatherjackets) (13 500 spp.)

The large, long-legged adults are very conspicuous. They are most abundant in tem-

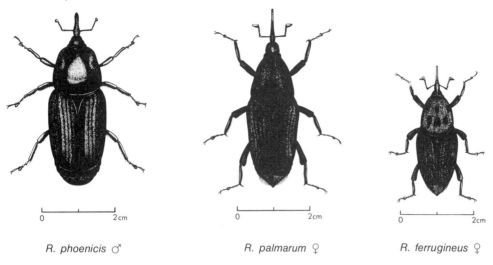

R. phoenicis ♂   R. palmarum ♀   R. ferrugineus ♀

Fig. 5.154 Palm Weevils (*Rhynchophorus* spp. – Col.; Curculionidae).

Fig. 5.155 Rice Weevil field infestation of maize cob (*Sitophilus oryzae* – Col.; Curculionidae); Alemaya, Ethiopia.

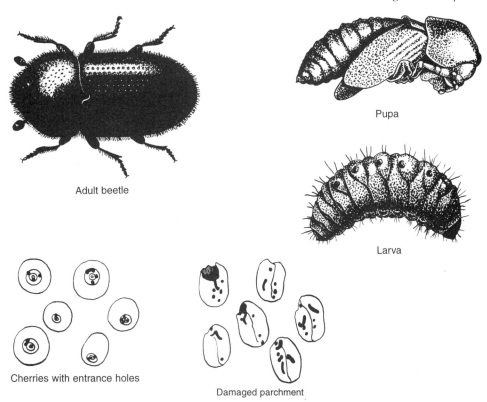

Adult beetle

Pupa

Larva

Cherries with entrance holes

Damaged parchment

Fig. 5.156 Coffee Berry-borer (*Hypothenemus hampei* – Col.; Scolytidae); Kenya.

perate regions, and their larvae are pests of grassland and will eat cereal seedlings and some root crops. They are most damaging to high-quality turf, such as golf courses and bowling greens. The main pest genus is *Tipula* (Fig. 5.158), a Holarctic genus.

(b) Family Cecidomyiidae (Gall Midges) (4500 spp.)

Almost all of these midges have phytophagous larvae and many induce galls on the host plant (hence the common name). The group is of great interest ecologically. Most species are quite host-specific; and leaves, flowers and fruits are more or less equally chosen as infestation sites. The tiny delicate adults are seldom seen but the galls

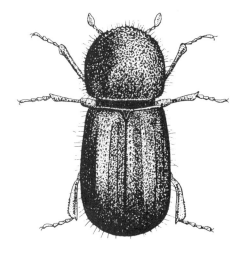

Fig. 5.157 Shot-hole-borer (*Xyleborus* sp. – Col.; Scolytidae); Kenya.

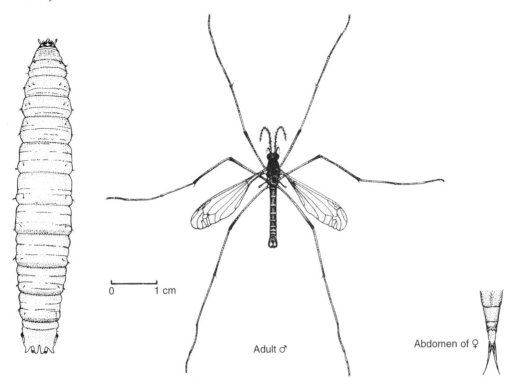

Adult ♂

Abdomen of ♀

Fig. 5.158 Crane Fly and larval Leatherjacket (*Ti*pula* sp. – Dipt.; Tipulidae); Cambridge, UK.

are usually conspicuous. Some of the more important pests are listed below:

- *Asphondylia* spp. (Sesame Gall Midge, etc.) – in sesame pods; India and East Africa (Fig. 5.159).
- *Contarinia* spp. (Blossom/Flower Midges) – many pest species in flowers and fruits of a very wide range of crops in all parts of the world.
- *Dasineura* spp. (Leaf Midges, etc.) – larvae fold or gall leaves of many plants; world-wide (Fig. 5.160).
- *Mayetiola destructor* (Hessian Fly) – larvae gall stems of wheat and other cereals and grasses; Holarctic.

A few other Nematocera are pests of cultivated plants, including *Bibio* spp. and *Sciara* spp., but these are seldom of importance.

(c) Family Syrphidae (Hover Flies, etc.) (5000 spp.)

The adults are most important as flower pollinators and the larvae as predators of aphids and other small insects, but a few are phytophagous and damage cultivated plants.

There are several species of *Eumerus* whose larvae infest bulbs of various types (flowers, onions, etc.) and some attack underground rhizomes and tubers (ginger, cassava, etc.). *Merodon* (Fig. 5.161) is a larger fly with larvae mostly found in *Narcissus* bulbs. *Mesogramina polita* is the Corn-feeding Syrphid Fly of the USA whose larvae eat pollen from maize flowers.

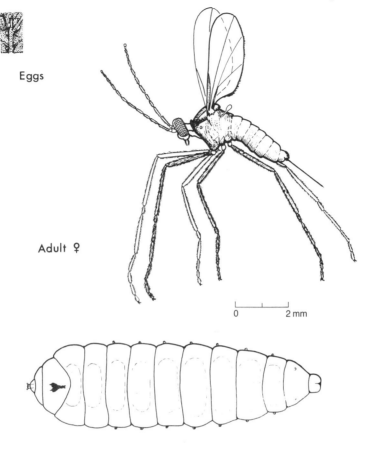

Eggs

Adult ♀

0   2 mm

Larva

Fig. 5.159 Sesame Gall Midge (*Asphondylia sesami* – Dipt.; Cecidomyiidae); adult and larva; Uganda.

(d) Family Ephydridae (Shore/Brine Flies) (1500 spp.)

Small, dark flies found in marshes, seashores, etc. The larvae of a few species make leaf mines, usually in aquatic or riparian plants, including Gramineae. They include:

- *Hydrellia* spp. (Leaf-miners) – rice, wheat and a wide range of pond plants have leaves mined by the larvae; worldwide (Fig. 5.162).
- *Scatella* spp. (Leaf-miners, etc.) – a few species are leaf-miners, including one in Chinese cabbage in south China.

(e) Family Psilidae (Carrot Fly, etc.) (170 spp.)

A small group but conspicuous in parts of the UK as locally serious pests of carrot, parsnip, celery and other Umbelliferae. The two main pest species are:

- *Psila nigricornis* (Chrysanthemum Stool-miner) – larvae mine stool of chrysanthemum, and sometimes root of carrot and lettuce; Europe, Canada.
- *Psila rosae* (Carrot Fly; Carrot Rust Fly) – a very serious pest of carrot, parsnip, celery, etc.; Europe, the USA and Canada (Fig. 5.163).

Fig. 5.160 Longan Gall Midge (*Gen.* and *sp. indet.* – Dipt.; Cecidomyiidae); south China.

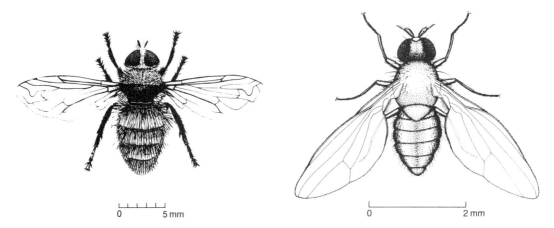

Fig. 5.161 Large Narcissus Fly (*Merodon equestris* – Dipt.; Syrphidae); Cambridge, UK.

Fig. 5.162 Rice Leaf-miner (*Hydrellia griseola* – Dipt.; Ephydridae).

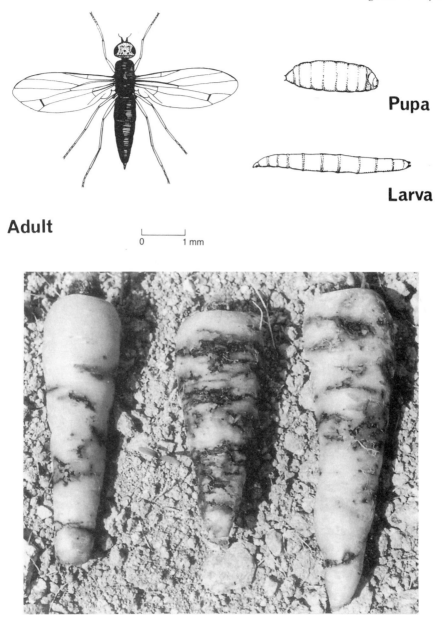

Pupa

Larva

Adult

0   1 mm

Fig. 5.163 Carrot Fly (*Psila rosae* – Dipt.; Psilidae); and damaged carrots; Cambridge, UK.

(f) Family Anthomyiidae (Root Flies, etc.) (1200 spp.)

A group with mixed habits, formerly included with the Muscidae; many larvae are saprophagous but some do attack growing plants and are pests. They include:

- *Delia* spp. – there are a dozen important pest species (formerly placed in *Hylemya*)

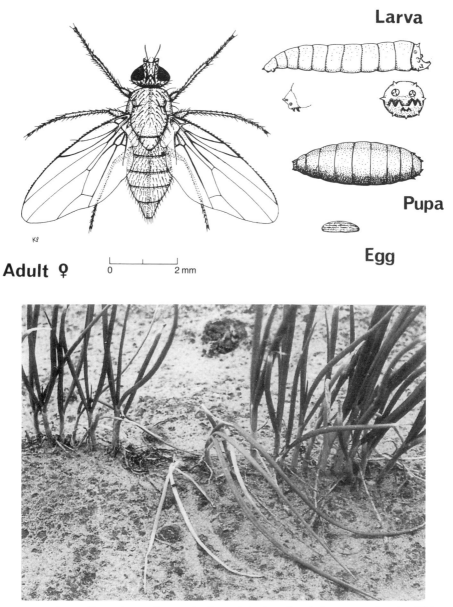

Fig. 5.164 Onion Fly (*Delia antiqua* – Dipt.; Anthomyiidae) and salad onions; Sandy, UK.

including Onion Fly (Fig. 5.164), Barley Fly, Bean Seed Fly and Cabbage Root Fly, some of which are Holarctic and widely distributed.

- *Pegomya* spp. – include Mangold Fly/Beet Leaf Miner, a common pest in Europe, Asia and now North America.

(g) Family Muscidae (House Flies; Shoot Flies, etc.) (3900 spp.)

Only a few members are truly phytophagous, but the genus *Atherigona* has a number of pest species causing 'dead-hearts' in cereal seedlings in parts of Africa and Asia. Rice, maize

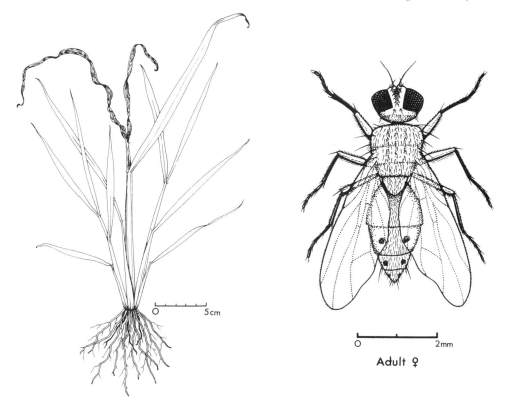

Adult ♀

Fig. 5.165 Rice Shoot Fly (*Atherigona oryzae* – Dipt.; Muscidae); Kenya; and typical shoot 'dead-heart'.

and sorghum are especially susceptible (Fig. 5.165).

(h) Family Agromyzidae (Leaf-miners, etc.)
(2000 spp.)

An interesting large group of small flies whose larvae mine in leaves, or sometimes twigs of woody plants. They attack all types of higher plants, worldwide. The female fly typically makes a series of feeding punctures on the upper surface of the leaf. This is usually only cosmetic but can be expensive, and there are records of seedlings being killed. One or two of the feeding punctures are usually used as oviposition sites. The larval tunnels in the leaf are usually of two types: a blotch mine as seen in holly leaves or a tunnel mine that starts narrow, becomes broader and terminates with a pupa on the lower side of the leaf (Fig. 5.166). The main damage is a loss of quality so that the plant becomes unsaleable, while the destruction of the leaf tissues can seriously reduce the rate of photosynthesis, with a consequent loss of yield.

Some of the more important pests include:

- *Agromyza* spp. (Cereal Leaf-miners, etc.) – on cereals and many herbaceous plants; worldwide.
- *Liriomyza* supp. (Vegetable Leaf-miners, etc.) – ona wide range of vegetables and ornamentals and some field crops; worldwide.

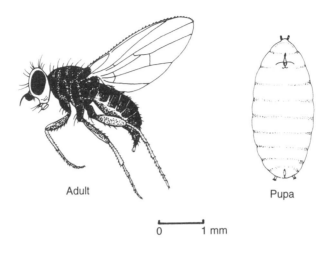

Adult

Pupa

0　　　1 mm

(a)

(b)

Fig. 5.166 (a) Pea (Vegetable) Leaf-miner (*Phytomyza horticola* – Dipt.; Agromyzidae); and (b) tunnelled leaf of Chinese cabbage showing pupae protruding from ventral surface; Hong Kong.

Fig. 5.167 Holly Leaf-miner (*Phytomyza ilicis* – Dipt.; Agromyzidae); Skegness, UK.

Fig. 5.168 Tea Leaf-miner (*Tropicomyia theae* – Dipt.; Agromyzidae); Mahé, Seychelles.

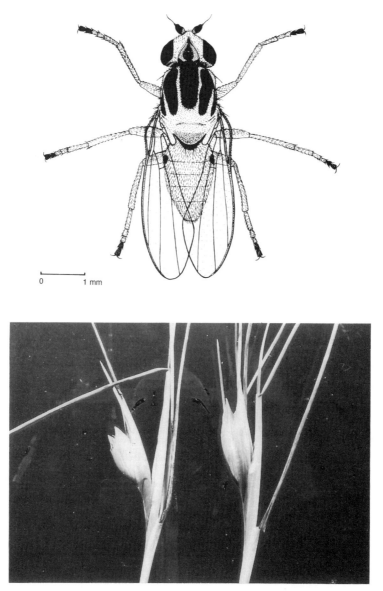

Fig. 5.169 Gout Fly (*Chlorops pumilions* – Dipt.; Chloropidae); and infested sea couch grass shoots; Skegness, UK.

- *Melanagromyza* spp. (Pulse Pod Flies, etc.) – in pods or stems of legumes; pantropical.
- *Ophiomyia phaseoli* (Bean Fly) – larvae bore seedling stems and kill plants; Africa, Asia and Australasia.
- *Phytomyza* (=*Chromatomyia*) spp. (Vegetable Leaf-miner, etc.) – several major pest species (Fig. 5.166); the Holly Leaf-miner (Fig. 5.167) is widespread in the UK.
- *Tropicomyia* spp. – several pest species; some are polyphagous and some mine pea leaves, while *T. theae* is only found in tea leaves (Fig. 5.168).

(i) Family Chloropidae (Gout/Frit Flies, etc.) (2000 spp.)

Small flies, most of which are phytophagous and feed in the stems of grasses and cereals. Some are common in wild grasses (Fig. 5.169) and a few are important pest species. The main pests include:

- *Chlorops* spp. (Gout Fly; Wheat/Rice Stem Maggot, etc.) – in stems of cereals and grasses; Europe, Asia and North America (Fig. 5.169).
- *Meromyza* spp. (Wheat Stem Maggot, etc.) – larvae found in wheat and grass stems; Europe, Japan and the USA.
- *Oscinella* spp. (Frit Fly; Cereal and Grass Flies) – several major pest species both in Europe and the USA.

(j) Family Tephritidae (Fruit Flies) (4500 spp.)

A large group, worldwide but most abundant in warmer regions. The fly is often brightly coloured, and the wings may be banded. The larvae are phytophagous and feed on or in growing plant tissues. Some live in flower heads (Compositae), while others make galls in stems, etc. The main agricultural pests are, however, the Fruit Flies whose larvae develop inside fruits of all types – both botanical and commercial. Adult flies feed on sugary food and are long-lived (five to six months); they fly strongly. Eggs are oviposited into the plant tissues, so the larvae live entirely inside the host plant tissues, but most leave when fully grown to pupate in soil. These pests are thus difficult to kill as larvae, and most control measures have to be applied against the adult flies, which is unusual.

The single most important pest is the Mediterranean Fruit Fly (*Ceratitis capitata*) which has the distinction of having an abbreviated English name – the Medfly (Fig. 5.70). It attacks a very wide range of commercial fruits native to the Mediterranean region; it has long been present in tropical Asia,

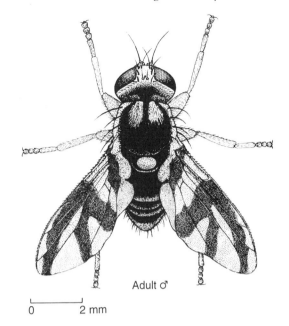

Adult ♂

0   2 mm

Fig. 5.170 Apple Maggot (*Rhagoletis pomonella* – Dipt.; Tephritidae); specimen ex Natural History Museum, UK.

Africa and Australasia; and is now found in Hawaii and South and Central America. So far it is thought that it has not been allowed to become established in the USA (although some authorities claim that there are some local populations now). There have been well-publicized regular accidental introductions into California, Texas and Florida, and then the State or Federal authorities have put the Fruit Fly eradication programme into action. To date, these actions have apparently kept the Medfly at bay, but at a considerable cost. However, the fruit industry in the southern USA is so valuable that the eradication programme costs, although high, are relatively insignificant.

The pest species fall mostly into a few large genera in different regions of the world, as shown below:

- *Anastrepha* spp. (150–200 spp.) (Central and South American Fruit Flies) – several polyphagous species found in South and

Central America and now the southern USA.

- *Ceratitis* spp. (100+ spp.) (Fruit Flies) – subtropical in the Mediterranean region and Africa, but *C. capitata* is more widespread now (Fig. 5.70).
- *Dacus* spp. (500 spp.; 30–40 spp. are pests) – a tropical group in Africa, Asia and the Pacific in a wide range of fruits; many pest species are polyphagous (Fig. 5.169). Wood (1989) suggests that many pest species of *Dacus* should be placed in the genus *Batrocera*.
- *Rhagoletis* spp. (50 spp.) (Temperate Fruit Flies) – found in both temperate and subtropical regions, these attack apple, cherry, walnut and blueberry, etc. (Fig. 5.170).

5.4.10 Order Lepidoptera (Moths and Butterflies) (120 000 spp.)

A very large and conspicuous group of insects, with very pronounced dimorphism between adults and larvae. The adults are almost entirely non-pests, and the damage to agricultural crops is invariably by the caterpillars (larval stages). There is overall tremendous diversity of habit among the caterpillars, but little morphological variation. They are all basically phytophagous and most feed in a concealed manner on the plant. A few species are poisonous or have urticating setae ('hairs'); these advertise their presence, being diurnal and sitting conspicuously on plant foliage. Generally, all parts of the plant body are attacked by feeding caterpillars but probably most caterpillars are 'leafworms'.

A number of very small moths, many of which are regarded as being rather primitive, are lumped together as the Micro-lepidoptera. Many have larvae that are leaf-miners of one sort or another. They may be quite conspicuous ecologically but most are not important economically (except for the Coffee Leaf-miners – Lyonetiidae).

Fig. 5.171 Bagworm (*Clania* sp. – Lep.; Psychidae) on palm frond; Malaysia (case length 30 mm).

(a) Family Psychidae (Bagworm Moths) (800 spp.)

A widespread group, most abundant in the tropics and often locally quite common. The larvae live inside bags made of silk, usually with pieces of twig or leaf material stuck on to the surface, and they feed on plant foliage. The largest bags measure up to 6 cm, but the smallest are 3–4 mm. Populations can be so high on some trees that defoliation

Fig. 5.172 Citrus Leaf-miner (*Phyllocnistis citrella* – Lep.; Phyllocnistidae) on grapefruit bush; south China.

results, and several pests in the Pacific region defoliate palms (oil and coconut). There are many small genera and it is difficult to generalize about pests, but a few more important pests are listed:

- *Clania* spp. – a large genus found on a wide range of cultivated plants in Africa and Asia (Fig. 5.171).
- *Crematopsyche pendula* (Oil Palm Bagworm) – found on oil palm in India, South East Asia and Australia.
- *Mahasena corbetti* (Coconut Case Caterpillar) – found on a wide range of crops in addition to coconut and oil palm in South East Asia.

(b) Family Phyllocnistidae (Leaf-miners) (50 spp.)

A very small group of leaf-miners; includes the Citrus Leaf-miner (*Phyllocnistis citrella*) (Fig. 5.172), which is very abundant in parts of tropical Asia and Africa, although damage levels are usually low.

(c) Family Lyonetiidae (Leaf-miners)

Quite a large group of very small moths, but heterogeneous. A few species of *Bedellia* (Sweet Potato/Cotton Leaf-miners), *Opogona* (Sugarcane Leaf-miners) and others are often recorded as being conspicuous, but actual damage is usually slight. However, the genus *Leucoptera* contains several species known as Coffee Leaf Blotch-miners (Fig. 5.173), which are major pests of coffee in Africa because attacked leaves are usually shed.

(d) Family Sesiidae (Clearwing Moths) (1000 spp.)

A large group, worldwide, with characteristic adults and larvae that bore in wood or vine

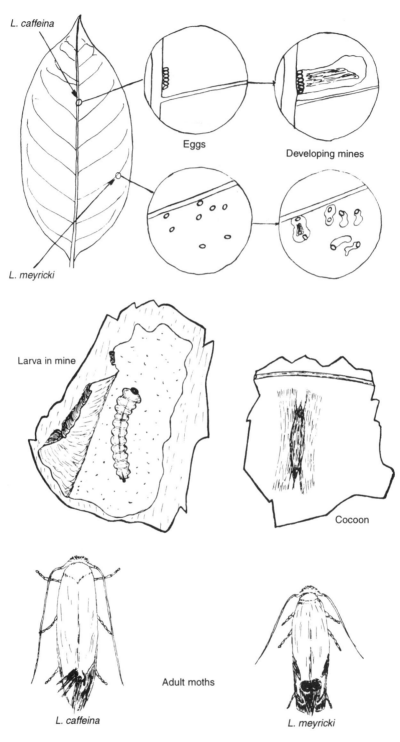

Fig. 5.173 Coffee Leaf Blotch-miners (*Leucoptera* spp.– Lep.; Lyonetiidae); Kenya.

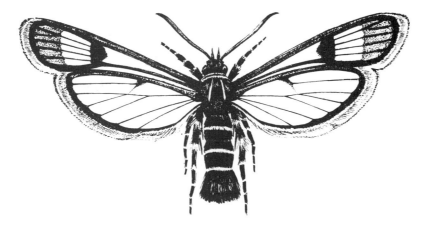

Fig. 5.174 Currant Clearwing Moth (*Synanthedon salmachus* – Lep.; Sesiidae); Cambridge, UK.

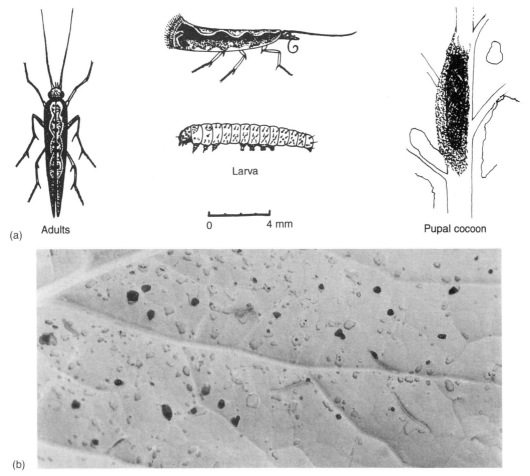

(a) Adults

Larva

0   4 mm

Pupal cocoon

(b)

Fig. 5.175 (a) Diamond-back Moth (*Plutella xylostella* – Lep.; Yponomeutidae); and (b) damaged cabbage leaf; south China.

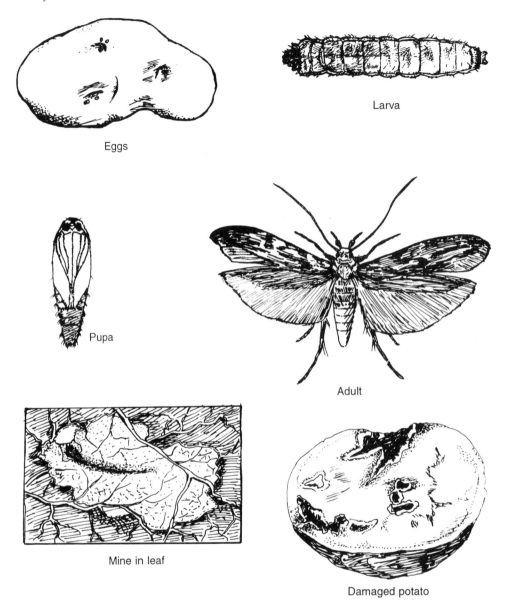

Eggs

Larva

Pupa

Adult

Mine in leaf

Damaged potato

Fig. 5.176 Potato Tuber Moth (*Phthorimaea operculella* – Lep.; Gelechiidae); Kenya.

stems (grape, cucurbits and sweet potato). Damage levels are not usually serious. The main pest genus agriculturally is *Synanthedon* and different species attack apple, strawberry, peach, currants and sweet potato (Fig. 5.174).

(e) Family Yponomeutidae (Small Ermine Moths, etc.) (800 spp.)

A small group with several interesting pest species in different parts of the world. The Diamond-back Moth (*Plutella xylostella*)

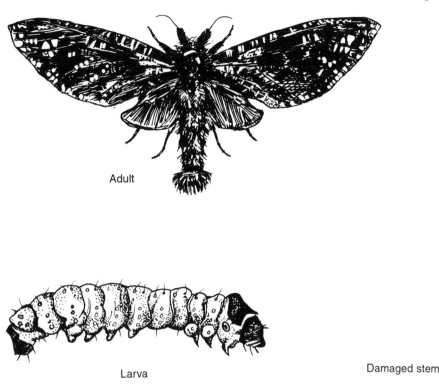

Adult

Larva

Damaged stem

Fig. 5.177 Cocoa Stem-borer (*Eulophonotus myrmeleon* – Lep.; Cossidae); Kenya.

(Fig. 5.175) is a serious pest of cultivated Cruciferae worldwide, but especially in tropical Asia where there can be 12–14 generations per year and in some areas pesticide resistance is very widespread.

(f) Family Gelechiidae (4000 spp.)

A large group with 400 genera, found worldwide, and with some serious crop pests. Most of the larvae bore inside plant tissues (stems, leaves, fruits, etc.), and some are well adapted as miners with the caterpillar emerging from the egg through the base directly into the host plant tissues. This practice makes Pink Bollworm a major pest of cotton and difficult to control. Measures have to be applied against the adult moths rather than the larvae.

Some of the more important crop pest species include:

- *Anarsia lineatella* (Peach Twig-borer) – larvae bore shoots and fruits of many stone fruits; Europe, Asia, North America.
- *Keiferia lycopersicella* (Tomato Pinworm) – on tomato; New World.
- *Pectinophora* spp. (Pink Bollworm, etc.) – bore cotton bolls; pantropical.
- *Phthorimaea operculella* (Potato Tuber Moth) – larvae mine leaves and bore stems and tubers of potato, etc.; pantropical (Fig. 5.176).
- *Scrobipalpa* spp. (Beet Moths, etc.) – larvae attack beets, etc.; Europe, Asia and North America.
- *Scrobipalpopsis solanivora* (South American Potato Tuber Moth) – major potato pest; Central and South America.

- *Sitotroga cerealella* (Angoumois Grain Moth) – a major stored grain pest although infestations usually start in the field on the ripening grain; cosmopolitan throughout the warmer parts of the world.

(g) Family Cossidae (Leopard/Goat Moths)

Mostly large moths; worldwide as a group but especially numerous in Australia. The larvae tunnel in wood and some in the stem of reeds. Some fruit trees are heavily attacked and may die. The main pests include:

- *Cossula magnifica* (Pecan Carpenterworm) – bores pecan, etc.; USA.
- *Eulophonotus myrmeleon* (Cocoa Stem-borer) – larvae bore stems of cocoa, coffee, cola, etc.; tropical Africa (Fig. 5.177).
- *Xyleutes capensis* (Castor Stem-borer) – larvae bore castor stems; East Africa. Other species found in other plants.
- *Zeuzera* spp. (Leopard Moths) – bore citrus, coffee, fruit trees, etc.; Europe, Asia and North America (Fig. 5.178).

(h) Family Limacodidae (Stinging and Slug Caterpillars)

Some important polyphagous defoliators, mostly on palms, bushes and trees. Some caterpillars are quite slug-like and may be called 'Jelly Grubs' (Fig. 5.179). Others are spiny with urticating bristles and can be painful to handle. Coffee, tea, cocoa, coconut and oil palm are all heavily attacked, each having about a dozen species recorded as pests worldwide.

A few major pests are listed below:

- *Niphadolepis* spp. (Jelly Grubs) – on coffee, tea, etc.; East Africa (Fig. 5.179).
- *Parasa* spp. (Stinging Caterpillars) – polyphagous on a wide range of Monocotyledoneae, coffee, citrus, etc.; tropical Africa and Asia (Fig. 5.180).
- *Setora nitens* (Nettle Caterpillars) –

polyphagous on palms, tea, coffee, cocoa, etc.; throughout South East Asia.
- *Thosea* spp. (Slug Caterpillars) – polyphagous on many crop plants; Africa and tropical Asia.

(i) Family Tortricidae (Tortrix Moths) (4000 spp.)

A large group of small moths, very important as pests of fruit trees in temperate regions. The family shows great diversity of habits, especially in relation to feeding and the damage they inflict. Many species produce silk, and the larvae web leaves together to construct shelters for feeding and pupation. Many species are economic pests: a few of the more important and diverse are listed here. A large number of species eat leaves, shoots and fruits on fruit trees and are collectively known as 'Fruit Tree Tortricids'.

- *Acleris* spp. (Fruit Tree Tortricids, etc.) – polyphagous; Europe, Asia and North America.
- *Archips* spp. (10+ spp.) (Fruit Tree Tortricids, etc.) – some polyphagous; web leaves and damage surface of fruitlets; Europe, Asia and North America (Fig. 5.181).
- *Cnephasia* spp. (Omnivorous Leaf-tier, etc.) – several species on different crops throughout Europe, Asia and North America.
- *Cryptophlebia* spp. (False Codling Moth, etc.) – polyphagous in bolls, fruits, nuts and legume pods; Africa and parts of Asia.
- *Cydia* spp. – there are several important agricultural pests in this genus including *C. molesta* (Oriental Fruit Moth), *C. nigricana* (Pea Moth) (Fig. 5.182) and *C. pomonella* (Codling Moth) (Fig. 5.183), most of which are found virtually worldwide. They are very damaging to their particular hosts.
- *Tortrix* spp. (Leaf-rollers, etc.) – larvae

0   1 cm

Fig. 5.178 Leopard Moth (*Zeuzera pyrina* – Lep.; Cossidae); Cambridge. Larval damage and protruding pupal exuvium.

Adult moth

Moth in natural position

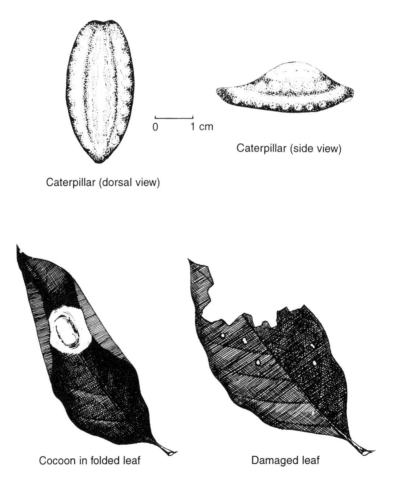

Caterpillar (dorsal view)

0   1 cm

Caterpillar (side view)

Cocoon in folded leaf

Damaged leaf

Fig. 5.179 Coffee Jelly Grub (*Niphadolepis alianta* – Lep.; Limacodidae); Kenya.

Adult moth

Adult moth, natural position

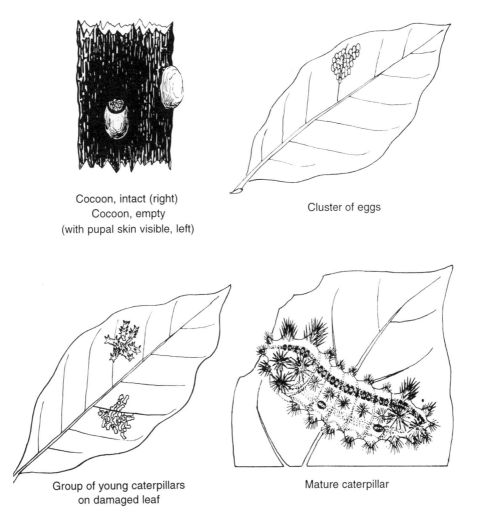

Cocoon, intact (right)
Cocoon, empty
(with pupal skin visible, left)

Cluster of eggs

Group of young caterpillars
on damaged leaf

Mature caterpillar

Fig. 5.180 Stinging Caterpillar (*Parasa vivida* – Lep.; Limacodidae); wingspan 38 mm; on coffee; Kenya.

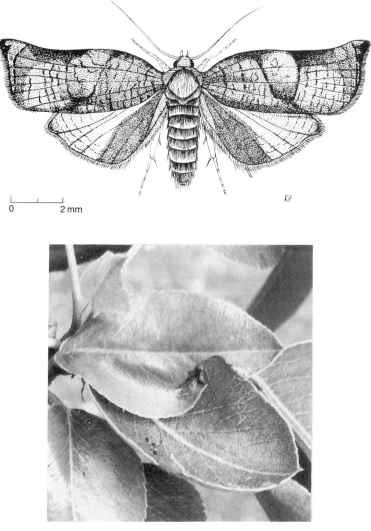

Fig. 5.181 Fruit Tree Tortrix (*Archips podana* – Lep.; Tortricidae), adult moth and larval leaf webbing; Skegness, UK.

attack leaves of fruit and nut trees, as well as coffee, cocoa, etc.; Europe, Asia and Africa.

There are many other genera involved but most have only a few pest species.

(j) Family Pyralidae (Snout Moths)

A very large group of medium-sized moths, found throughout the world. The group is divided into more than a dozen distinct subfamilies, most with different larval habits, some are adapted to aquatic lifestyles. One group is associated with Gramineae and there are several stalk-borers of cereal plants that are very damaging – rice, sugarcane, maize and sorghum suffer in particular. Most larvae produce silk and feed in concealment; leaf-rolling and leaf-folding using strands of silk are both widely practised.

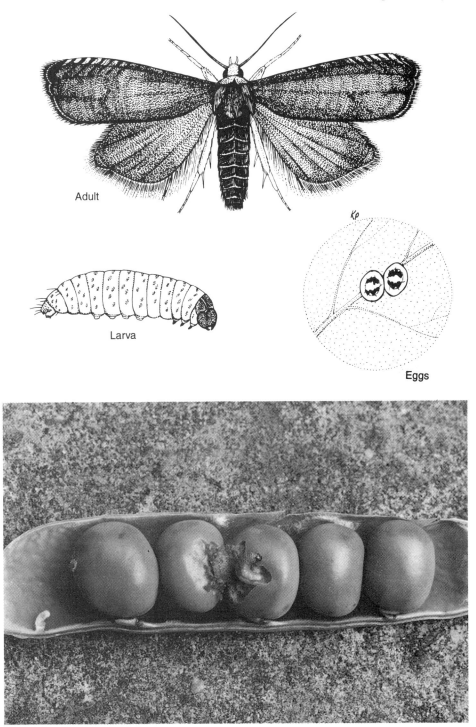

Adult

Larva

Eggs

(a)

(b)

Fig. 5.182 (a) Pea Moth (*Cydia nigricana* – Lep.; Tortricidae); and (b) larval damage in a pea pod; Cambridge and Skegness, UK.

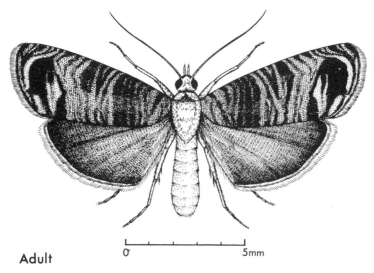

Adult

$\stackrel{\vdash}{0}$ 5mm

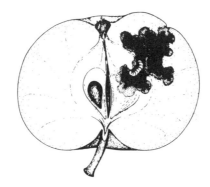

Larva inside fruit

Fig. 5.183 Codling Moth (*Cydia pomonella* – Lep.; Tortricidae); Cambridge, UK.

Some of the most important pests include:

- *Chilo* spp. (Rice Stalk-borers) – larvae bore stems of rice, sugarcane, etc. in tropical Asia and Africa (Fig. 5.184).
- *Crambus* spp. (Grass Moths; Sod Webworms) – pasture pests in Europe and North America.
- *Crocidolomia* spp. (Cabbage Cluster Caterpillars) – on Cruciferae; Africa, Asia and Australasia.
- *Diatraea* spp. (Sugarcane Stalk-borers, etc.) – sugarcane and cereals are attacked; USA and Central and South America.
- *Etiella* spp. (Pea Pod-borers) – larvae bore pods of legumes; worldwide in warm regions.

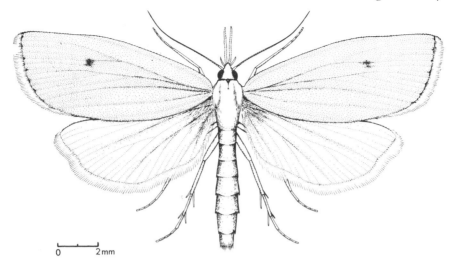

Adult

(a)

(b)

Fig. 5.184 (a) Spotted Stalk-borer (*Chilo partellus* – Lep.; Pyralidae); and (b) damage to sorghum stalks; Alemaya, Ethiopia.

- *Hellula* spp. (Cabbage Webworms) – on Cruciferae; Africa, South East Asia and Central and South America.
- *Ostrinia* spp. (Corn Borers) – larvae bore stem of maize and other cereals; Europe, Asia and eastern North America (Fig. 5.185).
- *Scirpophaga* spp. (Paddy Stem-borers, etc.) – in rice, sugarcane and grasses; tropical Asia and Australasia.

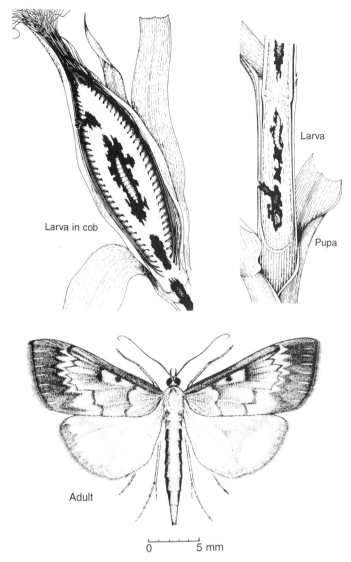

Larva in cob

Larva

Pupa

Adult

0   5 mm

Fig. 5.185 European Stalk-borer (*Ostrinia nubilalis* – Lep.; Pyralidae) and damage to maize.

Many other genera are important pests of particular crops, but these are too numerous to mention here.

(k) Family Lycaenidae (Blue Butterflies; Coppers; Hairstreaks)

A large group of small, brightly coloured butterflies, found worldwide. They are diurnal and active and usually conspicuous in flight, but cryptic at rest. Some larvae are carnivorous and feed on aphids, and others live with ants, but most are phytophagous. These feed on shoots, flowers and young fruits of *Citrus* and other fruits, and on legume pods. A few species are pests of some importance, including:

- *Chilades lajus* (Lime Blue Butterfly) – on *Citrus* and Rutaceae in tropical Asia (India to China).

Fig. 5.186 Pea Blue Butterfly (*Lampides boeticus* – Lep.; Pyralidae) and damage to maize. Top left, adult; top right, adult; bottom, larva. (wingspan 2 cm).

- *Lampides boeticus* (Pea Blue Butterfly) – larvae bore pods of pea and other legumes; cosmopolitan in the Old World (Fig. 5.186).
- *Virachola* spp. (Coffee Berry Butterfly, etc.) – one species bores berries of coffee in Africa (Fig. 5.187), others are polyphagous on many fruits and tree legumes; Africa and India.

(l) Family Pieridae (White Butterflies; Yellows; Orange-tips, etc.)

Distinctive white or yellow butterflies of

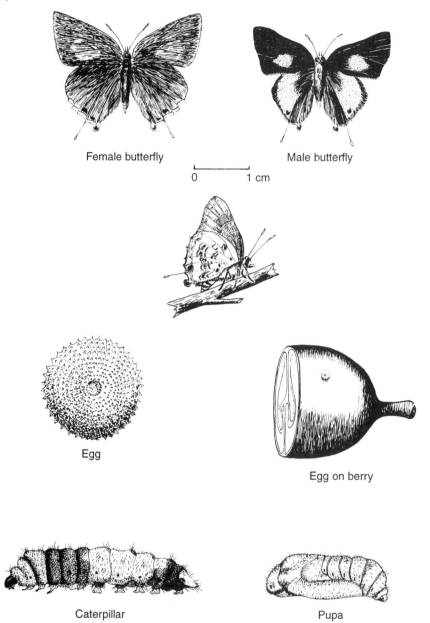

Female butterfly   Male butterfly

0   1 cm

Egg

Egg on berry

Caterpillar   Pupa

Fig. 5.187 Coffee Berry (Blue) Butterfly (*Virachola bimaculata* – Lep.; Lycaenidae); Kenya.

medium size, found worldwide. Several species are spectacular migrants in both Asia and North America.

Two of the main groups of host plants are Cruciferae and Leguminosae – the species of *Pieris* on Cruciferae are especially damaging.

Some of the more important pest genera include:

- *Colias* spp. (Clouded Yellows) – larvae feed on forage legumes; Europe, Asia and North America.

Fig. 5.188 Small White Butterfly (*Pieris canidia* – Lep.; Pieridae) on *Nasturtium*; Hong Kong. Top left, larva; top right, pupa; bottom left and right, adults.

- *Eurema* spp. (Grass Yellows) – feed on tree legumes in the Old World tropics.
- *Pieris* spp. (White Butterflies) – a north temperate genus with species extending into warmer regions. Hosts are Cruciferae and many are species of *Brassica*. Several major pests are known – both widespread and locally abundant (Fig. 5.188).

(m) Family Papilionidae (Swallowtail Butterflies) (600 spp.)

A tropical group, some very large in size, and most are of striking beauty. Only a few actually have the distinctive trailing 'tails'

from the hind wings. Some species, such as the spectacular Birdwings, are valuable as collectors' items. A few are pests of cultivated plants but the main pests are the Citrus Butterflies (*Papilio* spp. (10–20 spp.)) (Fig. 5.189), found worldwide on *Citrus* and other Rutaceae wherever they are grown.

(n) Family Hesperiidae (Skippers)

Stout-bodied insects that fly strongly; most are diurnal. The larvae feed mostly on Gramineae (and other monocotyledenous plants), and the leaf is usually rolled or

Fig. 5.189 Citrus Butterflies (*Papilio polytes* (left) and *P. demoleus* – Lep.; Papilionidae); Hong Kong. Top left, egg; top right, small larva; centre left, large larva; centre right, pupa.

folded to create a hiding-place for the soft-bodied caterpillar. Rice is attacked more often than any other crop plant. A few pest species are listed below:

- *Borbo* spp. (Swifts) – on cereals and grasses in South East Asia.
- *Erionota* spp. (2 spp.) (Banana Skippers) – on leaves of banana, and also on oil palm; India to South Japan (Fig. 5.190).
- *Panara* spp. (Rice Skippers) – on rice and other Gramineae; South East Asia (Fig. 5.191).
- *Telicota* spp. (Rice Skippers, etc.) – on rice, other Gramineae and coconut palm; India to China, Pacific Islands.

Fig. 5.190 Banana Skipper (*Erionota torus* – Lep.; Hesperiidae); south China (wingspan 65 mm). Top left, eggs; top middle, larva; top right, leaf damage; bottom left, pupa; bottom right, adult.

Fig. 5.191 Rice Skipper (*Panara guttata* – Lep.; Hesperiidae); south China (wingspan 32 mm).

(a)

(b)

(c)

(d)

Fig. 5.192 Magpie Moth (*Abraxas grossulariata* – Lep.; Geometridae); Cambridge, UK. (a) Adult resting on tree trunk; (b) adult; (c) larva; (d) pupa.

(o) Family Geometridae (Carpet Moths, Pugs; Loopers) (12 000 spp.)

A large group, ecologically dominating worldwide. Essentially a forest group so the few crop pests tend to attack fruit trees and forest shrubs (coffee, cocoa, hazel, etc.). The family includes many different genera and so it is difficult to generalize about crop pests.

One group has wingless females that emerge over the winter period and lay eggs in bark crevices on trees; the hatching larvae in early spring can defoliate fruit and nut trees. A small selection of crop pests includes:

- *Abraxas* spp. (Magpie Moths) – on currants, fruit and nut bushes; Europe and Asia (Fig. 5.192).

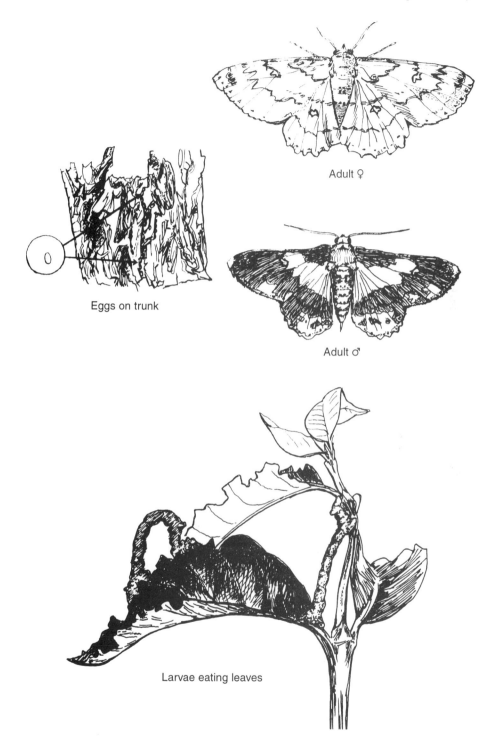

Adult ♀

Eggs on trunk

Adult ♂

Larvae eating leaves

Fig. 5.193 Giant (Coffee) Looper (*Ascotis selenaria* – Lep.; Geometridae) on coffee; Kenya.

- *Alsophila* spp. (March Moth; Fall Canker-worm) – on fruit trees; Europe, Asia, Canada and the USA.
- *Ascotis selenaria* (Giant Loopers) – polyphagous on coffee, citrus, sweet potato, soybean, etc.); Africa and Asia (Fig. 5.193).
- *Erannis* spp. (Winter Moths, etc.) – on deciduous fruit trees; Europe, Asia and the USA.
- *Hyposidra* spp. – polyphagous on coffee, cocoa, tea, cinchona, etc.; Africa, India, South East Asia and Australia.
- *Operophtera* spp. (Winter Moths) – on fruit and nut trees; Europe, Asia, Canada and the USA.

(p) Family Lasiocampidae (Eggars; Lappets; Tent Caterpillars) (1000 spp.)

Large, stout-bodied moths; caterpillars are large and bristly and most live gregariously inside a silken 'tent' in a bush or tree. The bristles can be painfully urticating. The large numbers of caterpillars can defoliate the host plant and several species can be serious pests:

- *Gastropacha* spp. (Lappet Moths) – on a wide range of bush and tree crops from India up to Japan.
- *Malacosoma* spp. (Tent Caterpillars) – larvae polyphagous on fruit trees in Europe, parts of Asia and North America.
- *Trabala* spp. (Oriental Lappet Moth, etc.) – on a wide range of fruit trees and bushes from India to China (Fig. 5.194).

(q) Family Saturniidae (Emperor Moths; Giant Silkmoths)

These largest of all Lepidoptera are mostly forest insects, although a few are recorded on cultivated fruit and nut trees and some other crop plants, mostly in tropical Asia and Africa. The following of these spectacular tropical insects can be classed as agricultural pests:

- *Actias selene* (Moon Moth) – on many fruit and nut trees in the area from India to southern China (Fig. 5.195).
- *Attacus atlas* (Atlas Moth) – on many different trees and some bushes in South East Asia (Fig. 5.196); the genus is pantropical, as half a dozen species.
- *Caligula* spp. – on fruit and nut trees and others; India to Indonesia.
- *Naudaurelia* spp. (Silkworms) – on a wide range of hosts in East and South Africa.

(r) Family Sphingidae (Hawk Moths; Hornworms) (1000 spp.)

Large, distinctive moths, worldwide but more abundant in the tropics. Some important pest species are cosmopolitan. The large caterpillars ('Hornworms' in the USA) are solitary, although there may be so many on a bush (e.g. coffee) that it is defoliated.

Some of the distinctive pest species include:

- *Acherontia* spp. (Death's Head Hawk Moths) – several species in Europe, Asia and Africa; mostly on Solanaceae, but some species are more polyphagous; a serious pest in the past on potato crops in the UK (Fig. 5.197).
- *Agrius* spp. (Convolvulus/Sweet Potato Hawk Moth) – one species in the New World and another in the Old World feeding on *Ipomoea* and *Convolvulus*, and occasionally other plants (Fig. 5.198).
- *Cephonodes hylas* (Coffee Hawk Moth) – on coffee, *Gardenia* and some other Rubiaceae; Africa, Asia and Australasia (Fig. 5.199).
- *Hyles* spp. (Striped Hawk Moth, etc.) – polyphagous and cosmopolitan.
- *Manduca* spp. (Tobacco/Tomato Hornworms) – these are the New World equivalent of *Acherontia* and feed on Solanaceae.

(a)

(b)

Fig. 5.194 Oriental Lappet Moth (*Trabala vishnou* – Lep.; Lasiocampidae); (a) adult female (wingspan 60 mm); and (b) larva on leaf of *Rhodomyrtus* (body length 60 mm); south China.

(a)

(b)

Fig. 5.195 Moon Moth (*Actias selene* – Lep.; Saturniidae); (a) adult male resting on tree trunk (wingspan 140 mm); (b) larva (70 mm) on twig; south China.

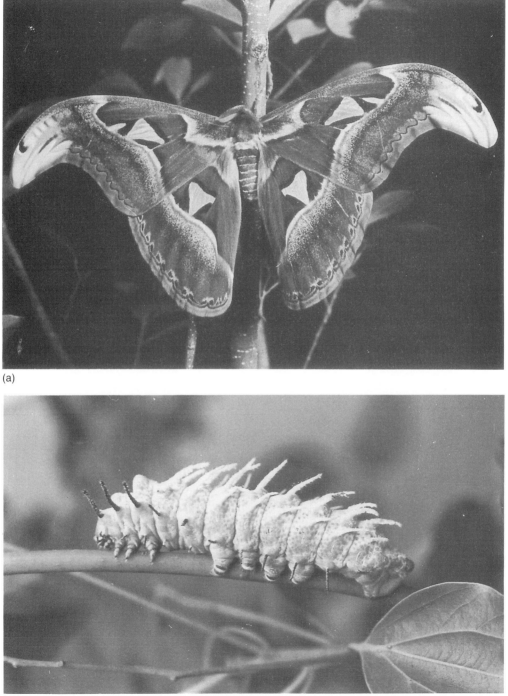

(a)

(b)

Fig. 5.196 Atlas Moth (*Attacus atlas* – Lep.; Saturniidae); (a) newly emerged adult male (wingspan 210 mm), and (b) larva (90 mm) on twig of camphor.

Fig. 5.197 Death's Head Hawk Moth (*Acherontia styx* – Lep.; Sphingidae); wingspan 100 mm; south China.

Fig. 5.198 Convolvulus/Sweet Potato Hawk Moth (*Agrius convolvuli* – Lep.; Sphingidae); wingspan 96 mm; south China.

- *Theretra* spp. (Grapevine Moths) – on grapevine; Europe and West Asia.

(s) Family Arctiidae (Tiger Moths; Woolly Bears) (10 000 spp.)

A large family of mostly brightly coloured moths; some aposematic in coloration and clearly poisonous to predators; many are diurnal. Some larvae are very bristly with urticating setae. Most feed on herbaceous plants and some larvae are polyphagous. A few of the more interesting species include:

- *Amsacta* spp. (Red Hairy Caterpillars) – polyphagous on a wide range of her-

(a)

(b)

Fig. 5.199 Coffee Hawk Moth (*Cephonodes hylas* – Lep.; Sphingidae); adult wingspan 65 mm; larva and pupa by F. Bascombe; south China.

Fig. 5.200 Garden Tiger Moth (*Arctia caja* – Lep.; Arctiidae); wingspan 75 mm; Gibraltar Point, UK.

baceous crops, including some Gramineae; India and South East Asia to Japan.

- *Arctia* spp. (Tiger Moths) – larvae polyphagous in low herbage; Europe and parts of Asia (Fig. 5.200).
- *Diacrisia* spp. (Buffs; Tiger Moths; Hairy Caterpillars) – many species on a wide range of hosts, many polyphagous; India, parts of Asia and Africa.
- *Spilosoma* spp. (Ermine Moths) – on some fruit trees; Europe and parts of Asia.
- *Tyria jacobaeae* (Cinnabar Moth) – already mentioned, this species feeds on the weed ragwort and is an important biocontrol agent.

(t) Family Noctuidae (Owlet Moths) (25 000 spp.)

The largest family of Lepidoptera and probably the most important economically. Most are medium sized, brownish moths, nocturnal in habits (with caterpillars also nocturnal) and which pupate in the soil. Most damage is done by leaf-eating, but some larvae specialize as 'fruitworms' and others as 'stem-borers', and a few are gregarious 'armyworms'. Some 14 separate subfamilies are recognized, but mostly on rather esoteric grounds. Some of the many important agricultural pests are listed below:

- *Agrotis* spp. (Cutworms, etc.) – 10–20 species are cutworms; polyphagous and cosmopolitan (Fig. 5.201).
- *Autographa gamma* (Silver-Y Moth) – larvae polyphagous semi-loopers; adults in Europe are migratory; Holarctic (Fig. 5.202).
- *Busseola fusca* (Maize Stalk-borer) – larvae bore stems of maize and sorghum in tropical Africa (Fig. 5.203).
- *Ceramica* spp. (Broom Moth; Zebra Caterpillar) – larvae polyphagous; Europe, Asia and North America.
- *Diparopsis* spp. (Red Bollworms) – cotton pests in Africa.
- *Earias* spp. (Spiny Bollworms) – on cotton in the Old World.
- *Euxoa* spp. (Darts; Cutworms) – many cutworm pests, mostly in the Northern Hemisphere.
- *Helicoverpa* spp. (African/American Bollworm; Corn Earworm) – major pests

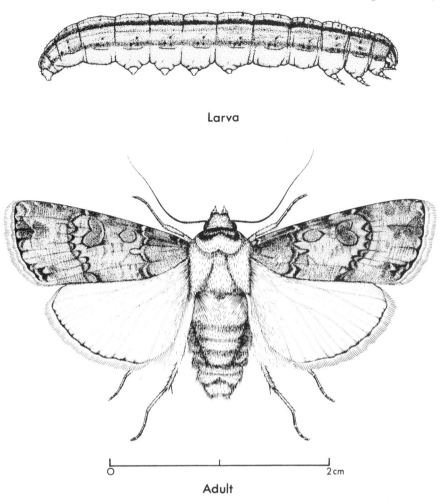

Larva

2 cm

Adult

Fig. 5.201 Turnip Moth/Common Cutworm (*Agrotis segetum* – Lep.; Noctuidae); Cambridge (MAFF), UK.

(4 spp.), polyphagous, worldwide (Fig. 5.204).

- *Heliothis* spp. – many species (80+) with a few as minor pests of many crops; worldwide.
- *Lacanobia* spp. (Tomato Moth, etc.) – polyphagous; worldwide.
- *Mamestra* spp. (Cabbage Moth, etc.) – polyphagous but prefers *Brassica*; Europe, Asia and North America.
- *Melanchra* spp. (Dot Moth; Zebra Caterpillar) – polyphagous in Europe, Asia and North America.
- *Mythimna* spp. (Cereal Armyworms, etc.) – polyphagous on cereals and other crops; genus is worldwide.
- *Noctua pronuba* (Large Yellow Underwing) – a common polyphagous cutworm; Palaearctic (Fig. 5.205).
- *Sesamia* spp. (Maize Stalk-borer, etc.) – larvae bore stems of maize, sorghum and other large Gramineae; Africa.

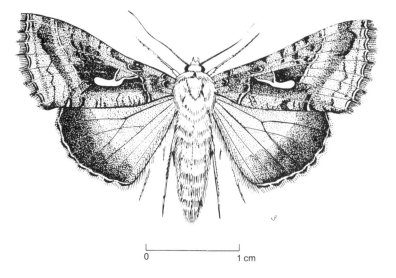

0   1 cm

Fig. 5.202 Silver-Y Moth (*Autographa gamma* – Lep.; Noctuidae); Hong Kong.

Fig. 5.203 Maize stem bored by larvae of *Busseola fusca* (Lep.; Noctuidae); Alemaya, Ethiopia.

- *Spodoptera* spp. (Armyworms/Leafworms) – a large tropical genus; some species strongly migratory; some show preference for Gramineae, some not so. Some very important pests.
- *Xestia* spp. – large genus, worldwide, many are polyphagous cutworms.

(u) Family Lymantriidae (Tussock Moths, etc.) (2500 spp.)

Medium-sized moths, some show sexual dimorphism; worldwide but most abundant in Old World tropics. Larvae are stout and bristly with urticating setae. Most are forestry pests but some are of importance on agricultural crops. Defoliation of trees and shrubs is the typical damage, although some attack Gramineae and others attack some herbaceous plants.

The family contains several very large genera, some of which include major pest species.

- *Dasychira* spp. (400 spp.) – many are polyphagous on a wide range of cultivated plants and crops in the Old World.

Fig. 5.204 Old World Bollworm (*Helicoverpa armigera* – Lep.; Noctuidae); Hong Kong.

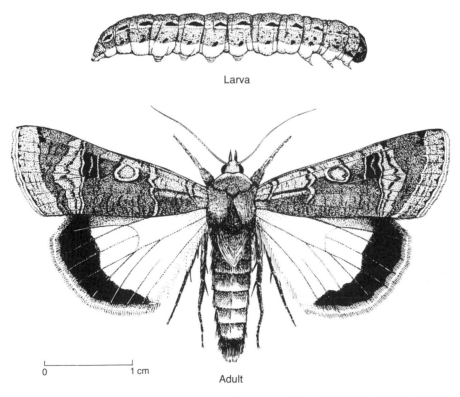

Larva

0   1 cm

Adult

Fig. 5.205 Large Yellow Underwing (*Noctua pronuba* – Lep.; Noctuidae); Cambridge, UK.

Fig. 5.206 Caterpillar of Yellow-tail Moth (*Euproctis similis* – Lep.; Lymantriidae) on leaf of *Rosa*; Gibraltar Point, UK.

- *Euproctis* spp. (600 spp.) – many crop pests, some polyphagous, in the Old World (Fig. 5.206).
- *Lymantria* spp. (150 spp.) – an Old World genus with a few species on fruit and nut trees.
- *Orygia* spp. (60 spp.) – Holarctic in distribution; the pests have been recorded on a very wide range of cultivated plants.

5.4.11 Order Hymenoptera (Sawflies, Ants, Bees and Wasps) (100 000 spp.)

A very large group of insects showing a tremendous diversity of size, shape and habits. Not many are pests of agricultural crops, although some are serious pests of trees. The sawflies as a group are phytophagous; most are leaf-eaters and can defoliate trees, but some are wood-borers, etc., and many other Hymenoptera are either parasitic or otherwise carnivorous. Some of the agricultural pests are reviewed below.

(a) Family Cephidae (Stem Sawflies)

A small group with apodous larvae which bore in stems of Gramineae, bush fruit and others. They include:

- *Cephus* spp. (Wheat Stem Sawflies) – larvae found in stems of wheat, other cereals and grasses; Europe, Asia and North America.

- *Hartiga* spp. (Blackberry Shoot Sawfly, etc.) – larvae bore canes of bush fruit and roses; Asia, the USA and Canada.

(b) Family Tenthredinidae (typical Sawflies) (4000 spp.)

This is the main group of sawflies and shows some diversity of habits. Most are found in temperate regions. Some are pests, but their damage is not usually serious. These include:

- *Allantus* spp. (Strawberry/Rose/Cherry Sawflies, etc.) – larvae eat leaves and may bore into shoots on woody hosts; Europe, Asia and the USA.
- *Athalia* spp. (Cabbage Sawflies) – on cultivated Cruciferae; Europe, Asia and Africa (Fig. 5.207).
- *Caliroa* spp. (Pear Slug Sawfly, etc.) – larvae skeletonize leaves of temperate fruit trees; Europe, Asia and the USA.
- *Croesus* spp. (Hazel Sawfly, etc.) – larvae are gregarious and eat leaves of hazel, chestnut, etc.; Europe, Asia and the USA (Fig. 5.208).
- *Hoplocampa* spp. (Fruit Tree Sawflies) – different species attack different fruit trees (Rosaceae); Europe, Asia and the USA (Fig. 5.209).
- *Nematus* spp. (Currant Sawflies) – several species on currant foliage worldwide.

(c) Family Formicidae (Ants) (15 000 spp.)

A most important group on various counts; worldwide; there are some important pest groups:

- *Acromyrmex* spp. and *Atta* spp. (Leaf-cutting Ants) – polyphagous leaf-cutters and fungus growers; southern USA and Central and South America; very serious pests.
- *Crematogaster* spp. (1000 spp.) (Cock-tail Ants) – these are pests of some agricultural importance in that they feed on honey-

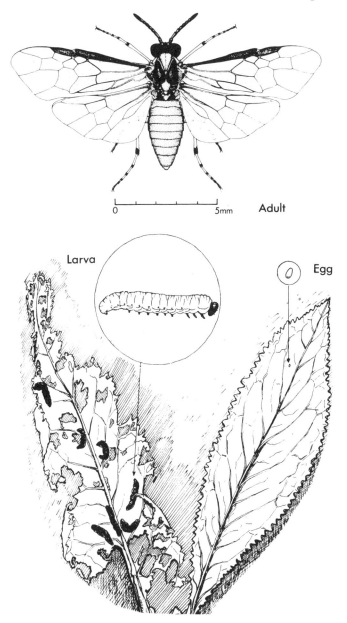

Fig. 5.207 Cabbage Sawfly (*Athalia rosae* – Hym.; Tenthredinidae); Skegness, UK.

dew and protect and 'farm' Coccoidea and Aphididae worldwide.

- *Macromischoides aculeatus* (Biting Ant) – this is one of several arboreal ant species (*Oecophylla* spp.) that are 'pests' because they allegedly attack field workers (this particular one in coffee plantations in Africa). However, personal experience suggests they are not really very aggressive!

Fig. 5.208 Hazel Sawfly (*Croesus septentrionalis* – Hym.; Tenthredinidae) larvae on leaf of hazel; Gibraltar Point, UK.

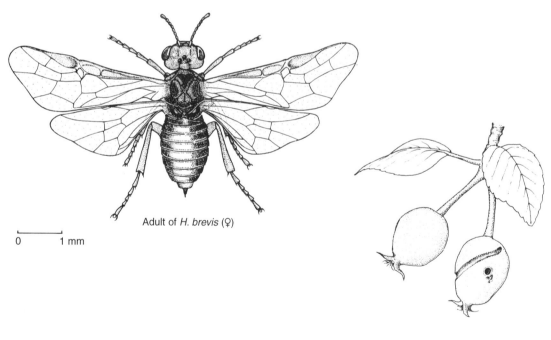

Adult of *H. brevis* (♀)

0   1 mm

Apple sawfly damage
to fruitlet

Fig. 5.209 Pear Sawfly (*Hoplocampa brevis* – Hym., Tenthredinidae); Cambridge, UK.

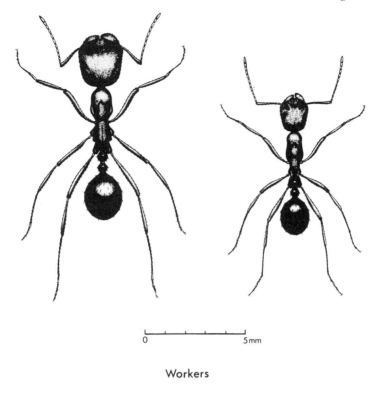

Workers

Fig. 5.210 Harvester Ant (*Messor barbarus* – Hym.; Formicidae); Kenya.

- *Messor* spp. and *Pheidole* spp. (Harvester Ants) – workers collect seeds of grasses and cereals in arid areas of Africa and Asia (Fig. 5.210).

(d) Family Vespidae (Social Wasps) (800 spp.)

These wasps are large in size and social in habits and kill their prey by stinging. If the nest is threatened, the workers are very fierce in their defence. Agriculturally the main pests are *Polistes* (Paper Wasps) (Fig. 5.211) which live in the tropics and nest in bushes (*Citrus*, etc.). *Vespa/Vespula* wasps are pests in that they pierce ripe fruits (grapes, etc.) for the sugary sap and so spoil them; they are also very aggressive in defence of their nest which may be in the orchard/plantation.

5.4.12 Class Arachnida, Order Acarina (Mites)

These tiny arachnids are mostly almost invisible to the unaided eye, but can be of considerable importance on many agricultural crops as they can kill the foliage.

(a) Family Eriophyidae (Gall/Rust/Blister Mites)

A large group of tiny worm-like mites, with only two pairs of reduced legs. Some species produce conspicuous and characteristic galls, usually on leaves; some of these galls are termed 'erinea'. These consist of a dorsal swelling on the leaf; the cup-like interior on the lower surface is filled with hairs or other epidermal projections which the mites live

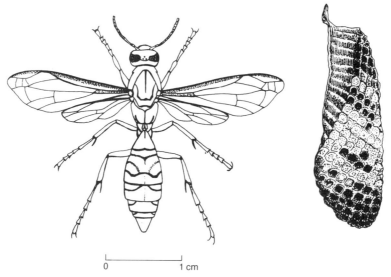

0   1 cm

Fig. 5.211 Paper Wasp worker and nest (*Polistes olivaceous* – Hym.; Vespidae); Hong Kong.

between. Distortion of older leaves is minimal, but on very young leaves it may be total. Some of the most frequently encountered pest species include:

- *Acalitus* spp. (Blister/Bud Gall Mites) – cotton, almond and plum are attacked and plants can be killed; Europe, Asia, the USA and Central and South America.
- *Aceria* spp. – several species attack fruit trees; they are pantropical as a group (Fig. 5.212).
- *Aculops* spp. (Rust Mites) – *citrus*, other fruit and Solanaceae attacked; virtually worldwide.
- *Eriophyes* spp. – many species, some of which should perhaps be placed in other genera (Wood, 1989); hosts include fruit and nut trees, tomato and also cereal crops; worldwide.
- *Phyllocoptella avellaneae* (Hazel Big-bud Mite) – swollen, unopened buds on hazel die; Europe, Asia and North America (Fig. 5.213). Other species attack currants, etc.
- *Phytoptus* spp. (Blister Mites, etc.) – *P. pyri* (the Pear Leaf Blister Mite) is found

throughout Europe, Asia and the USA (Fig. 5.214). *P. similis*, the Plum Leaf Bead Gall Mite, makes small pouch galls around the leaf edge on plum, damson and blackthorn.

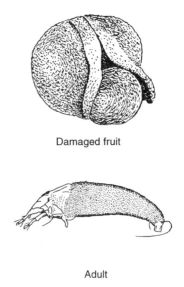

Damaged fruit

Adult

Fig. 5.212 Citrus Bud Mite (*Aceria sheldoni* – Acarina, Eriophyidae); Kenya.

(b) Family Tarsonemidae (Tarsonemid Mites)

Small mites, not so anatomically reduced as the eriophyids. A few species are crop pests – some are quite polyphagous on a wide range of plants:

- *Polyphagotarsonemus latus* (Broad Mite; Yellow Tea Mite) – polyphagous on a wide range of tropical crops (tea, coffee, tomato, etc.) and in greenhouses in temperate regions; cosmopolitan (Fig. 5.215).
- *Stenotarsonemus* spp. – several species specific to pineapple, sugarcane, oats and *Narcissus* bulbs, in all parts of the world.
- *Tarsonemus pallidus* (Strawberry/Cyclamen Mite) – on strawberry, cyclamen and watercress; Europe, Asia and the USA.

(c) Family Tetranychidae (Spider Mites)

By comparison, these are large plant mites – they can be seen with the unaided eye! The

Fig. 5.213 Hazel Big-bud Mite (*Phyllocoptella avellaneae* – Acarina, Eriophyidae); Alford, UK.

Fig. 5.214 Pear Leaf Blister Mite (*Phytoptus pyri* – Acarina; Eriophyidae); Skegness, UK.

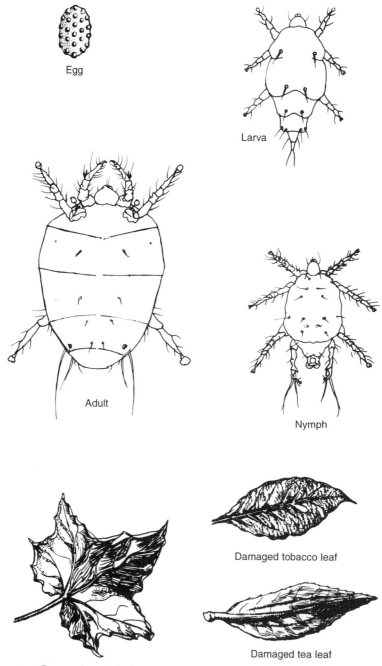

Egg

Larva

Adult

Nymph

Damaged tobacco leaf

Damaged tea leaf

Damaged cotton leaf

Fig. 5.215 Broad Mite (*Polyphagotarsonemus latus* – Arcarina; Tarsonemidae); Kenya.

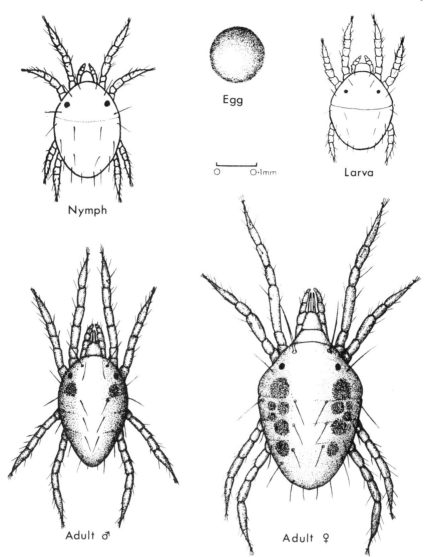

Fig. 5.216 Tropical Red Spider Mite (*Tetranychus cinnabarinus* – Acarina; Tetranychidae); Kenya.

body is rounded and the legs long, and large globular eggs (usually red) are laid on the leaf surface. There is a large number of agricultural pest species, a few of which are mentioned below:

- *Bryobia* spp. – a temperate group on fruit trees, cereals, clovers, etc. in all parts of the world.
- *Eotetranychus* spp. – on fruit and nut trees, grapevine, etc., in most parts of the world.
- *Eutetranychus* spp. – on fruit trees, cotton, cucurbits and other crops; pantropical.
- *Mononychellus tanajoa* (Green Cassava Mite) – an endemic minor pest of cassava in South America, recently introduced into Uganda and now spread throughout Central and West Africa. It is a very serious pest of cassava in Africa.

- *Oligonychus* spp. – a large genus (that looks like *Tetranychus*) with one group of species on fruit and nut trees worldwide, and another on Gramineae worldwide.
- *Panonychus* spp. (Fruit Tree Red Spider Mites, etc.) – recorded on a wide range of fruit trees, bushes and herbaceous plants; worldwide.
- *Tetranychus* spp. (Red Spider Mite, etc.) – a large genus with many species in all parts of the world. Some are tropical, some temperate; a few are oligophagous but most are polyphagous (Fig. 5.216).

(d) Family Tenuipalpidae (False Spider Mites)

A small group but with a few important pests of cultivated plants. Damage is usually slight: a bronzing of leaves and speckling on fruits. It includes *Brevipalpus* spp., several of which are polyphagous on tea, citrus, walnut, grapevine, coffee, rubber, olive and many ornamentals throughout the world.

5.5 FORESTRY PESTS

There is some overlap between forestry pests and agricultural pests – seedling pests in particular are much the same as those which attack fruit and nut trees.

Forestry on a worldwide basis presents several distinct aspects (see Table 5.8), and this has repercussions with regard to the pest situation. Before looking at the insects that are forestry pests, it is desirable to review the main aspects of forestry.

The primitive form of forestry is the felling of native wild trees, sometimes to clear land for cultivation and sometimes for the supply of timber or fuelwood/charcoal. Fell-

Table 5.8 Aspects of forestry and its products

Types of forestry (both temperate and tropical)	
Wild trees	for timber, wood
	for products
Cultivated trees	nursery beds for propagation
	forest/plantation
	agroforestry

Forestry products	
Timber – rough	building/scaffolding
– quality	furniture
Wood products	fuelwood
	charcoal
	plywood
	chips/sawdust
	paper pulp
Shade trees and ornamentals (specimen trees)	
Tree products	aromatic oils (camphor, etc.)
	resins (balsam, myrrh, etc.)
	wild rubber (*Hevea*, etc.)
	chicle in South America (*Sapodilla*)
	medicines (quinine, etc.)
	fruits (Durian, etc.)
	nuts (Brazil nuts, etc.)
	cork (from cork oaks)
	tannin from tree bark

ing for timber should be done selectively, taking only mature trees (and sometimes only a selection of these), thus allowing young trees to grow and seedlings to become established. Regrettably, history records the mass destruction of the remaining native oak forests in the UK and continental Europe in the 16th century, partly for timber for ship-building and later for charcoal for iron-smelting. The initial deforestation in the UK occurred in Saxon times to provide agricultural land and to remove habitat for wild animals (such as wolves and bears), for the villagers often lived in fear of large predators.

The essence of the science of forestry lies in sustainability – the long-term continuous production of trees and, sad to say, this has been a matter of concern only relatively recently. The time factor is critical, for in Europe native hardwoods (oak, beech, etc.) take up to 100 years to reach a good size; and worldwide many wild trees do not reach maturity (i.e. do not begin to bear fruit) until they are 15–20 years old. At the present time, there is growing international concern because in several areas of the world the last remaining areas of truly natural forest are being destructively felled. In parts of Malaysia and Thailand, the forest is being cleared for oil palm cultivation, but in much of Indonesia the felling is basically to harvest the tropical hardwoods for sale to Japan, Europe and North America. In South America the land is being used to make vast pastures for the mass rearing of beef cattle for the markets of North America. However, this should not be regarded as a Third World problem, for destructive (total) felling of native forest is being practised in parts of the north-western USA and in the taiga forests of Siberia.

In most areas, forest destruction results in extensive soil erosion. Forests usually grow in areas of high rainfall and forest soils are often shallow; if the vegetation cover is removed the rain often washes away the topsoil. In South East Asia, it is quite horrifying to see the streams running bright red in colour, for the water is laden with the distinctive red topsoil and looks like thick tomato soup!

The entire world has a sad history of deforestation. Forest is the natural climatic climax vegetation for more than half of the land surface of the Earth, but it is somewhat inimical to the development of human societies. Both trees and land as resources were initially regarded as inexhaustible and have been subject to continuous exploitation. Now that most of our natural forests have been destroyed, we are belatedly concerned, and most countries are now actively engaged in reafforestation programmes.

Modern forestry is concerned with reafforestation as much as immediate timber production. As Table 5.8 shows, there are many reasons for growing trees. The time factor is a problem as most trees are slow-growing, and the essential basis now is succession – young trees are required to replace those felled so that the forest as an ecological community continues. Nowadays, in many countries, the foresters are felling trees that were planted by humans, and these are being continuously replaced by seedlings.

Thus modern forestry consists of the rearing of seedlings in nurseries, followed by the establishment of a forest or plantation with the saplings. An important aspect of the science is now termed 'agroforestry'; this is the growing of trees as an adjunct to agriculture. The trees are grown for shade, wind protection and as sources of fuelwood and timber. The foliage is also a source of fodder for livestock.

Timber comes in two main categories: rough and quality material. The former is usually softwood (pine, eucalyptus, etc.) and is used for building-timbers (beams, rafters, etc.) and for scaffolding and shuttering (regarded as expendable in this context). Quality timber is usually hardwood, and the finest is used for furniture (teak, mahogany, rosewood, etc.) and also for expensive flooring, wall panels, doors and even boats. As a substitute for solid wood, these expensive woods are now being used widely as veneers for plywood.

Probably the most ancient use of wood is that of fuel, and in the developing world most cooking and heating still utilize wood. Since most countries suffer from ever-increasing overpopulation, it is no wonder that the shortage of fuelwood has become so acute. A more efficient form of domestic fuel is charcoal, which of course is a more extravagant use of wood. Recent work in the development of small domestic cookers is, however, producing equipment that can burn twigs and small pieces of wood quite effectively. This is leading to the more widespread use of legume trees/bushes such as *Leucaena*, the very rapid growth of which produces a useful annual biomass of small branches (and foliage for fodder). Once established, these plants can be coppiced annually indefinitely. Chips and sawdust are normal timber byproducts, but now are usually made into chipboard, compost, etc. Pulp for newsprint and other cheap forms of paper is a major commercial item in the modern world, and many quick-growing trees are used for this purpose; mostly conifers in the Northern Hemisphere, but some tropical species are now also used.

Shade trees and ornamentals are often very valuable, and their health is a major concern.

Most tree products which are collected are either sap or chemicals in the bark (or the bark itself); the tree is left alive and may be productive for many years (100 or more). A good example is the cork-oak 'forests' of northern Spain – they are really a form of wooded savannah of great ecological importance and of great antiquity. The agricultural equivalent can be seen in ancient Mediterranean olive groves.

With wild trees being used either for timber or for different tree products, insect pests are usually ignored. However, if the tree is damaged then the wood may be ruined or the product rendered useless. So insects may be causing quite serious losses and thus are economically important, but control measures are not deemed appropriate. However, cultivated trees can to some extent be protected. It is customary to rear seedlings in seed-beds. Direct seeding, including aerial seeding of inaccessible locations, usually has a low success rate. Planting acorns in a temperate forest may, however, be quite successful and can yield a high proportion of young plants (70–80% germination success is not uncommon). Current practice is to sow seeds densely in rather narrow seed-beds, which are irrigated and may often be protected. The established seedlings (10–20 cm) are later transferred singly to narrow (3–5 cm diameter) plastic (polythene) bags/tubes (pots) filled with potting soil; the soil column is usually 10–20 cm. At this stage a root dip may be given. The seedlings are then left exposed in the nursery, packed together, to grow and harden-off. At this time they can be given pesticide sprays, if warranted. Thus the seedling/sapling is planted out with the soil around the roots intact and roots well developed, and the plant suffers far less stress than in previous times when they were planted out with naked roots. One advantage with this technique is that pesticides and fertilizer can be incorporated in the potting soil, all of which will help the seedling to survive. Most trees are planted when 10–30 cm tall, and typically they do not get watered-in but are planted at a time when seasonal rain is expected. In Europe, the tree seedlings/saplings are usually given physical protection in the form of a plastic tube or something similar, but in the tropics the cost of this often cannot be justified. Forest and agroforestry tree seedlings on average have quite a low survival rate, although there is tremendous variation; a good season may see 50% survival, but if the rains fail there may be a total loss. With agroforestry especially, in the tropics and subtropics the single greatest hazard is browsing by ungulates, particularly domestic goats.

Seed-beds can be monitored for insect pests and treated with insecticides if required, as well as chemicals being added to the soil through polythene tubes, so the small seedlings can be protected. However, there can be problems with protected cultivation – at Alemaya (Ethiopia) the forestry seed-beds were covered with grass mats for physical protection, but under the mats lived large Field Crickets which ate the seedlings. Tree seedlings are susceptible to biting pests such as crickets, grasshoppers and termites, and to sap-sucking pests such as aphids, mealybugs and Cottony Cushion Scale. On a tree seedling the apical bud/shoot is especially vulnerable, for if damaged the plant habit can be transferred into a bush. Insect pests of seedlings are therefore of particular importance. Once the tree is well established, it can typically withstand quite a lot of pest damage, and many pests can be more or less ignored. The times at which tree species are especially vulnerable to pest attack are firstly as the seed germinates and secondly at field establishment as a young sapling.

Established trees can withstand leaf loss easily – up to 20% loss causes no problem, and neither does a similar loss of sap. However, a single beetle tunnel can severely lower the value of mahogany log. Traditionally, felled trees are stored as log stacks under rain shelters in the forest in order for the wood to season (slowly dry, shrink and harden) and during this time they are vulnerable to certain groups of insect pests. Nowadays in many countries timber is no longer seasoned, but instead is kiln-dried – quite rapidly and at a temperature lethal to insect larvae.

One feature of forests is important in relation to insect pest attack: natural temperate forests are characterized by having only a single dominant tree species over large tracts. Climatic and edaphic factors control which species dominates. North temperate regions with deep loamy soil have oak (*Quercus* spp., etc.) with calcareous soil beech (*Fagus* spp.); wetter, colder conditions have birch (*Betula* spp.) and the northern taiga has different species of conifers. These are natural monocultures, some of which are very extensive across Europe, Asia, the northern USA and Canada. Some insects have taken advantage of this and will have regular population irruptions which can cause widespread defoliation – such as Conifer Sawflies, several moth species and some beetles (Scolytidae). Most damage is done when European or Asian sawflies and moths accidentally become established in North America where, in the absence of their natural enemies, population explosions occur. Both deciduous temperate forest and northern evergreen conifers suffer regularly from these pests.

Tropical rainforest is characterized by having 10–20 co-dominant tree species and an overall tremendous diversity of species. Since most insect pests show a measure of host-specificity such forest is not susceptible to such insect pest outbreaks. In tropical rain forest, phytophagous insects have to search very diligently for their hosts.

When humans first became aware of the need for reafforestation, or at least for the planting of some trees, the choice was often for species that were quick-growing and tolerant of a wide range of conditions. Thus about 100 years ago widespread planting of *Eucalyptus* and *Acacia* throughout the tropics was started. In Africa the *Acacia* species (both native and Australian) grow in the deserts and dry regions, and the *Eucalyptus* where water is more available. In fact, in parts of tropical Asia, *Eucalyptus* trees are planted to drain swampland as some species are notoriously water-extravagant. The widespread planting of a few selected species like this can have major implications for insect pests with some host-specific pests suddenly becoming quite serious. These include the Eucalyptus Psyllid (see Fig. 5.225) which in Ethiopia can destroy the apical shoot of young trees (1–2 m) causing them to have a bushy habit. A similar problem is being shown with *Leucaena* in tropical Africa and Asia. This leguminous tree is being widely planted as a source of fuelwood and animal fodder, and it also covers and stabilizes bare soil to reduce soil erosion. However, the psyllid *Heteropsylla cubana* invaded South East Asia in 1984 from Central America and is devastating the *Leucaena*. It has spread throughout South East Asia and India and is now reported in both East and West Africa; losses of up to 50% have been recorded. With the development of small, cheap, efficient cooking-stoves that can effectively burn twigs and wood chips, it may be that future tree plantings for fuelwood may be leguminous trees that can be coppiced and the foliage used as livestock fodder.

In northern temperate regions replanting has usually been with quick-growing *Pinus* and other conifers. This softwood is very suitable for cheap paper such as newspaper and toilet paper, and it can also be used as building timber. Hence in many parts of Europe and North America native oak/beech/birch forest is being replaced by vast *Pinus* monocultures (and sometimes other conifer species). For heavy cropping the *Pinus* trees might be appropriate in a plantation, but where the forest has any amenity functions such regimented monoculture is not suitable. In most countries there is now real pressure to replant with a mixture of native hardwood trees together with some fast-growing exotic species.

In 1990 a published figure for forestry losses in the USA due to insect attack was 14 million board feet of timber per year.

One slight anomaly in the science of forestry is that bamboos are usually within its remit. This group is, of course, part of the Gramineae (subfamily Bambusinae) and so is technically a tropical grass!

Three major illustrated publications dealing with forestry pests are those by Johnson and Lyon (1976), Alford (1991) and Bevan (1987). A major publication in wood technology/preservation is the 1975 book by N.E. Hickin, entitled *The Insect Factor in Wood Decay*, and there are other relevant titles in 'The Rentokil Library'.

As mentioned previously, the pests of tree seedlings are essentially the same pests that attack seedlings of woody agricultural plants. There is also great similarity between the pests of forest trees (especially deciduous broad-leaved trees) and the pests of most fruit and nut trees. This, of course, is partly due to the fact that in ancient times the fruit and nut trees were part of the forest flora, at least at its edges. When considering insect pests of forestry in its broadest sense, it is clear that the following host categories have to be distinguished:

1. Seed-bed and seedling pests ⎫
2. Saplings in the nursery ⎬ Nursery stock.
3. Saplings at planting-out. ⎭
3. Saplings at planting-out.
4. Growing (healthy) trees.
5. Stressed/injured trees.
6. Dead trees/branches/stumps.
7. Logs (usually drying in forest); some dry, some damp.
8. Dead wood – both dry and damp (the latter usually with fungal attack).
9. Timber in storage or in use (poles, planks, beams, plywood, etc.).
10. Pests of fruits and seeds – important for natural regeneration.

The main groups of insects that are of concern to forestry are listed in Table 5.9.

Xylophagy – The Eating of Wood

Some of the most obviously damaging insect pests of forestry are the wood-borers that live in tree trunks and timber, leaving holes and tunnels sometimes as wide as 10–12 mm in diameter. This tunnelling can seriously reduce the productivity of the tree, as well as reducing the saleable value of the timber, especially of quality hardwoods. Some tun-

Table 5.9 Major insects and mites as forestry pests

Group	*Common name*
Order Collembola	Springtails
Order Orthoptera	
Family Tettigoniidae	Long-horned Grasshoppers; Bush 'Crickets'
Gryllidae	Field Crickets; Tree Crickets
Order Isoptera (Termites)	
Family Kalotermitidae	Dry-wood Termites
Rhinotermitidae	Wet-wood Termites
Termitidae	Bark-eating Termites
Order Hemiptera, suborder Homoptera	Plant Bugs
Family Cicadidae	Cicadas
Cicadellidae	Leafhoppers
Flattidae	Moth Bugs
Membracidae	Treehoppers
Psyllidae	Jumping Plant 'Lice'
Aphididae	Aphids; Plant 'Lice'
Margarodidae	Cottony Cushion Scale
Diaspididae	Hard/Armoured Scales
Order Hemiptera, suborder Heteroptera	(True Bugs)
Family Coreidae	Twig-wilters, etc.
Pentatomidae	Stink Bugs
Order Coleoptera (Beetles)	
Family Scarabaeidae	Chafers; White Grubs
Anobiidae	'Woodworm' Beetles
Bostrychidae	Black Borers; Auger Beetles
Buprestidae	Jewel Beetles; Flat-headed-borers
Cerambycidae	Longhorn Beetles
Curculionidae	Weevils; Pine Sawers
Scolytidae	Bark Beetles
Order Lepidoptera	Moths and butterflies
Family Gracillariidae, etc.	Micro-lepidoptera – Leaf-miners
Sesiidae	Clearwing Moths
Cossidae	Carpenter/Goat Moths
Metarbelidae	Wood-borer Moths
Tortricidae	Tortricids, etc.
Geometridae	Pugs/Carpets; Loopers
Lasiocampidae	Tent Caterpillars
Saturniidae	Emperor Moths
Notodontidae	Processionary Caterpillars, etc.
Noctuidae	Owlet Moths; Cutworms, etc.
Lymantriidae	Tussock Moths
Order Hymenoptera	Sawflies, Ants, Wasps, etc.
Family Siricidae	Woodwasps
Cimbicidae	Birch Sawfly, etc.
Diprionidae	Conifer Sawflies
Tenthredinidae	Sawflies
Formicidae	Ants
Torymidae	Seed Chalcids
Class Arachnida, order Acarina (Mites)	
Family Eriophyidae	Gall/Rust Mites
Tetranychidae	Spider Mites

nels become infected with fungi and wood-staining can result.

With most wood-boring insects, the rate of larval growth is slow – some larger species in temperate regions take up to three to four years for full development. Some timber beetles can emerge from furniture and domestic timbers (beams, doors, etc.) many years after harvest – several instances have been reliably recorded of up to 15–25 years. One of the main reasons for such slow development is the poor nutritive value of wood as a diet.

The structure of a tree trunk consists of several very distinct layers, as follows:

1. Bark – outer dead layer of cork;
   – thin inner living layer (including phloem tubes for food conduction).
2. Cambium – lying between bark and wood.
3. Wood – sapwood of living cells, including xylem vessels for water conduction;
   – heartwood of dead lignified cells.
4. Pith – central core, either hollow or composed of special large, thin-walled cells; sometimes in trees greatly reduced.

The insects that are broadly regarded as being wood-eaters feed in several different ways:

1. Sap-drinkers – xylem vessels carry mostly water;
   – phloem tubes carry dissolved foods.
2. Wood-chewers – feed on cell (sap) contents.
3. Wood-digesters – cellulases secreted in intestine;
   – micro-organisms in intestine;
   – fungus (mycelium)-eaters;   Ambrosia Beetles; fungus gardens (see below).

Wood consists of a mixture of smallish box-like parenchyma cells (used for storage) and elongate tracheid cells used for the transportation of water (xylem vessels) or food materials such as sugars (sieve tubes). The cell walls are composed of cellulose and hemi-celluloses (complex polysaccharides which can be broken down into glucose and other basic sugars), with lignin (often about 25%) being incorporated for strength. The cell content (sap) is mainly water but with a range of dissolved substances. These include carbohydrates (usually in the form of simple sugars but may also include starches); proteins are present either in solution or as granules. Fats and oils may also be present – usually as globules. Several secondary compounds may be present, such as pigments, glycosides, alkaloids, tannins, resins and minerals, but most of these are not usually important as nutrients.

Nutritionally, the phloem-tappers (some Hemiptera) and the inner bark-borers (some Buprestidae, some Cerambycidae) do the best. Xylem-feeders (especially Cicadidae) have to excrete enormous quantities of surplus water. Sapwood-chewers imbibe cell sap, but a lot of chewing is needed (like 'eating' sugarcane – one chews for ages, swallows the juice and then has to spit out a mouthful of chewed woody debris). So far as wood is concerned, it appears that a few insects (mostly termites) secrete intestinal cellulases and can digest cellulose directly but most of the 'true wood-eaters' rely on fungi and bacteria which can produce cellulases that degrade cellulose down to starch and simple sugars. Some of the insects house these bacteria and fungi as intestinal micro-organisms (the Kalotermitidae have a rectal pouch), while others inoculate their tunnels with fungi (Ambrosia Beetles, etc.) and eat the mycelium. Still others, like the termites (Termitidae), construct 'fungus gardens' and eat part of the mycelium (bromatia). Some of the insects that feed on rotten wood (Lucanidae, Passalidae, and some Scarabaeidae, etc.) are probably relying on the fungi and mycelium for most of their food intake, as do many saprophagous insects.

Living wood contains as much as 30% water (of dry weight). Air-drying in temperate regions reduces the water content to 15–20% (while in Egypt it gets as low as 6%); and indoors furniture and domestic timbers are usually in equilibrium with atmospheric moisture at about 6–7%. Final moisture level in dried wood is always directly dependent upon ambient conditions, and fluctuation is normal.

Looking at the Insecta and Acarina in more detail, the following outlines the genera and species most damaging to forestry, along with some common, but less damaging, pests.

5.5.1 Order Collembola (Springtails)

The globular-shaped bodies of the Sminthuridae show that these are soil surface-dwellers. *Bouretiella* has been recorded damaging *Pinus* seedlings and other conifers in Europe and North America. The subterranean *Onychiurus* attacks seedling roots, and several species can be pests.

5.5.2 Order Orthoptera (Crickets, Grasshoppers, etc.) (17 000 spp.)

(a) Family Tettigoniidae (Long-horned Grasshoppers; Bush 'Crickets', etc.)

These nocturnal, arboreal insects are basically forest-dwellers. The females have a distinctive ovipositor – those in which it is long and straight usually lay eggs in the soil, but those with a short, stout, often curved ovipositor usually cut a slit in a twig (or sometimes a leaf edge) and lay eggs in the slit in a short row. The twig is often killed and a favoured host can have many dead twigs in the canopy (with conspicuous dead brown leaves). They

are seldom recorded as major forestry pests. Some members of the group are called Bush 'Crickets', and these eat the leaves of some bushes as well as tall herbs.

(b) Family Gryllidae (Crickets; Tree Crickets) (2300 spp.)

As mentioned, the Field Crickets (*Acheta*, *Brachytrupes* spp.; Fig. 5.88) can be very damaging to seedlings (especially if the seed-beds are covered) and also in greenhouses in temperate countries. Several species of *Oecanthus* (Tree Crickets) are pests of bushes and trees in Canada and the USA, causing damage both by leaf-eating and oviposition.

(c) Family Gryllotalpidae (Mole Crickets) (50 spp.)

These subterranean burrowing insects are worldwide and can be damaging in seed-beds and for newly planted seedlings. *Gryllotalpa* (Fig. 5.89) is the Old World genus and *Scapteriscus* is found throughout the USA.

5.5.3 Order Isoptera (Termites) (1900 spp.)

This tropical group is renowned for its ability to use wood as food (either directly or indirectly) and for its spectacular nests. Some species are basically grassland inhabitants (including arid grassland), where they have subterranean nests, often with a large protruding nest mound. Others are essentially forest species and many nests are in the form of cartons stuck on to the trunk or branches of trees; alternatively the small nest is actually inside the tree trunk or branch. So far as tree hosts are concerned, most damage consists of bark-gnawing; the wood removed is usually dead, although some can feed on live wood and can hollow out the heartwood of large trees. Seedlings and saplings are very vulnerable to termite attack, both below and above ground level.

Some of the more important pests of forestry include:

(a) Family Kalotermitidae (Dry-wood Termites) (250 spp.)

These are forest insects adapted to living inside dead tree branches *in situ*. They have an intestinal microflora, especially in their large rectal pouch, which can degrade cellulose, and they can use metabolic water. Thus they can live inside and eat quite dry wood successfully – in the Forestry Laboratory outside Kampala, Uganda, there was a small colony of *Cryptotermes* in a dry piece of wood inside a large glass jar that had lived there for three years. Three of the main genera are *Cryptotermes* (see Fig. 5.43), *Kalotermes* and *Neotermes*. Plantation teak in Java is seriously damaged by *Neotermes* – the winged adults find cracks in the bark or dead branches and start colonies, the workers later penetrating into the living trunk. Infested trees that are girdled are likely to have a dramatically increased *Neotermes* infestation.

(b) Family Rhinotermitidae (Wet-wood Termites, etc.)

These are also forest species, which live in damp, dead tree-stumps and in fallen logs. They are very serious pests in the tropics: Although they do not usually attack living trees, bridges, buildings, posts and other wooden structures in a forest are at risk and some damage may occur to log poles in the presence of moisture. There are 45 species of *Coptotermes* in South East Asia and Australasia, at least ten of which are recorded as being destructive to forest trees in the tropics; the workers excavate galleries in the living wood, often deep in the heartwood.

In Australia *Eucalyptus* seems to be particularly susceptible to *Coptotermes* attack; the whole of the heartwood of the trunk can be eaten out. Mahogany and pines in Central America can suffer similar damage from another *Coptotermes* species. Initial infestation is usually of dead, damp wood, but once the colony is established the workers are apparently able to attack 'living' wood, and they tunnel the heartwood of the tree. (Of course the heartwood of large trees cannot really be regarded as being alive – the cells are so heavily lignified.)

(c) Family Termitidae (Bark-eating Termites, etc.) (1300 spp.)

In the tropics the large nest mounds (see Fig. 5.93) are a very conspicuous element of local scenery, as they are also in parts of Australia. Here, there might be more than 10–20 mounds per hectare; some mounds as tall as 2–4 m (or even more!).

Most of these termites are ground-oriented in that the nest is subterranean, and foraging is either underground (where they eat roots and stems) or above ground, protected by a sheet of chewed wood and soil. The soil sheet can be built over a tree trunk to a height of several metres (or more) (Fig. 5.217) and the workers strip the bark underneath. The soil sheet offers physical protection for the foraging workers.

On the other hand, some of these termites are true forest species and live entirely arboreal lives centred around a carton nest (Fig. 5.218) either stuck on to the side of a tree trunk or else in the tree canopy. They gnaw away the tree trunk. Palms in tropical Asia are often particularly favoured host 'trees'.

Members of the Termitidae are very important pests of tree seedlings and saplings, for they will eat the roots and the stems and are often very destructive in seed-beds. They can also be very damaging so far as natural regeneration of forests is concerned, making colonies based on dead tree stumps and cutting down natural seedlings.

The Nasutitermitinae is a very large subfamily with 53 genera; according to Harris (1971) it is found in tropical Asia and Africa, but probably best developed in South and

Fig. 5.217 Tree trunk attacked by termites under a sheath of soil and wood fragments (Isoptera, Termitidae); Alemaya, Ethiopia.

Central America. Most are arboreal forest species, and some are recorded as very destructive to forest trees. Some of the subterranean species (Macrotermitinae) also live in forests, especially at the edges.

Some of the most important genera and species of termites in relation to forestry are listed below:

- *Macrotermes* (Bark-eating Termites; Subterranean Termites; Mound Termites, etc.) (10 spp.) – very abundant throughout Africa

and tropical Asia; quite important general and seedling pests, but not particularly damaging to established trees.

- *Microcerotermes* (Live Wood-eating Termites) – a group showing great diversity of habits, some nesting underground and others making carton nests on tree trunks or living in dead branches. The genus is pantropical and the workers can apparently penetrate living wood quite easily. They attack seedlings and saplings as well as mature trees.

Fig. 5.218 Termite carton nest stuck on to the side of a coconut palm; Malaya.

- *Microtermes* (Seedling Termites) – these have small nests underground but can be very abundant in parts of Asia and Africa and are particularly serious pests at times in seed-beds.
- *Nasutitermes* – tropical forest species that are quite aboreal and make carton nests on palm and tree trunks and in the canopy. Probably most damaging as forestry pests

in South America, where more than 100 host trees are recorded.
- *Odontotermes* (Subterranean Termites) – many species are recorded attacking plants of economic importance, as well as scavenging litter from the forest floor. They are most damaging as pests of seed-beds, seedlings and saplings, being basically subterranean insects.

5.5.4 Order Hemiptera (true Bugs), suborder Homoptera (Plant Bugs)

As a group these are not so important on forest trees as they are on agricultural crops, but some are quite common and can be damaging. Seedlings can be destroyed in the same manner as agricultural seedlings, but mature trees suffer little. Damage is usually done by the feeding action and sometimes a chemical reaction to the bug saliva, but in some cases the bug oviposits into twigs which are usually killed.

The bugs as vectors of plant diseases (viruses especially) are not so important in forestry as in agriculture. The groups most encountered in forestry are as follows:

(a) Family Aphrophoridae (Froghoppers; Spittle Bugs) (800 spp.)

Most are found on herbaceous plants but some prefer trees as their feeding host, and spittle masses can be seen in the foliage. Some of the nymphs are quite host-specific, but other species are polyphagous. Damage is sometimes done by feeding adults, but more usually by the nymphs – the spittle mass is unsightly on ornamentals. Some of the genera found on trees include:

- *Aphrophora* spp. – found on alder, willow, poplar, ash, etc. in Europe and North America.
- *Ptyelus* (Yellow Froghopper) – a widespread genus on many different trees and other plants in Asia and Africa.

(b) Family Cicadidae (Cicadas) (4000 spp.)

Basically a forest group, with *Mogannia* as one of the few species on Gramineae. Trees of all types and sizes are used as hosts, and generally there are few signs of host-specificity shown. (There probably **is** host-specificity, but equally probably it is broad and the 'deciding' factors could be many.) The group is widespread throughout the tropics and subtropics. Males 'sing' very loudly – they have a pair of special organs at the base of the abdomen, a vibrating membrane in a sound-box and (usually) a resonating chamber, all hidden beneath the large ventral tympanum. In many tropical regions there is a succession of cicada species that 'sing' one after the other (with some overlap) so that there is virtually always a cicada song to be heard. A few species are quite small (20 mm) but many are 20–40 mm long, and the Large Brown Cicada of Malaysia measures 110 mm from its head to its wing tips.

Most are xylem-feeders and imbibe vast quantities of sap, excreting streams of water in thin jets as they feed. They occur almost exclusively on tree branches and narrow trunks.

The life cycle is strange in that the females lay eggs in twigs split by the ovipositor, and the hatching nymphs fall to the ground where they burrow into the soil. The nymph lives a subterranean life, sucking sap from plant roots, and development is a very lengthy process which is generally measured in years. The American Periodical Cicada (*Magicicada septemdecim*) spends 17 years as a nymph in the northern, and 13 years in the southern, states, and the adults of each generation emerge simultaneously in vast numbers.

Oviposition often kills the twig, and in the USA considerable damage is done to various forest trees – the dead patches in the canopy are very conspicuous.

In Hong Kong the commonest cicada is *Cryptotympana*, a large brown insect (Fig. 5.219) to be seen in large numbers on the branches of camphor, tung, *Dalbergia*, *Paulowina* and other forest trees (Fig. 5.220).

(c) Family Membracidae (Treehoppers) (2500 spp.)

These small brown bugs are characterized by being gregarious and have anatomical

Fig. 5.219 Large Brown Cicada of Hong Kong (*Cryptotympana* sp. – Hem.; Cicadidae); body length 33 mm.

Fig. 5.220 Large Brown Cicadas (*Cryptotympana* sp. – Hem.; Cicadidae) feeding on a branch of *Paulowina* in Hong Kong.

projections of the thorax – usually spines. The group is found in tropical forests, and the eggs are laid in slit twigs. Infestations are regularly seen, but damage is usually slight.

(d) Family Cicadellidae (= Jassidae) (Leafhoppers) (8500 spp.)

A very large and important group, found worldwide. Many species occur on Gramineae and low-growing herbaceous plants, with an equally large group to be found on woody hosts – both bushes and trees (Fig. 5.221). Most are 4–8 mm in length but smaller and larger (tropical) species occur. The spines on the hind tibiae are characteristic of the group. Most are phloem-feeders, but the Typhlocybinae usually feed in the mesophyll tissues of leaves. The forest and shrub species usually sit on the underside of leaves (Fig. 5.222) –

Fig. 5.221 Large Brown Leafhopper, found on *Tristania* trees in south China.

Fig. 5.222 Rose Leafhoppers (*Edwardsiana* sp. – Hem.; Cicadellidae) on underside of rose leaf; Skegness, UK.

symptoms of their infestation include some leaf edge curling and the dorsal lamina has a speckled silvery appearance (Fig. 5.223). Many Cicadellidae are quite host-specific, but some of the crop pests are notoriously polyphagous. In temperate regions many leafhoppers overwinter as adults in the foliage and on warm days may be quite active.

Many leafhoppers are conspicuous and locally abundant – they can be damaging to young plants, but large trees suffer no discernible damage even though infestations can be high. Foliage discoloration is the main effect on deciduous trees and thus can be of importance on ornamentals (but, of course, the leaves are eventually shed).

Some of the more important tree-inhabiting leafhoppers include:

- *Alebra* spp. – on oak, elm, lime, horse chestnut, sycamore and maple, etc. in Europe. In the USA *A. albostriella* is the Maple Leafhopper.

Fig. 5.224 Large Brown Leafhopper (*Bothrogonia* sp. – Hem.; Cicadellidae); body length 14 mm.

Fig. 5.223 Typical symptoms of leafhopper infestation – upper surface of leaf silvered and flecked (*Edwardsiana* sp. – Hem.; Cicadellidae); Skegness, UK.

- *Bothrogonia* spp. (Large Brown Leafhoppers) – in South East Asia to be found on foliage of *Tristania* in local forests (Fig. 5.224).
- *Edwardsiana* spp. (Rose Leafhopper, etc.) – found on Rosaceae, beech and hornbeam (Fig. 5.222); the species on sycamore in the UK has very distinctively marked nymphs; occur in Europe, Asia and North and South America.
- *Empoasca* spp. (Green Leafhoppers, etc.) – a cosmopolitan genus with many species and many recorded hosts; in the UK common on Rosaceae (*Rosa* and *Rubus* are used as overwintering hosts).
- *Erythroneura* spp. (Fruit Tree Leafhoppers, etc.) – on a wide range of fruit trees and other hosts; cosmopolitan.
- *Fagocyba cruenta* (Beech Leafhopper) – widely distributed on beech, oak, sycamore, etc.; specimen trees can be disfigured.
- *Typhlocyba* spp. – a cosmopolitan genus whose species feed on leaves of trees, including beech, hornbeam, elm, birch and *Nothophagus*.
- *Zygina* spp. (Blue Leafhopper, etc.) – on fruit trees and other trees in Europe and Asia.

(e) Family Psyllidae (Jumping Plant 'Lice'; Suckers) (1300 spp.)

A very interesting group, worldwide, on many different hosts, and sometimes regarded as six separate families. The

Fig. 5.225 Eucalyptus Psyllid (*Ctenarytaina eucalypti* – Hem.; Psyllidae) infesting shoot of *E. globulus*; Alemaya, Ethiopia.

nymphs have a characteristic flattened shape and are often called 'Suckers'. The nymphs of some species produce waxy filaments; infestations can be very conspicuous and infested shoots can be killed. Tree seedlings/saplings can be converted into bushes by shoot destruction. Pantropically the genus *Ficus* (*Chlorophora* and other Moraceae) is especially favoured as a host. A few species are now very serious pests on some species of *Eucalyptus* and *Leucaena* trees; host-specificity is the rule, often strikingly so.

Some of the more important species include:

- *Ctenarytaina eucalypti* (Eucalyptus Psyllid) – infests and may kill shoots of *Eucalyptus* (Blue Gums), native to Australia but now introduced into Europe, Asia and Africa (Fig. 5.225). At Alemaya (Ethiopia) only *E. globulus* was seen infested; particularly damaging to young trees.
- *Euphyllura olivina* (Olive Psyllid) – on twigs of *Olea*; Mediterranean and East Africa (Fig. 5.229).

- *Heteropsylla cubana* (Leucaena Psyllid) – from Central America, now established in India and parts of Africa. It is proving very damaging to planted *Leucaena*.
- *Homotoma* spp. (Fig Psyllids) – on Moraceae in the Mediterranean region, Africa, and parts of Asia, but no foliage deformation.
- *Macrohomotoma* spp. (Fig-shoot Psyllid, etc.) – infest shoots of *Ficus microcarpa* (Chinese Banyan), etc. (Fig. 5.226).
- *Megatrioza* spp. (Syzygium Psyllid, etc.) – nymphs cause leaf-pits on *Syzygium* leaves (Fig. 5.227), throughout South East Asia.
- *Mesohomotoma* spp. – one species occurs on Malvaceae in East Africa and tropical Asia and another on Stereuliaceae in West and Central Africa.
- *Pachpsylla* spp. (Hackberry Psyllids) – on Hackberry trees in the USA.
- *Pauropsylla* spp. (Fig-leaf Psyllids) – on various species of *Ficus* in tropical Asia; *P. udei* makes spectacular leaf galls (Fig. 5.228).
- *Psylla* spp. (Alder Sucker, etc.) – many species on temperate woody hosts (mostly

Fig. 5.226 Fig-shoot Psyllid (*Macrohomotoma striata* – Hem.; Psyllidae) found on *Ficus microcarpa* throughout South East Asia; inset of adult resting on a leaf; Hong Kong.

Fig. 5.227 Syzygium Psyllid (*Megatrioza vitiensis* – Hem.; Psyllidae) causing leaf-pits; Malaya.

(a)

(b)

Fig. 5.228 *Pauropsylla udei* (Fig-leaf Gall Psyllid) (Hem.; Psyllidae) galls on *Ficus variegata* var. *chlorocarpa*; (a) some opened; and (b) some leaves completely covered with galls; Hong Kong.

Fig. 5.229 Olive Psyllid.

Fig. 5.230 Camphor Psyllid (*Trioza camphora* – Hem.; Psyllidae) showing leaf-pits; Hong Kong.

trees), including ash, alder, box, hawthorn and other Rosaceae (such as apple and pear), also *Schefflera* in South China and *Acacia* in Australia. Nymphs are often covered with wax (as in Fig. 5.229).

- *Trioza* spp. – this is a large group found on both herbaceous and woody plants; some are host-specific, but a few are polyphagous; most make leaf-pits (e.g. the Camphor Psyllid) (Fig. 5.230). Oak, Rutaceae and some Lauraceae are favoured woody hosts in Europe and the USA.

The Spondyliaspinae contain some 250 species in 14 genera and are mostly confined to *Eucalyptus* trees in Australia.

(f) Family Aleyrodidae (Whiteflies, etc.) (1156+ spp.)

A widespread group but most abundant in the tropics; these tiny, winged bugs have scale-like nymphs usually found on the underside of leaves. Gymnosperms are not used as hosts, and only a few occur on grasses. Palms are regular hosts and whitefly infestations can be very heavy. Tree families that are extensively used include Betulaceae, Fagaceae (especially *Quercus*), Lauraceae,

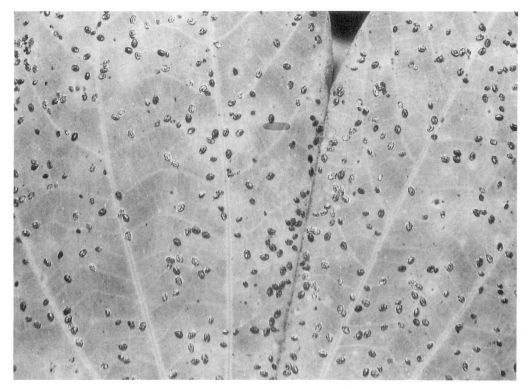

Fig. 5.231 Blackfly (*Aleurolobus marlatti* – Hem.; Aleyrodidae) nymphs on upper surface of *Bauhinia* leaf; Hong Kong.

Leguminosae, Moracea, Myrtaceae, Rutaceae and Ulmaceae (Mound and Halsey, 1978). The genera *Acer* and *Salix* are both used by many whiteflies. Worldwide most species are probably to be found on herbaceous dicotyledenous plants, but a good number are on trees.

Some of the more important genera to be found on trees are included below.

- *Aleurocanthus* – polyphagous and pantropical.
- *Aleurodicus* – on palms, mango, etc.; pantropical.
- *Aleurolobus* – on Moraceae and other trees in tropical Asia (Fig. 5.231).
- *Aleyrodes* – some on palms, bamboos, *Ficus* and other trees; cosmopolitan.
- *Dialeurodes* – on Rutaceae, Moraceae, etc.; pantropical (Fig. 5.232).

- *Pealius* spp. (Fig Blackflies, etc.) – on Moraceae and *Quercus*; worldwide.
- *Siphoninus* – on a wide range of trees (Rosaceae, ash, olive, etc.); Palaearctic and India.
- *Trialeurodes* – recorded on a wide range of hosts in the tropics and greenhouses in temperate regions, probably most damaging to seedlings and saplings.

Some plants are attacked by a large number of Aleyrodidae, worldwide. Mound and Halsey (1978) list 81 species on *Ficus*, 36 on *Quercus*, 34 on *Cocos* and 18 on *Bambusa*.

(g) Family Aphididae (Aphids or Greenfly, etc.) (3500 spp.)

A large, ecologically dominating group; worldwide but best developed in northern

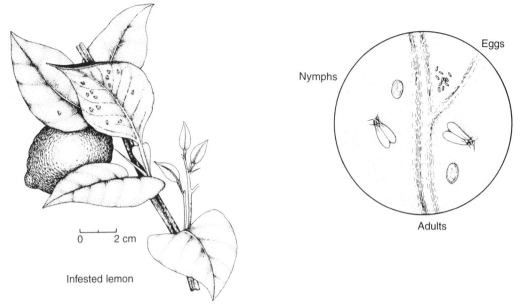

Fig. 5.232 Citrus Whitefly (*Dialeurodes citri* – Hem.; Aleyrodidae); Kenya.

temperate regions. All groups of higher plants are used as hosts, more or less equally although fewer are found on Gramineae.

From the viewpoint of forestry, there is a group of aphids confined to various tree species that are pests of some importance. A second group of polyphagous pest species is damaging to seedlings and young plants.

Taxonomically, this family could be split into half a dozen separate families, but it is in many ways more convenient to keep the group together. However, the Lachninae are large dark aphids to be found in the foliage of gymnosperms and some other trees, and the Callaphidinae and Chaitophoridae are both largely tree-inhabiting species.

The aphids that can be particularly damaging to tree seedlings are the polyphagous species listed below:

- *Aphis craccivora* (Groundnut Aphid) – polyphagous but prefers legumes; cosmopolitan.
- *Aphis gossypii* (Cotton Aphid) – polyphagous in warmer regions.

- *Myzus persicae* (Peach-Potato Aphid) – polyphagous and worldwide.
- *Toxoptera aurantii* (Black Citrus Aphid) – polyphagous in warmer regions (Fig. 5.233).

One reason for the importance of aphids as pests of cultivated plants is that many are vectors of viruses and other disease-causing organisms. These are, however, usually most serious on herbaceous plants.

The most important pests of trees include:

- *Cavariella* spp. (Willow Aphids) – on willows; Holarctic.
- *Chaitophorus* spp. (Willow Leaf Aphids) – on *Salix* and poplars; Europe.
- *Cinara* spp. (Black Pine Aphids, etc.) – 17 species in the UK, on pines, spruce, larch, cypress, juniper, etc.; widespread in Europe, Asia, Africa and the USA.
- *Drepanosiphum platanoides* (Sycamore Aphid) – only on sycamore; Europe and North America.

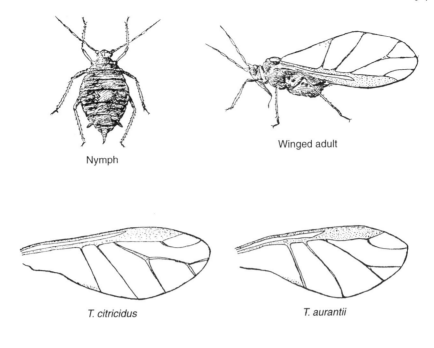

Nymph

Winged adult

T. citricidus

T. aurantii

(a)

(b)

Fig. 5.233 (a) Black Citrus Aphid (*Toxoptera aurantii* – Hem.; Aphididae); (b) aphid under leaf of *Ficus microcarpa*; Hong Kong.

Fig. 5.234 Beech Woolly Aphid (*Phyllaphis fagi* – Hem.; Aphididae); Skegness, UK.

- *Elatobium* spp. (Spruce Aphids) – on *Picea* in North America and now Europe.
- *Euceraphis* spp. (Birch Aphids) – on *Betula*; Europe.
- *Eulachnus* spp. (Pine Aphids) – on *Pinus* spp.; Europe, Africa and the USA.
- *Greenidea* spp. (Fig Aphids) – on Moraceae; South East Asia to China.
- *Lachnus* spp. (Oak Aphids) – on *Quercus*; Europe.
- *Myzocallis* spp. – on *Quercus*, hazel, sweet chestnut; Europe and North America.
- *Myzus* spp. – on Rosaceae, privet, etc.; cosmopolitan.
- *Periphyllus* spp. (Maple Aphids) – on *Acer* in Europe and the USA.
- *Phyllaphis fagi* (Beech Woolly Aphid) – on beech; Europe and North America (Fig. 5.234).
- *Pterocomma* spp. (Willow Aphids) – on twigs of *Salix*; Europe.
- *Rhopalosiphum* spp. – on *Prunus*, *Malus* and other Rosaceae; cosmopolitan.
- *Toxoptera* spp. (Black Citrus Aphids) – on many hosts in warmer parts of the world.
- *Tuberolachnus salignus* (Large Willow Aphid) – on stems and branches of *Salix* in dense colonies; Europe and the USA.

Exotic conifer plantations in Southern and Eastern Africa are being seriously damaged by a complex of aphid species from Europe (*Cinara cupressi* – the Cypress Aphid; *Eulachnus rileyi* – the Pine Needle Aphid and *Pineus* sp. – the Pine Woolly Aphid). An international effort is being made to find predators and parasites that might be introduced.

(h) Family Pemphigidae (Woolly Aphids)

These are distinguished by having no long abdominal siphunculi; the openings are small lateral slits. The nymphs produce large quantities of wax, mostly as long filaments, and some species produce galls on the host foliage. Many species overwinter on trees in

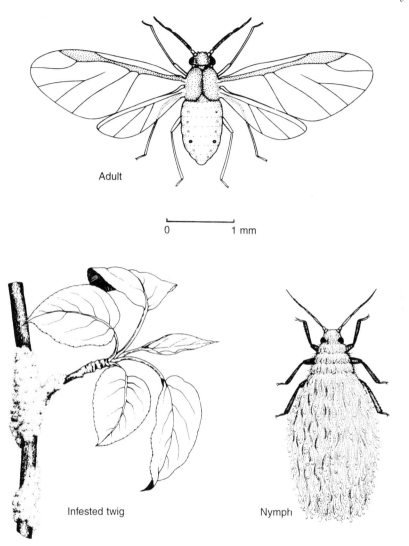

Adult

0   1 mm

Infested twig   Nymph

Fig. 5.235 Apple Woolly Aphid (*Eriosoma lanigerum* – Hem.; Pemphigidae) on *Malus* spp.; Kenya.

temperate regions and spend the summer on herbaceous plants, sometimes on the roots.

Some important genera include:

- *Astegopteryx* spp. (Palm Woolly Aphids) – on Palmae; South East Asia.
- *Cerataphis* spp. (Palm Woolly Aphids) – on Palmae; pantropical.
- *Eriosoma* spp. (Apple/Elm Woolly Aphids, etc.) – some species on *Malus* and *Pyrus*,

others on oaks and elm trees; Holarctic (Fig. 5.235).

- *Forda* spp. – make leaf galls on *Pistacia* in the Near East.
- *Pemphigus* spp. (Root Aphids) – overwinter on poplars, as leaf or petiole galls (Fig. 5.108), and during summer on roots of lettuce, carrot, etc.
- *Pseudoregma bambusicola* (Bamboo Woolly Aphid) – encrust stems and shoots of

(a)

(b)

Fig. 5.236 Bamboo Woolly Aphid (*Pseudoregma bambusicola* – Hem.; Pemphigidae); Hong Kong; (a) encrusting stem; and (b) young shoot.

Fig. 5.237 Pineapple-gall Adelges (*Adelges viridis* – Hem.; Adelgidae) on Norway spruce; Skegness, UK.

some larger bamboos; South East Asia and China (Fig. 5.236).

(i) Family Phylloxeridae (Phylloxeras)

A small group with reduced wing venation and short antennae, no siphunculi and a complex life cycle. The Grape Phylloxera is a very serious pest of viticulture. One species is common in the UK on the leaves of English Oak, and others (*Phylloxera* spp.) on oaks in South Africa and Asia.

(j) Family Adelgidae

Adults also reduced, and life cycle very complex using *Picea* (spruces) as primary host, and *Pinius*, *Larix* or *Abies* as secondary host. However, a few species live only on the secondary host. The family includes:

- *Adelges* spp. (Adelges) – seven species attack conifers in the UK. The Pineapple-gall Adelges (Fig. 5.237) is probably the

most common, and spruces grown as Christmas trees can be rendered unmarketable.

- *Pineus* spp. (Pine Adelges) – some native to Europe, others to North America on *Pinus*; found as waxy clumps at the base of the needles on young shoots.

(k) Family Margarodidae (Fluted Scales)

The female, with a slightly reduced body, carries a large, fluted, white waxy egg-sac posteriorly and is very distinctive. Infestations can be heavy and young trees can have their trunk and branches completely covered. Shoots and small branches can be killed and saplings seriously debilitated; seedlings can also be killed.

The species found on trees include:

- *Drosicha* spp. (Giant 'Mealybugs') – on a wide range of trees; India and South East Asia.

Fig. 5.238 Cottony Cushion Scale (*Icerya purchasi* – Hem.; Margarodidae) infesting the trunk of a young (2 inch diameter) *Celtis* tree; Alemaya, Ethiopia.

- *Icerya* spp. (Cottony Cushion Scale, etc.) – polyphagous on trees, shrubs and some palms, etc.; Old World Tropics (Fig. 5.238).
- *Matsucoccus* spp. (Pine Scales) – on *Pinus*; Europe and Asia.

(l) Family Orthezidae

A small group, mostly Holarctic in distribution; females with ovisac are quite mobile. The only species of particular importance is *Orthezia insignis* (Jacaranda Bug), possibly pantropical and found on many trees and shrubs. It is common in Kenya and Uganda.

(m) Family Lacciferidae (Lac Insects)

Includes the Indian Lac Insect, *Laccifera* and the Forest Lac Insects, *Tachardina*; the female body is covered with a thick resinous scale and they encrust the twigs of trees and woody shrubs.

(n) Family Pseudococcidae (Mealybugs)

A tropical group of insects, reduced anatomically and covered with a waxy layer, usually white. Most infest young shoots or leaves, but some encrust roots underground. Several species can be damaging to seedlings and saplings, which may even be killed. Established trees are seldom attacked.

The tropical nursery pests include the common and widespread:

- *Planococcus* spp. (Root Mealybugs, etc.) – polyphagous and pantropical.
- *Pseudococcus* spp. (Mealybugs) – polyphagous and cosmopolitan; a few temperate species in Europe attack conifers.

On established trees there can be infestations of *Planococcus* and *Pseudococcus*, especially on Palmae in the tropics. However, the most damaging species is probably *Maconellicoccus hirsutus* (Hibiscus Mealybug) which in Hong Kong has been seen to kill shoots of *Celtis* and other trees and shrubs. It is recorded in Eastern Africa and tropical Asia.

(o) Family Eriococcidae (Woolly Scales)

This group is more or less confined to temperate forest trees. It includes:

- *Cryptococcus* spp. (Felted Beech Scale, etc.) – on *Fagus* an important pest in Europe and now North America. Occurs on trunks and branches.
- *Eriococcus* spp. – on various conifers, and *Eucalyptus* in Australia; other trees and woody shrubs; worldwide (Fig. 5.239).
- *Gossyparia spuria* (Elm Scale) – on *Ulmus*; Europe.
- *Kermes* spp. (Oak Scale, etc.) – on oaks; Palaearctic and the USA.

Fig. 5.239 Juniper Scale (*Eriococcus juniperus* – Hem.; Eriococcidae) on juniper bush; Alemaya, Ethiopia.

- *Pseudochermes fraxini* (Ash Scale) – on *Fraxinus* in Europe.

(p) Family Coccidae (Soft Scales)

A very large and widespread group, most abundant in warmer regions but not uncommon in northern temperate countries. In temperate regions they are mostly found on woody hosts; on most hosts the infestations are usually light and so little damage is done. Honey-dew production is often copious and sooty mould is typically found on foliage, especially on evergreens. Seedlings and saplings can suffer set-back, but this is seldom serious. Ants may be found in attendance.

The main forestry pests include:

- *Ceroplastes* spp. (Waxy Scales) – mostly tropical and subtropical; can be abundant on *Ficus* and *Acacia* (Fig. 5.240) but seldom damaging.
- *Coccus* spp. (Soft Green Scales, etc.) – polyphagous and cosmopolitan, most damaging to seedlings, but on some tree leaves (*Laurus*, etc.).
- *Eulecanium tiliae* (Nut Scale) – on a wide range of trees and shrubs, including elm, oak, lime, horse chestnut and alder, from India to Europe and also in North America.
- *Gascardia* spp. (White Waxy Scales) – pantropical and polyphagous on a wide range of trees and shrubs.
- *Lecanium* spp. (Soft Scales) – in North America several species are major pests on a wide range of trees.
- *Parthenolecanium* spp. (Brown Scales, etc.) – subtropical and temperate species, found in both hemispheres on a wide range of woody hosts, especially Rosaceae. One species is confined to yew.
- *Pulvinaria* spp. (Woolly Scales) – temperate species in Europe, Asia and North America on a very wide range of woody hosts (including elm, lime, maple, sycamore, horse chestnut, willow and birch) (Fig. 5.241).

Fig. 5.240 Giant (Orange) Waxy Scale (*Ceroplastes* sp. – Hem.; Coccidae) on *Acacia albida* (15 mm diameter); Alemaya, Ethiopia.

Fig. 5.241 Woolly Scale (*Pulvinaria* sp. – Hem.; Coccidae) on stem of grapevine; Skegness, UK.

• *Saissetia* spp. (Black/Helmet Scale, etc.) – cosmopolitan in warmer regions. Some species polyphagous on a wide range of woody hosts (Fig. 5.242).

(q) Family Diaspididae (Hard or Armoured Scales)

A large and important group worldwide; many are pests. The hosts are mostly woody plants (trees and shrubs), and twigs and branches are the usual infestation sites. Some trees (usually small) may be entirely covered with encrusting scales and the tree can be killed. A few species are capable of serious damage. Females are either oviparous or viviparous, and eggs or young are deposited under the 'scale'. The female body then shrinks and dies leaving the scale cavity filled with offspring. In the UK, eggs typically overwinter under the dead female scale. Young trees can be killed by some species, mature trees seldom suffer serious damage. Classification within this group is somewhat confusing and many different names (especially genera) have been, and are still being, used in different parts of the world. Some of the more important pests include:

• *Aonidiella* spp. (Red Scale, etc.) – mostly fruit pests, but can be abundant on Palmae and some forest trees; pantropical.
• *Aspidiotus* spp. (Palm Scales, etc.) – the genus is cosmopolitan, and several species

Fig. 5.242 Black (Olive) Scale (*Saissetia nigra* – Hem.; Coccidae) on twig of *Ficus microcarpa*; Hong Kong.

are serious pests on palms, rubber trees, tung, Moraceae and others.

- *Chionaspis* spp. (Snow Scales, etc.) – a cosmopolitan genus found on trees, both deciduous and evergreen. Especially common on willows, alder, ash poplar and other forest trees and ornamentals.
- *Chrysomphalus* spp. (Purple Scale, etc.) – worldwide several species occur on a wide range of trees and shrubs. They can be very damaging.
- *Lepidosaphes* spp. (Mussel/Oystershell Scales) – cosmopolitan, with *L. ulmi* common in northern temperate regions. It occurs on a wide range of woody hosts including many tree species (130 host species are recorded); some small trees can be totally covered with encrusting scales (Fig. 5.243) and the tree may die.
- *Quadraspidiotus perniciosus* (San José Scale) – this cosmopolitan (although not found in the UK) pest is very serious on deciduous trees and shrubs – trees can be killed and more than 700 hosts are recorded. Other species attack birch, ash, poplar and other trees.

5.5.5 Order Hemiptera, Suborder Heteroptera (Animal/Plant Bugs)

These are the larger, active bugs with toxic saliva that causes death and necrosis of tissues – feeding sites are typically seen as a series of brown spots.

The more important forestry pests are as follows:

(a) Family Tingidae (Lace Bugs) (1800 spp.)

Small flattened bugs with a lacy appearance to be found on the foliage of a wide range of plants, both herbaceous and woody. Only

Fig. 5.243 Mussel Scale (*Lepidosaphes ulmi* – Hem.; Diaspididae) encrusting branch of *Pyrus*; Skegness, UK.

occasionally do they cause serious damage. Some ornamental and shade trees suffer foliage disfiguration by pests such as:

- *Corythuca* spp. (Lace Bugs) – in North America and Europe leaves of sycamore and plane are damaged, as well as some other trees (elm, oak, willow, poplar, alder, etc.) by a total of some 27 different species.

- *Teleonemia* spp. – in the Old World tropics several tree species are attacked, especially wild olive in Africa (Fig. 5.244).

(b) Family Miridae (= Capsidae) (Capsid and Plant Bugs) (6000 spp.)

This large group of small flattened bugs is most serious as pests of herbaceous plants, but they do damage trees to some extent. In North America a few species are pests of trees:

- *Lygus lineolaris* (Tarnished Plant Bug) – on a range of native trees from Mexico, through the USA to Canada, causing tattering of foliage.
- *Plagiognathus albatus* (Sycamore Plant Bug) – on sycamore and *Platanus* trees in the eastern half of the USA and Canada.
- *Tropidosteptes* spp. (Ash Plant Bugs) – three species are found on the leaves of the two species of ash in the USA.

(c) Family Coreidae (Brown Bugs; Twig-wilters) (2000 spp.)

These large brown bugs are mostly found on woody hosts, but not many on trees. The saliva is very toxic and fruits will fall prematurely; the twigs are often killed. The species damaging to trees include:

- *Anoplocnemis* spp. (Giant Twig-wilters) – in Africa and Asia these are polyphagous on a wide range of trees.
- *Leptoglossus* spp. (Leaf-footed Plant Bugs) – the genus is worldwide (pantropical) and polyphagous; in the USA native pecan trees and *Pinus* are attacked; the fruits are destroyed.
- *Mictis* spp. (Giant Twig-wilters) – on some ornamental trees in Asia and Australia (Fig. 5.245).
- *Notobitus meleagris* (Bamboo Bug) – on some of the large species of bamboo in southern China.

Fig. 5.244 Wild olive leaves showing feeding punctures by Olive Lacebug (*Teleonemia australis* – Hem.; Tingidae); Alemaya, Ethiopia.

(d) Family Pentatomidae (Shield and Stink Bugs)

These are pests of herbaceous plants and some shrubs (such as *Camellia*) and a few fruit trees (e.g. *Citrus*). In the UK the Hawthorn Stink Bug (*Acanthosoma haemorrhoidale*) is native to mixed deciduous woods and also occurs on oak, birch, alder, poplars and hazel. In Europe it is reported on hornbeam.

There are records of some shield bugs and stink bugs feeding on shoots of *Pinus, Acacia, Chlorophora* and other forest trees in East Africa, but they are not serious pests. In Hong Kong the native forest tree, *Mallotus paniculatus*, has been found to harbour large numbers of *Cantao ocellatus*, mostly in mating pairs (Fig. 5.246). In the USA *Tetyra bipuncta* is the Shieldback Pine Seedbug whose feeding destroys the seeds in pine cones.

5.5.6 Order Thysanoptera (Thrips) (5000 spp.)

These tiny insects, with their characteristic reduced strap-like wings, are mostly pests of herbaceous plants and cereal crops. Most species inhabit flowers or the surface of young leaves, and their feeding causes surface scarification of the tissues. A few species are encountered during the practice of forestry but they seldom cause serious damage. These species include:

(a) Family Thripidae (Thrips) (2000 spp.)

- *Drepanothrips reuteri* (Grapevine Thrips) – these are recorded on oak trees in the UK.
- *Frankliniella* spp. (Flower Thrips) – a large genus found worldwide on many different plants. Herbaceous plants are preferred as

(a)

(b)

Fig. 5.245 (a) Giant Twig-wilter (*Mictis* sp. – Hem.; Coreidae; body length 20 mm); (b) feeding on and killing a shoot of *Cratoxylon ligustrinum*; Hong Kong.

Fig. 5.246 Shield Bug (*Cantao ocellatus* – Hem.; Pentatomidae); body length 17 mm; from *Mallotus paniculatus*; Hong Kong.

young trees of *Pinus* in Kenya and on *Ficus*, *Acer*, palms and other trees in the USA. This is a leaf-feeder.

- *Taeniothrips inconsequens* (Pear Thrips) – mainly a pest in flowers of fruit trees in Europe and the USA but now a pest of sugar maple in eastern North America.
- *Thrips australis* (Gum Tree Thrips) – in flowers of *Eucalyptus* trees in Australia.
- *Thrips fuscipennis* (Rose Thrips) – in addition to *Rosa*, this pest is found on the young leaves of several tree species in Europe.

(b) Family Phlaeothripidae (Leaf-rolling Thrips, etc.) (c. 2000 spp.)

These thrips usually inhabit leaves of trees, often causing leaf-rolling or leaf distortion; damage is usually cosmetic rather than destructive but on (specimen) trees can be quite disfiguring. The family includes:

- *Gigantothrips* spp. (20+) (Giant Fig Thrips) – leaf infestation occurs on many species of *Ficus* in tropical Asia and Africa, but no distortion takes place.
- *Gynaikothrips ficorum* (Banyan Leaf-rolling Thrips) – these fold and redden the young leaves of *Ficus retusa* in the USA and *F.*

hosts but the pests will spread to adjacent woody hosts (USA).

- *Heliothrips haemorrhoidalis* (Black Tea Thrips, etc.) – polyphagous and cosmopolitan but most abundant in warmer regions. Recorded as abundant on

Fig. 5.247 Banyan Leaf-rolling Thrips (*Gynaikothrips ficorum* – Thy.; Phlaeothripidae) on *Ficus microcarpa*; Hong Kong.

microcarpa in tropical Asia. They can spoil the appearance of ornamental trees and can render young trees unsaleable (Fig. 5.247). A total of 80 species are known, ten of which attack *Ficus* species worldwide.

- *Haplothrips* spp. – 300 species are known worldwide, which infest a wide range of plants, including some trees.
- *Liothrips* spp. (200) (Leaf-rolling Thrips) – attack many trees worldwide, including olive, oak, figs, willow and hickory.

5.5.7 Order Coleoptera (Beetles) (330 000+ spp.)

This is both the largest order of insects and also the most important as pests of forestry (Imms (1960) recognizes 95 families). Adults have biting and chewing mouthparts and may damage both wood and foliage. There may or may not be definite ecological separation between adults and larvae. The larvae are typically white, legless or with functional thoracic legs, and well-developed biting mandibles. They may live in the soil and eat plant roots; bore in wood, eat foliage or live inside fruits and seeds. The group includes many important predators that kill other insects – both as adults and as larvae.

A large number of wood-boring beetles are recorded on trees and timber worldwide. Many trees that are heavily infested by beetle borers are clearly sickly, some being moribund or on the point of death. Careful examination of the trees almost invariably begs the question, 'is the tree dying because of the beetle infestation, or were the beetles attracted to a sickly/dying tree?'. To be accurate, it is usual that this question cannot be answered. Recent work in California suggests, however, that any tree that is stressed is more vulnerable to insect attack, and even a low level of water stress can be sufficient to make a tree vulnerable. In a healthy tree, boring beetle larvae are usually overwhelmed by a copious sap flow and the damaged plant tissues rapidly 'heal' (i.e. regrow). However, there is evidence that in a few cases, a beetle is able to attack, and indeed show preference for, a healthy tree (some Cerambycidae), whereas in a number of cases it is clear that only dying trees are chosen as hosts (some Scolytidae). Some beetles only infest stacked timber in the forest (i.e. dead wood). However, it seems likely that the majority of wood-borers choose host trees that are in some way stressed.

There are so many beetle pests of forestry that only a selection can be considered here. The main groups include:

(a) Family Passalidae (Sugar Beetles) (500 spp.)

These largish black (or brown) flattened beetles with parallel-sided bodies are found in many tropical forests, especially under loose and rotten bark on dead trees. Larvae are said to live in rotten wood on the forest floor, although they are almost never found. Strictly speaking, they are not pests but will often be encountered. *Didimus* and *Pentalobus* are common in East Africa.

(b) Family Lucanidae (Stag Beetles) (750 spp.)

These large black or brown beetles are forest species. The male often has very prominent mandibles, as well illustrated by the temperate *Lucanus*. The larvae live in rotten wood, so they are really only secondary pests. The genus *Prosopocoilus* (Fig. 5.248) is found throughout tropical Asia and Africa; the adults feed on nectar or sap exuding from tree trunk wounds. *Nigidius* is another common African genus.

(c) Family Scarabaeidae (Chafers/White Grubs; Cockchafers/June Beetles, etc.) (17 000 spp.)

There are six large and well-defined sub-families with somewhat different habits, but it is more appropriate here to view the group collectively. The larvae are 'scarabaeiform'

Fig. 5.248 Male Stag Beetle from south China (*Prosopocoilus biplagiasus* – Col.; Lucanidae) (body length 30 mm).

with a thick, fleshy, white, C-shaped body. They live in either soil with a high organic content, especially in rubbish tips and compost heaps, or in rotting wood. Worldwide many species are pasture pests, and these species can be damaging in seed-beds and to young saplings. Adults are often large, brightly coloured or black beetles with a strong rounded body, and most are nocturnal in habit. Rose Chafers feed on flowers, and many trees (*Acacia*, etc.) in blossom may attract dozens of beetles, with the result that seed production can be greatly reduced.

In the tropical rain forests there are some very spectacular beetles to be seen in the forest canopy, some of which are the largest insects in the world. The adults feed on flowers and young foliage, and the larvae are reported to live in rotting wood on the forest floor. Damage is usually slight, but the beetles are conspicuous by their presence. In Africa the Goliath Beetle (Fig. 5.126) can be found, while the Atlas Beetle (Fig. 5.249) is abundant in South East Asia,

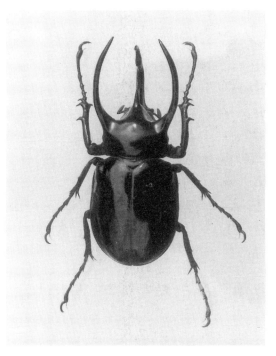

Fig. 5.249 Atlas Beetle (*Chalcosoma atlas* – Col.; Scarabaeidae), body length, including horns, 80 mm; Malaya.

usually associated with palms. *Dynastes* are the giant Hercules Beetles of Central and South America.

Some of the more important forestry pests include:

- *Anomala* spp. (Flower Beetles) (100+ spp.) – adults eat foliage, larvae in soil eat roots of young trees.
- *Melolontha* spp. (Cockchafers) – larvae are polyphagous root-eaters and feed on roots of seedlings and saplings; Palaearctic.
- *Oryctes* spp. (40+ spp.) (Rhinoceros Beetles) – adults gnaw growing point of palms in Asia and Africa; larvae live in rotten palm trunks.
- *Pachnoda* spp. (Flower Chafers) – in Africa and Arabia the adults eat the flowers of many trees and plants, including *Acacia*.
- *Popillia* spp. (Japanese Beetle, etc.) – adults

polyphagous, eat flowers and foliage; East Asia and North America.

- *Protaetia* spp. (Rose Chafers, etc.) – adults destroy tree flowers in Asia.
- *Serica* spp. (Brown Chafers) – larvae attack seedlings and saplings (roots) in Europe and Asia.
- *Xylotrupes gideon* (Unicorn/Elephant Beetle) – coconut palms and some trees have foliage eaten by the adult beetles in Asia and Australasia.

(d) Family Buprestidae (Jewel Beetles; Flat-headed-borers) (11 500 spp.)

These are nearly all timber-borers – the majority are tropical, but some can be found in temperate regions. The larvae all bore in plant tissues, but the adults cause no direct damage except for a little bark-gnawing; they are active and fly readily. Some Buprestidae can attack healthy trees but most are only successful in weakened trees. Larvae have a characteristic large, flattened thorax (hence the common name), and they do not extrude frass as they tunnel – the tunnel space behind the borer is filled with tightly packed frass (see Fig. 5.250), and there are no frass holes in the tree trunk or branch. Development is slow – in the tropics larvae can mature in a year, but in Europe two years are usually required. Within the group most larvae do not bore deeply: most feed just under the bark, and pupation occurs close to the surface. Many species prefer dying trees or logs as host, so in general these beetles may be regularly encountered. However, the damage they cause to forest trees is seldom serious.

This family is unusual in that there are a few genera with large numbers of species. Some of the more important forest pests include the following:

- *Agrilus* spp. (700 spp.) – larvae bore a wide range of shrubs and trees, worldwide (Fig. 5.131); most are small in size (1–2 cm). Many leguminous trees in Africa are used as hosts.

Fig. 5.250 (a) Pine Buprestid (*Chalcophora japonica* – Col.; Buprestidae), body length 35 mm; and (b) pine plank bored by larva, showing frass-filled tunnel; Hong Kong.

- *Chalcophora* spp. (Pine Buprestids) – quite large in size (3–4 cm); larvae bore trunk of *Pinus* and some other conifers in Asia and Europe (Fig. 5.250).
- *Chrysobothris* spp. (300 spp.) – on a wide range of hosts (trees and shrubs), mostly in Africa and India.
- *Stigmodera* spp. (400 spp.) – on a wide range of tree hosts in Australia.
- *Trachys* spp. (Buprestid Leaf-miners) – larvae mine leaves on many woody hosts in Europe and most of Asia. They are also recorded on oaks, beech, elm, willows, etc. in the USA.

Newly planted trees are especially at risk from Buprestidae for they can be seriously debilitated. Adult Jewel Beetles are often recorded emerging from timber in houses or other buildings that have been in place for years. The piece of pine wood in Fig. 5.250 was taken from a three-year-old door in Hong Kong – two adults emerged. In Australia, it is reported that an adult beetle emerged from a domestic timber 25 years old.

(e) Family Elateridae (Click Beetles; Wireworms) (7000 spp.)

A widespread group whose larvae live in soil, eat plant roots and bore tubers. They are most damaging to herbaceous plants and root crops. A number of species are recorded as damaging roots of tree seedlings and nursery stock. A typical polyphagous soil pest is *Athous niger*, the Garden Wireworm (Fig. 5.251). Probably the commonest genus is *Agriotes* (Common Click Beetle/Wireworm), found throughout Europe, Asia and North America. Click beetles are often found in tropical forests, for a few have larvae that feed on decaying wood (*Melanotus*), and a number have larvae that are predacious on the immature stages of other wood-borers. Thus these beetles are collected from logs and timber quite regularly but they are never present as primary pests.

(f) Family Anobiidae (Furniture/Timber Beetles) (1100 spp.)

This group is found in timbers in domestic premises, and in the forest in cut logs,

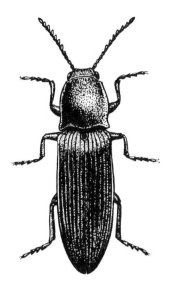

Fig. 5.251 Garden Click Beetle/Wireworm (*Athous niger* – Col.; Elateridae); length 12 mm; Boston, UK.

ring-barked trees and dead branches. The larvae bore in wood, and most species appear to be associated with fungal decay. They are small in size and usually occur in dense infestations, so that a piece of damaged wood will usually have many emergence holes (Fig. 5.252).

The two commonest and best known species are probably:

- *Anobium punctatum* (Furniture Beetle) – a widespread species of European origin, it is only to be found in dead wood (usually softwood, both deciduous and evergreen) Fig. 5.253).
- *Xestobium rufovillosum* (Deathwatch Beetle) – a temperate species found in dead wood (mostly structural timbers and usually hardwoods (oak, etc.)), and where fungal decay is present. Recorded throughout Europe, North Africa and the USA.

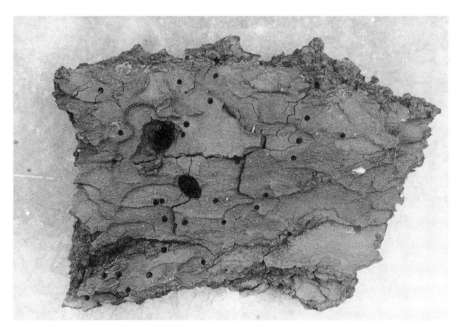

Fig. 5.252 Pine tree bark with Timber Beetle emergence holes (plus two Cerambycid holes); Friskney, UK.

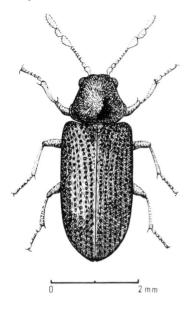

Fig. 5.253 Furniture Beetle (*Anobium punctatum* – Col.; Anobiidae); Hong Kong.

Fig. 5.254 Black Borer (*Bostrychopsis parallela* – Col.; Bostrychidae) from bamboo; Hong Kong.

(g) Family Bostrychidae (Black Borers/Auger Beetles, etc.) (430 spp.)

These distinctive black or brown beetles, with their cylindrical body shape and large hooded and toothed prothorax shielding the reflexed head, are well known in the tropics as pests of trees and wood. The boring is done by the adult beetles, and the female makes a breeding gallery where the adults shelter and feed and the young develop. The genus *Apate*, which includes several major agricultural pests, is apparently atypical in that it attacks living (but usually unhealthy) trees. Most Bostrychidae bore in dry sapwood, especially in hardwood trees, and they tend to be polyphagous. Some of the smaller species seem to prefer dead branches and twigs; in Africa they especially attack trees belonging to the Mimosaceae (*Acacia*, *Albizia*, etc.).

Some of the more important forestry pests include:

- *Apate* spp. (Black Borers) – usually attack unhealthy trees in Africa and South America (Fig. 5.133).
- *Bostrychopsis* spp. – attacks many different trees and bamboos. Recorded in Africa, Asia and Australia (Fig. 5.254).
- *Dinoderus* spp. (Small Bamboo-borers) – bore bamboos, rattan palm, *Albizia* and various trees; South East Asia, Australia and Africa.
- *Epicerastes* spp. – in *Celtis* logs in Uganda.
- *Sinoxylon* spp. – bore in bamboos, Acacias and many other leguminous trees, and other trees in Africa, the Mediterranean, India and South East Asia.
- *Xyloperthella* spp. – some major pests that attack sapwood and logs of *Celtis*, *Chlorophora*, *Albizia* and many other trees, in tropical Africa.
- *Xyloperthodes* spp. – another African genus, recorded in logs, poles and timber of *Acacia*, *Eucalyptus* and *Chlorophora*. (In

Africa there is quite a large number of small genera also recorded attacking this series of valuable timber trees.)

(h) Family Lyctidae (Powder-post Beetles) (70 spp.)

Small beetles, with larvae boring sapwood (mostly of hardwood species) worldwide. The genus *Lyctus* (Fig. 5.255) has several species widely distributed; others are found in Africa and Asia. Freshly sawn wood is often a preferred host. Typical damage consists of the sapwood being eaten away (leaving a fine powder behind) under a thin skin (veneer) of wood on the outside. In the residual veneer are the emergence holes made by the adult beetles. Of the 12 genera in this family a few seem to be true forest species and are found in girdled trees rather than timber. *Minthea* is another genus of importance in Africa, with species recorded in both girdled forest trees and in sawn timbers.

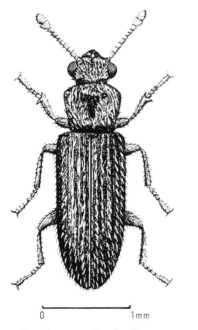

0   1mm

Fig. 5.255 Powder-post Beetle (*Lyctus* sp. – Col.; Lyctidae).

(i) Family Lymexylidae (Timber Beetles) (40 spp.)

A small group, mostly tropical, of elongate soft-bodied beetles whose larvae tunnel in wood of trees or palms. The two main genera are:

- *Atractocerus* – the adult has a long, slender and soft body, with tiny elytra but large hind wings with which they fly in tropical forests – the author caught several in the Budongo Forest in northern Uganda in 1968. The larvae are very long and thin, and reported to bore in dying trees or newly dead sappy logs. The tunnel walls are usually infected with fungi. These beetles are forestry pests in Africa, India, South East Asia and Australia.
- *Melittoma* – this is not so elongate (Fig. 5.256). *M. insulare* is the Coconut Palm-borer in the Seychelles and Madagascar; *M. sericum* is the Chestnut Timberworm of the USA, and another species is found in Australia.

(j) Family Cerambycidae (Longhorn Beetles; Round-headed-borers) (20 000 spp.)

This large and ecologically dominating group of beetles is probably the single most important family of forestry pests. This is in part because many species are large insects, and also because the group is abundant worldwide, with some making deep tunnels in tree trunks, spoiling quality timber. Adults of many species are attracted to lights at night, but they are also to be found on the tree from which they emerged. The larvae all bore in plant material – some of the smallest are found in herbaceous stems or tunnelling in ripe seeds (cotton and coffee), but most inhabit tree trunks and branches. Typically, the larva makes a series of holes from which frass is extruded as it tunnels. With some species, however, frass is hidden in crevices or under bark and is not extruded. However, the tunnel is kept clear, as distinct from

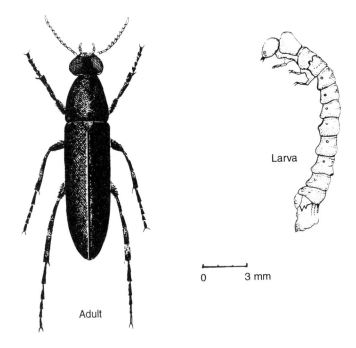

Larva

Adult

0   3 mm

Fig. 5.256 Coconut Palm-borer (*Melittoma insulare* – Col.; Lymexylidae); Seychelles.

the Buprestidae. Larval development can be completed in a few months for some small species, but the larger ones take one to three years. Pupation takes place within a cell just under the tree bark, at the end of a tunnel. The adult emergence hole is circular in shape, as distinct from the oval hole of the Buprestidae. Adults live for one to three months and many feed on tree bark or young foliage, others on flowers.

Some species prefer timber or dead trees as hosts, but others clearly prefer living, healthy trees.

The group is divided into three large subfamilies and two smaller ones, and some entomologists regard these as being of family status. The vast number of species to be found in forests and affecting trees grown for timber or ornamental purposes worldwide makes it impossible to mention all but a few of the more important genera here:

Subfamily Cerambycinae

The second largest group; adults are variable but most are diurnal; antennae are often very long and hind legs are the longest. A few species bore living trees but many appear to prefer dead wood. Some of the forest species include:

- *Chlorophorus* spp. (Bamboo-borer, etc.) – bore bamboos in Asia and the USA (Fig. 5.257), *Acacia* and other trees in Africa; usually dying trees.
- *Clytus arietis* – one of the commonest British longhorns, recorded on oak, chestnut, sycamore, beech, birch, *Ficus*, *Salix*, etc.; found throughout Europe.
- *Hylotrupes bajulus* (House Longhorn) – native to Europe and Asia, this species prefers pine, *Picea* and *Abies*, but in parts of Europe poplar, oak, alder and other deciduous trees are attacked. The sapwood

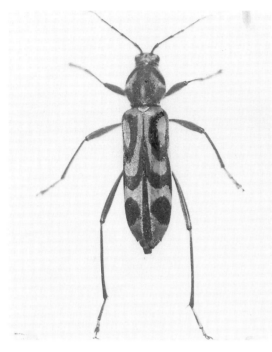

Fig. 5.258 Eucalyptus Longhorn (*Acanthophorus maculatus* – Col., Prioninae); Alemaya, Ethiopia (body length 85 mm).

Fig. 5.257 Bamboo Longhorn (*Chlorophorus annularis* – Col.; Cerambycinae); body length 14 mm; Hong Kong.

of dry timber is bored first, and later the larvae penetrate the heartwood. In the 'wild' these beetles are found in tree stumps and dead trees.

- *Oemida* spp. – major pests of living *Cupressus* and dead trees, both deciduous and evergreen; East Africa.
- *Phoracantha* spp. (Eucalyptus Longhorn, etc.) – a minor pest in Australia but serious pest of *Eucalyptus* trees in the USA and parts of Asia.
- *Xystrocera* spp. – attack logs of leguminous trees and dead trees in Africa and South East Asia, but *Celtis* and other trees may also be damaged.

Subfamily Prioninae

A small group of flattened and elongate beetles; most are brown in colour, with short antennae and protruding head. The larvae

usually infest dead, dry wood, which is often decaying, but a few species feed in apparently sound wood. A few common species are referred to below:

- *Acanthophorus* spp. – large brown species, polyphagous, and usually reared in dead trees, especially those starting to decay; a wide range of host trees is recorded in Africa and India (Fig. 5.258).
- *Macrotoma* spp. – recorded in dead *Celtis*, *Acacia* and many other trees in East Africa.
- *Mallodon (Stenodontes) downesi* – a common African species recorded on a wide range of logs and dead trees.

Subfamily Lamiinae

The largest group, showing great diversity, but usually with long antennae – those of the males being longer than those of the females. The head is strongly deflexed.

(a)   (b)

Fig. 5.259 (a) Jackfruit Longhorn (*Apriona germarii* – Col.; Lamiinae); and (b) larva from *Ficus* tree; south China (beetle length 48 mm).

Many pest species are recorded. Many prefer dead or dying host trees, but an equal number attack living and apparently healthy hosts. A few of the more common and important pest species include:

- *Acalolepta* spp. – larvae bore teak and other trees in India and South East Asia.
- *Acanthocinus* spp. (Pine-borers) – major pests of *Pinus* in the USA. Medium-sized beetles (*c.* 2 cm) with very long antennae.
- *Apriona* spp. (Jackfruit Longhorn, etc.) – larvae bore living trees of *Artocarpus*, *Ficus*, etc.; India, South East Asia and China (Fig. 5.259).
- *Batocera* spp. (Spotted Longhorn Beetles) – some eight species, large in size, bore in *Sapium*, walnut, Moraceae and many other trees in India, South East Asia and China (Fig. 5.260).
- *Glycobius speciosus* (Sugar Maple-borer) – a yellow and black wasp-like beetle, native to North America and a major pest of sugar maples, which can even be killed.
- *Monochamus* spp. (Pine Sawyers, etc.) –

bore *Pinus* in Europe, Asia, the USA and Canada, and other trees in Africa.
- *Petrognatha gigas* (Giant Fig Longhorn) – larvae bore in living trees of *Ficus natalensis* and sometimes *Chlorophora* in East Africa, and other *Ficus* trees in West Africa (Fig. 5.261).
- *Saperda* spp. – a large genus of small beetles, some of which bore in healthy trees (willow, alder, poplar and hawthorn), while others prefer stressed, dying or dead trees of elm or hickory; most bore in small branches or twigs.
- *Tetroplum gabrieli* (Larch Longhorn) – major pest of larches in Europe.

(k) Family Bruchidae (Seed Beetles; Bruchids) (1300 spp.)

These small beetles have larvae that feed on legume seeds usually inside the pods. They are important to forestry in that they destroy the seeds of leguminous trees (*sensu lato*) which has a very adverse effect on natural regeneration. Natural regeneration of *Acacia* trees in Africa and tropical Asia is very

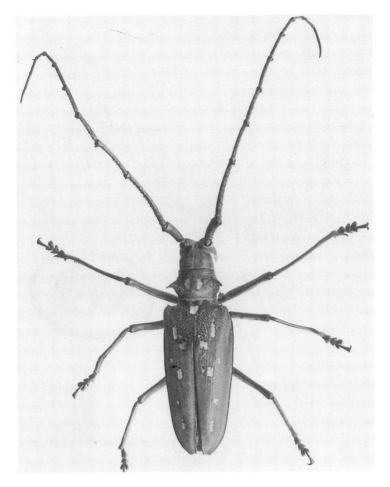

Fig. 5.260 Giant Red Spotted Longhorn (*Batocera rufomaculata* – Col.; Lamiinae); body length 70 mm; from *Sapium* tree; south China.

important in arid areas. Figure 5.262 shows a seed pod of *Acacia contorta* from the Sahara (Lybia) with ten seeds and with six bruchid emergence holes. Leguminous trees are being used in agroforestry and land reclamation projects in tropical Asia and Africa in ever-increasing quantities, for their multipurpose qualities are proving most valuable: they are used to stabilize the soil, enrich the soil, provide fuelwood for cooking, foliage for livestock feed and sometimes seeds as food (pulses). In Europe, shrubs such as lupin are being used to reclaim mine dumps and the like. Most bruchids infest legume pods in the wild, but some can also continue to develop in seed stores. The main pests belong to the genera:

- *Acanthoscelides* (300 spp.) – mainly a New World genus but a few species are pantropical or African.
- *Bruchus* – several species regularly recorded as pests; worldwide, including one species on *Acacia albida* pods in East Africa.
- *Caryedon* (Groundnut-borer, etc.) – the usual hosts are groundnut in West Africa, and tree legumes (*Tamarindus, Bauhinia,*

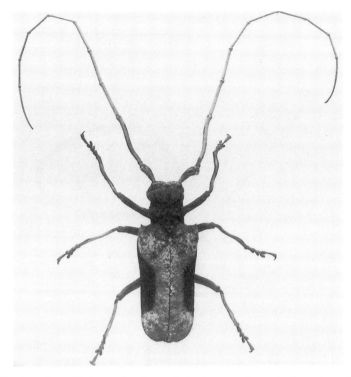

Fig. 5.261 Giant Fig Longhorn (*Petrognatha gigas* – Col.; Lamiinae); male; body length 65 mm; from *Ficus natalensis*; Uganda.

Fig. 5.262 Seed pod of *Acacia contorta* from the Sahara Desert (Libya) bored by Bruchid Beetles.

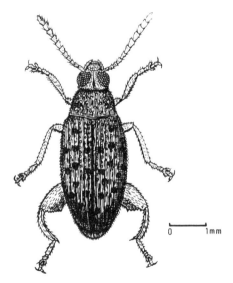

Fig. 5.263 Groundnut-borer (*Caryedon serratus* – Col.; Bruchidae); Kenya.

etc.) in India and East Africa (Fig. 5.263).

(l) Family Chrysomelidae (Leaf Beetles) (20 000 spp.)

A very abundant and widespread group of leaf-eating beetles with a large number of genera and species. They show great diversity of habits and 10–15 different sub-families are recognized. Many species are pests of cultivated plants but only a few are of importance to forest trees. Some of these can be injurious to seedlings and young plants. Specimen and shade trees and nurs-ery stock are vulnerable, because the leaf beetles damage foliage and spoil the appear-ance of the tree.

The species most likely to be encountered on forest trees and nursery stock include:

- *Aphthona* spp. (Flea Beetles) – adults hole leaves of many trees and plants in Europe, Asia and Africa (Fig. 5.264).
- *Chalcoides* spp. (Poplar/Willow Flea Beetles) – on poplars and willows in Europe.
- *Chrysomela* spp. (Poplar Leaf Beetle, etc.) – on poplars, *Salix* and alder; Europe.
- *Colasposoma* spp. (Acacia Leaf Beetle, etc.) – on wattle and other trees in Africa and tropical Asia.
- *Dactylispa* spp. (Hispid Beetles) – on leaves of *Sterculia* and other trees in Africa.
- *Galerucella* spp. (Willow/Elm Leaf Beetle, etc.) – hosts include willows, elms and horse chestnut; found throughout Europe.

Fig. 5.264 Damage to leaves of *Mallotus apelta* by adult Flea Beetles (*Aphthona* sp. – Col.; Chrysomelidae); Hong Kong.

Fig. 5.265 Large Yellow Flea Beetles (*Podontia lutea* – Col.; Chrysomelidae) on leaves of *Spondias*; south China (body length 15 mm).

- *Lochmaea* spp. – *Salix*, birch, poplars and hawthorn are attacked; on hawthorn the larvae of *L. crataegi* develop inside fruits; Europe.
- *Peploptera* spp. – on *Acacia* trees in East Africa.
- *Phyllodecta* spp. (Willow/Poplar Leaf Beetle, etc.) – on willows and poplars throughout most of Europe.
- *Podontia* spp. (Large Yellow Flea Beetles) – both larvae and adults eat foliage of *Spondias*; South East Asia (Fig. 5.265).
- *Promecotheca* spp. (Palm Leaf-miners) – larvae mine leaves of Palmae in South East Asia and the Pacific region.

(m) Family Anthribidae (Fungus Weevils) (2400 spp.)

A tropical group of sturdy little beetles with antennae clubbed but not elbowed, and snout short and broad. They are forest species associated with dead wood and fungi, often under loose bark; a few species bore in tropical lianas and a few in stored seeds. The group is best represented in the Indo-Malayan region.

A few species of note are listed below:

- *Araecerus fasciculatus* (Nutmeg Weevil, etc.) – widely recorded damaging seeds in storage (Fig. 5.266) (*Cassia, Maesopsis* and other tree legumes) in India and Africa. Other species attack ripe legumes in India and other ripe seeds in parts of Asia.
- *Epicerastes* spp. – in logs of *Celtis* in Uganda.
- *Mecocerus* spp. – common in dead trees and logs of many different species; East Africa.
- *Phloeobius* spp. – in dead *Acacia, Albiza, Celtis*, etc., with the species *P.*

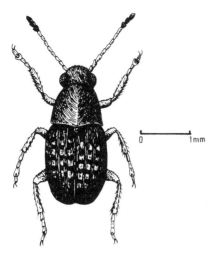

Fig. 5.266 Nutmeg Weevil (*Araecerus fasciculatus* – Col.; Anthribidae); Kenya.

podicalis common in large dead lianas; East Africa.

(n) Family Brenthidae (1300 spp.)

A tropical group of weevils (*sensu lato*) found in forests and associated with damp, dead wood. Some larvae are predatory, and some are thought to be damaging to logs. Adults are small and brown with long antennae and an elongate narrow body. Some adults are also predatory, usually on other wood-boring beetles.

A few species are quite common in some regions:

- *Anisognathus* spp. – common in logs of many tree species in East Africa.
- *Cerobates* spp. – on *Ficus*, *Brachystegia*, *Celtis* and other logs, and also under the bark of trees; East Africa.
- *Pseudocephalus* spp. – in *Celtis*, *Chlorophora*, *Ficus* and other forest trees, also found under the bark; East Africa.

A couple of dozen other different genera are recorded in East Africa, but are not thought to be of importance as pests.

(o) Family Attelabidae (=Rhynchitidae) (300 spp.)

A small group of weevils recently separated from the Curculionidae. They are characterized by being small, with an elongate 'snout' bearing clubbed but not geniculate antennae with a short scape. Some are important tree pests because the female weevils make leaf-rolls on several common temperate trees. The female beetle cuts the leaf lamina to start the roll and lays an egg inside where the larva feeds. On forest trees these pests are inconsequential, but on shade or ornamental trees or nursery stock their damage can be of importance. The few noteworthy species which attack trees include:

- *Apoderus* spp. (Hazel Leaf-roller Weevil, etc.) – these attack hazel, beech, alder, birch and hornbeam in Europe, and mango, litchi, etc. in India and South East Asia.
- *Attelabus* spp. (Oak Leaf-roller Weevil, etc.) – on oaks, hazel, alder, chestnut, etc.; throughout much of Europe.
- *Byctiscus* spp. (Poplar Leaf-roller Weevil, etc.) – roll leaves of *Populus*, hazel, elm, birch, alder and willows; Europe.
- *Deporaus* spp. (Birch/Maple Leaf-roller Weevil, etc.) – on birch, *Acer*, oak, beech, hazel, hornbeam, etc.; Europe. On mango in India.
- *Rhynchites* spp. (Apple Fruit Rhynchites, etc.) – twig-cutters sever shoots of oaks and Rosaceae after oviposition; fruit-eaters lay eggs in fruit of Rosaceae; Europe.

(p) Family Curculionidae (Weevils proper) (60 000 spp.)

A very large group of beetles with their distinctive clubbed and elbowed antennae. Many have the front of the head elongated into a long and narrow 'snout' or rostrum with terminal mouthparts; others have a short and stout rostrum. Typically, the female uses the long snout to make deep feeding holes into which she may lay a single egg; the larva lives inside the host plant tissues where it is usually safe from predation. Most of the pest species are agricultural pests, but some are forestry pests, eating leaf foliage and seeds or boring into the wood of the tree. Others are serious pests of seedlings and nursery stock and eat the roots in the soil. Because of the diversity of lifestyles it may be worth viewing weevil pests according to how they damage the tree hosts:

1. Seedling pests
   (a) Leaf-eating – adults of *Phyllobius*, *Systates* and *Strophosomus*.
   (b) Root-eating – larvae of *Phyllobius*, *Otiorhynchus* and *Tanymecus*.
2. Tree pests
   (a) Foliage
   • Leaf-eaters – adults of *Phyllobius*, *Dermatodes*, *Hypomeces*, *Magdalis*, *Polydrusus*, *Systates* and *Nematocerus*.
   • Leaf-rollers – adults and larvae of *Goniopterus*.
   • Leaf-miners – *Rhynchaenus*.
   (b) Fruits
   • Nut-borers – *Curculio*.
   • Fruit-borers – *Conotrachelus*.
   (c) Roots
   • Root-borers – *Hylobius* larvae.
   • Root-eaters – larvae of *Phyllobius* and *Otiorhynchus*.
   (d) Shoots
   • Bamboo shoots – *Cyrtotrachelus* larvae.
   • Palm Weevils – *Rhynchophorus* larvae.
   • Shoot Weevils – *Pissodes*, *Barypeithes* and *Alcidodes*.
   (e) Stems/trunks/branches
   • Bark-eating – larvae of *Magdalis*.
   • Living wood – Bamboo Weevil (*Cyrtotrachelus*).
     – Palm Weevils (*Diocalandra*).
     – *Mecocorynus* larvae and *Sipalinus*.
   • Dead wood/timber – *Mecopus* and *Stemoscelis* (with fungus).

The more important weevil pests of forestry are as follows:

• *Alcidodes* spp. (Shoot Weevils, etc.) – adults girdle seedling stem; larvae bore inside stem; Africa.
• *Barypeithes* spp. (Broad-nosed Weevils) – adults feed on shoots and seedlings of *Pinus*, oak, birch, etc.; Europe.
• *Conotrachelus* spp. (Fruit Curculios) – fruit and nut trees attacked; USA.
• *Curculio* spp. (10+) (Nut/Acorn Weevils) – hazelnut, chestnuts and acorns eaten by larvae; some make galls on oaks, hazel, birch, etc.; Europe and the USA (Fig. 5.267).
• *Cyrtotrachelus longimanus* (Bamboo Weevil) – larvae feed on apical shoots of some large bamboos; southern China (Fig. 5.268).
• *Goniopterus scutellatus* (Eucalyptus Weevil) – larvae and adults roll Eucalyptus leaves; can defoliate trees; Australia, Asia and Africa.
• *Hylobius* spp. (Pine Root Weevils) – larvae eat roots of *Pinus*; Japan and the USA.
• *Magdalis* spp. (Bark Weevils) – adults eat leaves, larvae bore bark; Europe and the USA.
• *Mecocorynus* spp. (Cashew Weevils, etc.) – larvae bore branches of various trees; East Africa.
• *Mecopus* spp. – larvae found in dead trees and logs on many different tree species; East Africa.
• *Myllocerus* spp. (15 spp.) (Grey Weevils) – adults eat leaves, larvae in soil eat roots of many different species; tropical Asia (India to Japan).
• *Nematocerus* spp. (Shiny Cereal Weevils) – adults eat leaves, larvae eat roots – tree seedlings are mostly attacked; tropical Africa.
• *Otiorhynchus* spp. (6+ spp.) – adults eat leaves of many plants and larvae in soil eat roots; Europe, Asia and North America.
• *Phyllobius* spp. (10+ spp.) (Leaf Weevils) – adults eat leaf edges and larvae in the soil eat roots; more serious as nursery pests; Europe, Asia and the USA (Fig. 5.269).
• *Pissodes* spp. (Conifer Shoot Weevils) –

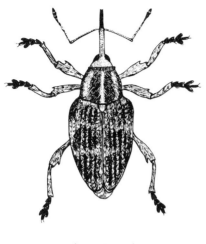

0　　　　　　2 mm

(a)

(b)

Fig. 5.267 (a) Acorn Weevil (*Curculio* sp. – Col.; Curculionidae); and (b) holed acorn; Alford, UK.

larvae bore and kill apical shoots of *Pinus* and other conifers; Europe, Asia and the USA.

- *Polydrusus* spp. (Leaf Weevils) – adults eat leaves of deciduous trees; Europe and the USA.
- *Rhynchaenus* spp. (Leaf-mining Weevils) – larvae mine leaves of beech, birch, alder, poplars, elm and oak; Europe, Asia and the USA.
- *Rhyncophorus* spp. (Palm Weevils) – larvae bore crown of palms in the tropics; pantropical.

- *Scolytoproctus* spp. – larvae bore logs of *Celtis*, *Morus* and many other trees in East Africa.
- *Sipalinus* spp. (Pine-borer, etc.) – larvae bore trunk of *Pinus* in China and Japan; in other trees (*Celtis*, *Erythrina*, etc.) in Africa.
- *Stenoscelis* spp. – in dead trees, logs and planks of a wide range of softwoods and hardwoods, usually associated with fungal decay in Africa.
- *Strophosomus* spp. (Nut Leaf Weevils) – adults feed on leaves and shoots of a wide range of trees and shrubs; damaging to nursery stock; Europe.
- *Systates* spp. (20+ spp.) (Systates Weevils) – adults have been reported defoliating many different trees (*Acacia*, *Eucalyptus*, *Pinus*, etc.); Africa.
- *Tanymecus* spp. (Surface Weevils) – adults attack foliage, larvae in soil eat roots; may be very damaging to seedlings; Asia.

(q) Family Scolytidae (Bark Beetles; Ambrosia Beetles)

A very important group of forestry pests – small in size but very numerous. Many are gregarious and many carry fungi. The adult beetles are the primary tunnellers and make breeding galleries in the wood under the bark of a wide range of trees where the eggs are laid. The larvae then make small lateral galleries where they live and feed (Fig. 5.270). The larvae of some species feed on the sugars and starches present in the sapwood; other species cultivate fungi and the larvae feed on the fungal mycelium – these are the 'Ambrosia' Beetles. Some species are renowned for the effective pheromones produced by the adults; these are used to ensure aggregation in large numbers on the stressed or sickly forest trees that make suitable breeding hosts. One drawback for the beetles is that this habit makes them very susceptible to pheromone traps placed in

Fig. 5.268 Bamboo Weevil (*Cyrtotrachelus longimanus* – Col.; Curculionidae) and damage; south China. Top left, damage; top right, larva; bottom left and right, adult female.

the forest. They are generally more important in temperate forests, where considerable damage can be done to both conifers and broad-leaved trees; in tropical forests they are usually regarded only as minor pests. One example of their importance as forestry pests is the recent devastation in the UK by Dutch Elm Disease – the fungus carried by the *Scolytus* beetles killed virtually all the English elm trees in the greater part of the UK, starting at the south coast of England. There are two subfamilies: these are regarded by some authorities as being separate families, and Imms (1960) regards them as subfamilies within the Curculionidae.

Some of the more important and widespread pests include:

Subfamily Scolytinae

- *Blastophagus* spp. (=*Tomicus*) (Pine Shoot Beetles) – attack *Pinus*, firs and spruces; Europe, Asia and North America.
- *Conophthorus* spp. (Pine Cone Beetles) – destroy seeds in pine cones; USA and Canada.

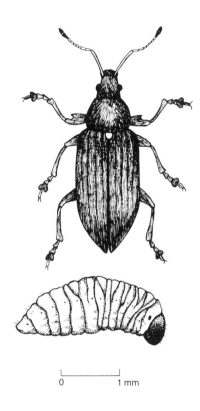

(a)

(b)

Fig. 5.269 Common Leaf Weevil (*Phyllobius pyri* – Col.; Curculionidae) and adult feeding damage to hazelnut leaves; Gibraltar Point, UK.

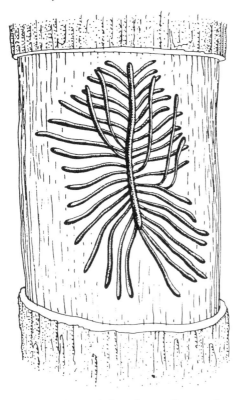

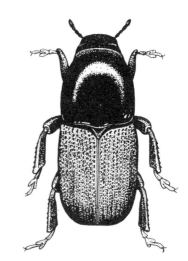

(a)

Fig. 5.270 Bark Beetle breeding gallery under tree bark (*Scolytus* sp. – Col.; Scolytidae); Alford, UK.

- *Dendroctonus* spp. (Pine Beetles) – attack *Pinus*; USA and Central America.
- *Dryocoetes* spp. (Spruce/Birch Bark Beetles) – attack various temperate trees; Japan and the USA.
- *Hylastes* spp. (Pine Bark Beetles) – in *Pinus*, spruce and larch trunks; Europe and Asia.
- *Hylesinus* spp. (2 spp.) (Ash Bark Beetles) – bore ash in the UK, *Olea* in Europe and *Celtis* in Africa.
- *Ips* spp. (Pine/Larch Beetles) – attack various conifers; Europe, Asia and North America.
- *Polygraphus* spp. (Fir Bark Beetles) – attack conifers; Asia and Africa.
- *Scolytus* spp. (Bark Beetles) – many species; attack a wide range of fruit and

(b)

Fig. 5.271 (a) Elm Bark Beetle (*Scolytus scolytus* – Col.; Scolytinae); and (b) dead English elm tree; Skegness, UK.

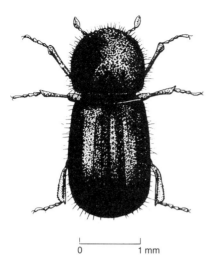

```
0                    1 mm
```

Fig. 5.272 Shot-hole-borer (*Xyleborus* sp. – Col.; Scolytinae); Kenya.

forest trees throughout Europe, Asia and North America. *S. scolytus* is the Larger Elm Bark Beetle and is the main vector of Dutch Elm Disease (Fig. 5.271).

- *Stephanoderes* spp. – bore seeds of many trees (and coffee) and also bore twigs and branches of *Acacia*, *Pinus* and other trees; Africa.
- *Xyleborus* spp. (Shot-hole-borers) – the genus is polyphagous and cosmopolitan; some species are primary pests and can infest living trees, others prefer moribund trees. Some species are Ambrosia Beetles and carry fungi to infect the breeding gallery as larval food (Fig. 5.272).
- *Xylosandrus* (Black Twig-borers) – some species primary pests, others secondary, on a wide range of trees; pantropical.

Subfamily Platypodinae (Ambrosia Beetles and Pinhole Borers)

These beetles make breeding galleries in weakened trees and freshly felled logs, and the tunnel is stained by fungus reared for the larvae. Sapwood is most often damaged but tunnels may also penetrate the heartwood.

Most species appear to be polyphagous, but hardwoods are preferred as hosts. Important genera include:

- *Crossotarsus* spp. – on native trees; India.
- *Doliopygus* spp. – in logs of a wide range of forest trees (broad leaved) in East Africa.
- *Platypus* spp. – in dying trees and logs of a very wide range of trees; India, Africa and Europe. One species is common in oak in continental Europe and southern England.

5.5.8 Order Diptera (Flies)

This group is not really associated with forestry. There are some flies with larvae that feed in fruits (Tephritidae, Drosophilidae) or on rotting wood or bark (Stratiomyiidae). A few species of Asilidae have larvae that prey on other insect larvae that live in wood (*Xylocopa*, etc.). However, the main group of forest flies are the Tachinidae that parasitize caterpillars feeding on the tree foliage – several dozen species are recorded in most larger countries or regions.

Two groups of flies are regularly found associated with leaves, and many species are recorded with forest trees. The Cecidomyiidae are Gall Midges and make galls on the leaves of many trees (oak, lime, elm, willow, sycamore, etc.) including shoot galls. The Agromyzidae are leaf-miners and the larvae make either tunnel or blotch mines in the leaves of a wide range of woody plants. Both groups are, however, regarded as being very minor and unimportant pests of forestry, although of great interest ecologically.

5.5.9 Order Lepidoptera (Moths and Butterflies) (120 000 spp.)

Adult Lepidoptera cause no damage to trees; any such damage is done by the larvae (caterpillars) during their feeding activities.

Worldwide the main damage to trees by caterpillars is consumption of the leaves. Thus seedlings can be defoliated by a few caterpillars, and mature trees can be defoliated by a very large population. Caterpillars also cause damage to trees in other ways, as summarized below:

1. Cutworms and soil pests
   Cutworms are the larvae of certain Noctuidae, that live in and on the soil and eat plant roots and gnaw seedling stems. Swift Moth (Hepialidae) larvae also live in the soil and eat roots. Seed-beds can be devastated by these pests.

2. Leafworms
   This is a very broad and general term for the caterpillars that eat leaves, damaging the lamina in various ways. Dozens of families fall into this category. Small amounts of damage can be ignored, for most trees over-produce leaves as foliage grazing and browsing is a vital part of most terrestrial food webs. Many shrubs and trees can actually lose half of their leaf cover before a discernible loss of yield is recorded. Typically any serious leaf loss is quickly restored by compensatory growth. Thus out of the vast range of caterpillar species that feed on leaves of trees, only a few have the capacity for appearing in such large numbers that they can cause serious defoliation. The others are really only of academic interest most of the time. However, it must be remembered that any unsightly foliage damage can reduce the value of nursery stock and can be undesirable in specimen or amenity trees. The leafworms include:

   - Bagworms (Psychidae) (Fig. 5.273). This is a specialized group of leaf-eaters that live inside a protective case of silk and plant fragments. In the tropics, bagworm populations are recorded defoliating palms and various trees.
   - Leaf-miners. Many of the Microlepidoptera have caterpillars that live between the two epidermal leaf surfaces where they make tunnel or blotch mines, which cause leaf distortion, discoloration and mortality.
   - Leaf-folders/rollers. There are not many leaf-folders/rollers on forest trees, other than a few Pyralidae and some Tortricoidea. The leaves are fastened by silken threads produced by the mouths of the larvae.

3. Budworms (Tortricoidea)
   The larvae of many tortricids feed inside the buds of mostly herbaceous plants, but some trees are also attacked. The destruction of a growing point on a tree can be serious, but in a forest is seldom so provided the tree is mature. A somewhat similar result occurs from damage caused by shoot-boring caterpillars (see below).

4. Flower-feeders
   A few Pyralidae and some Tortricidae eat flowers of a number of tree species and will reduce seed-set.

5. Fruitworms
   Agriculturally, these species, which damage fruits and seed pods, are very important, but only a few cause damage to forest trees. Seed pods of leguminous trees are bored by some Tortricidae and Pyralidae.

6. Stem-borers
   - Shoot-borers. A well-known British species causing conspicuous damage (i.e. a six-inch dead pine shoot) is the Pine-shoot-borer (*Rhyaciona*) (Tortricidae (Fig. 5.274). Several species of Yponomeutidae kill shoots of juniper and *Thuja*, and some Pyralidae (*Hypsipylla*, etc.) bore shoots of mahogany and other forest trees in Africa and Asia.
   - Bark-borers/eaters (Metarbelidae, some Pyralidae); *Mussidia* spp. (Pyralide) bore the bark of mahogany and *Ficus* in Africa.
   - Woody trunk/branch-borers (Sesiidae, Cossidae, Metarbelidae). In addition to

Fig. 5.273 Bagworms (*Clania* sp. – Lep.; Psychidae) on *Thuja* twig; cases 25–30 mm long; Hong Kong.

Fig. 5.274 Pine-Shoot-borer damage (*Rhyaciona* sp. – Lep.; Tortricidae); Gibraltar Point, UK.

these three families, which are all borers in plant bodies, there are a few Pyralidae that bore into tree trunks and cause considerable damage. In Australia some Hepialidae bore woody roots and tree trunks.

The main groups of Lepidoptera that are encountered as forestry pests are listed below:

(a) Family Eriocraniidae (Leaf-miners)

A small group of tiny primitive moths whose larvae mine leaves of Angiosperms and the seeds of some Gymnosperms. The larvae of *Eriocrania* spp. mine leaves of *Betula* and oak in Europe.

(b) Family Hepialidae (Swift Moths) (3000 spp.)

A widespread and primitive group whose larvae either live in soil eating plant roots or (as in Australia) bore woody roots and tree trunks. The caterpillars are white with a brown head capsule, and as soil pests can be very damaging to seedlings and saplings whose roots are eaten. Some of the more damaging pests include:

- *Abantiades* spp. (11 spp.) – larvae in soil eat roots of *Eucalyptus* in Australia.
- *Aeratus* spp. (16 spp.) (Trunk-borers) – larvae bore trunks of *Eucalyptus*, *Acacia*, etc. in Australia.
- *Endoclita* spp. (Trunk-borers) – larvae bore trunks of many shrubs and trees; India and South East Asia.
- *Hepialus* spp. (Ghost Moths) – larvae in soil eat plant roots; can be severe nursery pests in Europe and Asia.
- *Palpifer* spp. (Tree-borers) – larvae attack various trees in Japan and East Asia.
- *Sthenopsis* spp. (Tree-borers) – larvae bore in trees of many species in the USA.

(c) Family Nepticulidae (Pygmy Moths)

Tiny moths whose larvae mine leaves. The main genus is *Stigmella* and different species are found on oak, beech, willow and birch. The group is worldwide; in the UK there are 70 species of *Stigmella*, mostly quite host-specific.

Several other families, collectively known as the 'Micro-lepidoptera', have larvae that mine leaves of temperate forest trees, but they are not normally of pest status.

(d) Family Psychidae (Bagworm Moths) (800 spp.)

Most bagworms live in the tropics, especially the large species, and some show preference for palms as hosts. Infestations are often large enough to cause tree defoliation – generally large numbers of eggs are laid (up to 2000 per female). Important pest species include:

- *Acanthopsyche* spp. – recorded defoliating *Acacia* and *Pinus* in East Africa and India; also on *Delonix* in Africa and Asia.
- *Clania* spp. – a large genus of polyphagous species recorded on a wide range of trees in Asia and Africa (Fig. 5.275).
- *Cryptothelea* spp. – polyphagous on many trees in Africa and Asia.
- *Eumeta* spp. – plantations of *Acacia*, *Eucalyptus* and *Cupressus* defoliated; also on *Ichinus*; East Africa.
- *Kotochalia junodi* (Wattle Bagworm) – major pest of *Acacia*, *Eucalyptus*, *Cupressus*, *Delonix*, etc.; Africa.
- *Oiketus* spp. – on many different trees in the USA.
- *Pteroma* spp. – on leguminous trees and others; India and South East Asia.

(e) Family Sesiidae (Clearwing Moths) (1000 spp.)

A large family, worldwide in distribution. Adults are characterized by having most of the wing without scales and a yellow/black

Fig. 5.275 Palm Tree Bagworm (*Clania* sp.– Lep.; Psychidae); case 25 mm; Hong Kong.

coloration that makes them resemble large wasps. Larvae bore in plant tissues, mostly in the trunks of temperate trees. The tree trunk can be so extensively bored that the tree dies. The main sign of infestation is a pile of frass at the base of the tree, followed by emergence holes from which the pupal exuvium protrudes. Adult moths will sit on the tree trunk on the morning of their emergence, often causing alarm because of their wasp/hornet-like appearance; adults are diurnal.

The main species attacking trees include:

- *Conopida* sp. (Camphor Clearwing) – larvae bore camphor trees; southern China.
- *Ramosia* spp. (Bark-borers) – larvae bore bark of sycamores and oak; USA.
- *Sesia* spp. (Hornet Clearwings) – larvae bore trunks of oak, poplar, willow, birch,

Fig. 5.276 Hornet Clearwing Moth (*Sesia apiformis* – Lep.; Sesiidae); ex poplar tree trunk; Cambridge, UK (body length 22 mm).

(a)

(b)

Fig. 5.277 (a) Section of poplar tree trunk showing Clearwing larval tunnels (*Sesia* sp. – Lep.; Sesiidae); Skegness, UK; (b) Hornet Clearwing larvae on poplar trunk; Skegness, UK.

etc.; Europe, Asia and the USA (Figs 5.276 and 5.277).

- *Synanthedon* spp. (Fruit Tree-borers, etc.) – a wide range of fruit and nut trees bored; Europe, Asia and the USA.

(f) Family Yponomeutidae (Small Ermine Moths, etc.) (800 spp.)

A small group, but some species may be locally very abundant. Quite a few are silvery little moths to be seen sitting on tree foliage. Some larvae are gregarious leaf-eaters, often living within an extensive silken web. Others are solitary shoot-borers. The more important pests include:

- *Argyresthia* spp. (Shoot Moths) – larvae bore shoots of juniper, *Cupressus*, *Thuja* and rowan; Europe and North America.
- *Prays fraxinella* (Ash Bud Moths) – larvae bore shoots of ash; Europe.
- *Yponomeuta* spp. (Small Ermine Moths) – gregarious larvae web foliage and defoliate many trees and shrubs; Europe, Asia, Australia, Africa and North America (Figs 5.278 and 5.279).

(g) Family Coleophoridae (Case-bearer Moths) (400 spp.)

Small, narrow-winged moths, whose larvae start as leaf-miners but then move into small portable cases made of silk. Most belong to the large genus *Coleophora*, and several species damage leaves of larch, birch, alder and hazel in Europe, Asia and North America.

(h) Family Gelechiidae (4000 spp.)
A large and widespread group with only a few forestry pests:

- *Anacampsis* spp. – larvae web leaves of birch, poplars and willows; Europe.
- *Dichomeris* spp. (Juniper Webworm, etc.) – larvae web foliage of junipers, oak and various fruit trees; USA and Europe.

(i) Family Cossidae (Goat/Leopard Moths; Carpenter-worms)

A worldwide group of large moths with tapering wings; nocturnal and fast-flying.

Fig. 5.278 Small Ermine Moth (*Yponomeuta* sp. – Lep.; Yponomeutidae); Skegness, UK.

Fig. 5.279 Larval web of Small Ermine Moth on laurel bush; Skegness, UK.

The larvae bore in plant stems – most in trunks of trees and branches – and in woody shrubs. Many are serious pests of forest trees, for the larval tunnelling can be very extensive; upon emergence of the adult, the pupal exuvium is left extruded. The tunnelling larva usually make a series of frass holes and sometimes there is copious sap exudation.

Some important pests include:

- *Cossus* spp. (Goat Moths) – polyphagous tree-borers in oak, willow, ash, birch, etc.; Europe, Asia and Africa.
- *Prionoxystus* spp. (Carpenter-worms) – larvae bore trunks of many hardwoods; USA and Canada.
- *Xyleutes* spp. – many species (69 in Australia alone), boring trunks of many trees and shrubs; Africa, Asia and Australasia (Fig. 5.280).
- *Zeuzera* spp. (Leopard Moths) – polyphagous tree/shrub stem-borers; Europe, Asia, Africa and North America (Fig. 5.281).

(j) Family Metarbelidae (Wood/Bark-borers)

A small group found in tropical Asia and Africa, where they can be locally very abundant. The larvae are polyphagous and feed on a very wide range of trees. The larvae feed on the bark of the tree at night, under a tubular web of silk and frass. Each larva has a refuge hole deep in the heartwood, usually at a fork in the trunk, where it spends the day and where it will pupate. Small trees may be girdled, and large trees may have 15–30 larvae per trunk – clearly harmful. The hosts appear to be healthy, living trees. The four main pest genera are:

- *Indarbela* – found on a wide range of host trees from tropical Asia to North East Africa (Figs 5.282 and 5.283).
- *Metarbela* – on many different forest trees; Africa.
- *Salagena* – also on many different woody shrubs and trees in tropical Africa.
- *Tetragra* – on trees in South Africa.

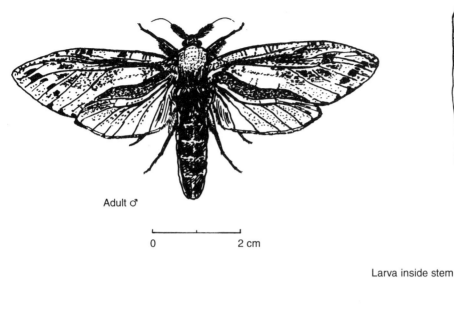

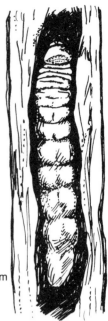

Adult ♂

0   2 cm

Larva inside stem

Fig. 5.280 Castor Stem-borer (*Xyleutes capensis* – Lep.; Cossidae); Kenya.

(k) Family Limacodidae (=Cochlididae) (Stinging and Slug Caterpillars)

A widespread tropical family found mostly on Palmae, which can be defoliated by the feeding larvae. Pupation usually takes place inside a hard, round cocoon stuck on to the tree trunk. One group of larvae is brightly coloured green and yellow, and is conspicuous with body protrusions (scoli) bearing sharp spines, often with urticating properties (i.e. 'stinging caterpillars'). The other group of larvae is slug-like with reduced legs and a smooth body – hence the common name of 'Slug Caterpillars' or 'Jelly Grubs'. It is essentially a tropical forest group, and some of the common genera include:

- *Cheromettia* (Gelatine Caterpillars) – polyphagous; South East Asia.
- *Parasa* (Stinging Caterpillars) – polyphagous on a wide range of shrubs and trees; tropical Asia and Africa (Fig. 5.284).
- *Setora* (Nettle Caterpillars) – polyphagous; South East Asia.
- *Thosea* (Slug and Nettle Caterpillars) – on herbaceous and woody hosts in Africa and tropical Asia to China.

(l) Family Tortricidae (Tortrix Moths) (4000 spp.)

A large, widespread and very important pest group, essentially (oak) forest insects and very damaging to fruit trees in temperate regions. They are also important forestry pests in the Holarctic region. Physically they are small in size, although the adults have broad wings. Some are polyphagous but others are quite host-specific: usually the leaf-eaters are in the former category and pod/fruit-borers in the latter. An important group of pests are known collectively as the

(a)

(b)

Fig. 5.281 (a) Leopard Moth/Red Coffee-borer (*Zeuzera coffeae* – Lep.; Cossidae) (body length 23 mm); and (b) pupal exuvium protruding from larval tunnel in tree branch; south China.

Fig. 5.282 Adult Wood-borer-moth (*Indarbela* sp. – Lep.; Metarbelidae); body length 20 mm; Hong Kong.

Fig. 5.283 Larval feeding tube of *Indarbela* sp. (Lep.; Metarbelidae) on *Acacia* tree trunk; south China.

(a)

(b)

Fig. 5.284 (a) Stinging Caterpillar (*Parasa lepida* – Lep.; Limacodidae) on leaf of camphor(?) and (b) pinned adult; Hong Kong.

'Fruit Tree Tortricids', but an equally important group are found on forest trees.

Because of this diversity, it is worth while looking at the group from the point of view of its habits.

1. Shoot/stem-borers
   - Pine-shoot Borers *Rhyaciona* and *Epiblema*.

2. Fruit-borers
   - *Cydia* and *Cryptophlebia*.
3. Seed pod-borers
   - *Cydia* and *Cryptophlebia*.
4. Bud-borers
   - *Cydia* (Fig Bud-borer); *Gretchena* (Pecan Bud-borer) and *Spilonota* (Bud Moths), etc.
5. Leaf-folders/tiers (larvae use silk to tie leaf edges together)/webbers
   - the majority of species are placed here.
6. Dead-leaf-eaters
   - *Clepsis* in Europe on hawthorn, etc.
   - Several species in Australia on *Eucalyptus* leaves.

Some of the more important forestry pests include:

- *Acleris* spp. (Spruce Budworms, etc.) – on spruce, larch, etc.; USA and Canada; also on oak, beech, sycamore, poplar, willow, etc. in Europe.
- *Adoxophyes* spp. – larvae web leaves and eat buds of birch, willow, alder, poplar and other trees; Europe, Asia and Africa.
- *Ancylis* spp. (Leaf-folders) – larvae on leaves of oak, beech, elm and birch; Europe.
- *Archips* spp. (Fruit Tree Tortricids, Oak Webworm, etc.) – larvae web foliage of many fruit trees and some forest trees, including *Pinus*, oak, birch, elm, alder, ash and lime; Europe, Asia, Africa and North America.
- *Argyrotaenia* spp. (Leaf-rollers; Pine Tube Moth) – on many host trees in Europe and Asia; common on *Pinus* in the USA and Canada.
- *Choristoneura* spp. – polyphagous leaf-rollers on fruit and forest trees in Europe and Asia; also Spruce Budworm in the USA, the most important conifer defoliator in North America.
- *Cnephasia* spp. (Ominivorous Leaf Tiers) – mostly pests of herbaceous plants, but sometimes damaging to young conifers;

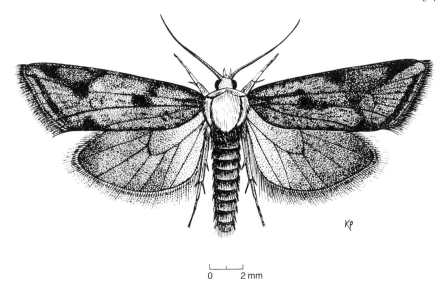

Fig. 5.285 Omnivorous Leaf Tier (*Cnephasia longana* – Lep.; Tortricidae) adult moth; Cambridge, UK.

Europe, Asia and parts of North America (Fig. 5.285).

- *Croesia* spp. (Leaf Tiers) – on maple, sycamore and trees of Rosaceae in Europe; two species are major pests of oak in the USA.
- *Cryptophlebia* spp. – polyphagous borers of fruits and nuts and seed pods of leguminous trees; Africa and Asia.
- *Cydia* spp. (Bud Moths, etc.) – many fruit pests; a few bore buds and cones of trees of *Ficus*, *Pinus*, etc., in Asia and parts of Europe.
- *Ditula* spp. – larvae web foliage of many different shrubs and trees, including beech, juniper, larch, pine and yew; Europe and North America.
- *Epinotia* spp. – larvae roll leaves of birch, *Salix*, alder and hazel; Europe.
- *Hedya* spp. – recorded on ash, alder and *Sorbus*; East Asia and North America.
- *Olethreutes* spp. – several species attack plants in different ways; the trees attacked include birch, larch, *Salix* and spruces; Europe and North America.
- *Orthotaenia undulana* (Leaf Tier) – a common pest of birch, alder, elm, juniper, *Acer*, *Salix* and *Pinus*; Europe and North America.
- *Pandemis* spp. – polyphagous leaf-rollers on alder, birch, elm, lime, oak, *Acer*, *Salix*, etc.; Europe, Asia and North America.
- *Petrova resinella* (Pine Resin-gall Moth) – larva makes a gall in the shoot of *Pinus*; Europe.
- *Ptycholoma* spp. – oak forest species in Europe and Asia recorded on oak, *Acer*, *Salix*, poplars and various conifers.
- *Rhyacionia* spp. (Pine-shoot Moths) – larvae bore shoots of *Pinus*; Europe, Asia and North America (Fig. 5.274).
- *Spilonota* spp. (42 spp.) (Bud Moths/Leaf Tiers, etc.) – larvae eat leaves and bore buds of oaks, alder, *Larix*, hazel, etc.; some are leaf-tiers on *Eucalyptus* in Australia; found also in Europe, Asia and North America.

(m) Family Pyralidae (Snout Moths)

A large and very important group agriculturally, but only a few are forest pests. A few species have larvae that eat leaves and a very few have larvae that bore in the trunks and branches of trees. These include:

- *Hypsipylla* spp. (Shoot Moths) – larvae bore shoots of forest trees; Africa, India, South East Asia and Australia. *H. robusta* is the Mahogany-shoot Borer of Africa. The Mahogany-shoot Borer of Brazil is a serious pest in that it causes shoot-tip split and prevents proper trunk formation – the split trunk is useless commercially.
- *Mussidia* spp. – some larvae bore pods of legume trees, others tunnel bark of mahogany trees and *Ficus* in Africa.

(n) Superfamily Papilionoidea (Butterflies)

These are not generally forest insects, although a few (Satyrinae) browns live on forest grasses and others on the foliage of bamboos. Some of the Swallowtails (Papilionidae) feed on tree foliage or on the foliage of some vines, but they are seldom regarded as pests. Several Skippers (Hespiriidae) are common on the foliage of palms in tropica Asia and Africa.

(o) Family Geometridae (Carpets/Pugs; Loopers/Cankerworms) (12 000 spp.)

A large group, basically forest insects, which can be quite dominating ecologically with many small genera. The larvae have a reduced number of prolegs and a long body, and they walk by 'looping'. They are leaf-eaters, cryptic and mostly nocturnal, usually occurring singly in the foliage. Although many species will be found on leaves of forest trees, very few are to be regarded as important forestry pests. The adults are mostly small and slender with large wings, held horizontally at rest. An interesting group in the forests of northern temperate regions is termed 'Winter Moths'. Here, the non-feeding adults emerge over the winter period and the wingless females lay

eggs on the shoots of trees; these hatch at bud-burst in the spring and the larvae feed on the young leaves.

A selection of the more important pest species includes:

- *Alsophila* spp. (March Moths; Fall Cankerworms) – on deciduous trees in Europe, Asia and North America (Fig. 5.286).
- *Ascotis selenaria* (Giant Looper) – polyphagous on a range of deciduous trees and shrubs; Africa and Asia (Fig. 5.287).
- *Biston* spp. (Peppered Moth, etc.) – defoliate temperate deciduous trees; Europe, Asia and North America.
- *Bupalus piniaria* (Pine Looper Moth) – a common and widespread pest of *Pinus* in the UK; damaging to forests and especially to conifers in nurseries. Several conifers are attacked, including *Larix*, *Abies*, *Picea* and *Pseudotsuga*; widespread in Europe.
- *Erannis* spp. (Winter Moths, etc.) – on a wide range of deciduous trees; Europe, Asia, the USA and Canada.
- *Hyposidra* spp. – on a wide range of shrubs and trees; Africa, Asia and Australia.
- *Neocleora* spp. – on *Eucalyptus* and other trees; Africa.
- *Operophtera* spp. (Winter Moths) – on fruit, nut and deciduous forest trees; Europe, Asia, the USA and Canada (Fig. 5.288).

In the UK, several dozen species are recorded as forestry pests, each usually in a different genus; for good colour illustrations see Alford (1991).

(p) Family Lasiocampidae (Eggars, Lappets; Tent Caterpillars) (1000 spp.)

These stout-bodied, large, bristly moths have large bristly caterpillars that eat the foliage of a wide range of shrubs and trees. Their large size and protective urticating bristles make them quite serious forest defoliators, especially the gregarious species. Females of some species lay up to 2000 eggs.

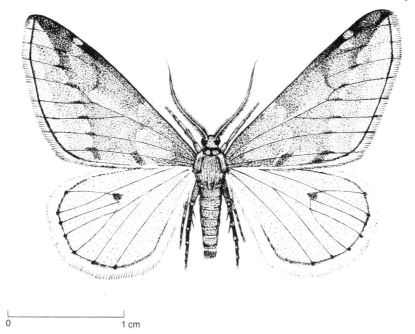

0   1 cm

Fig. 5.286 March Moth (*Alsophila aescularia* – Lep.; Geometridae) adult male; Cambridge, UK.

Fig. 5.287 Giant Looper Caterpillar (probably *Ascotis selenaria* – Lep.; Geometridae) (body length 48 mm); Hong Kong.

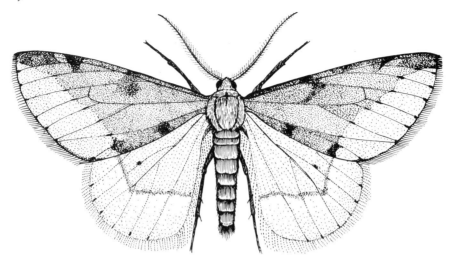

Fig. 5.288 Winter Moth, adult male (*Operophtera brumata* – Lep.; Geometridae); Cambridge, UK.

Some of the more important forestry pests include:

- *Bombycopsis* spp. – larvae on *Eucalyptus* and *Pinus*; East Africa.
- *Dendrolemus* spp. (Pine/Oak Tent Caterpillars) – on many different forest trees, both evergreen and deciduous; Europe, Asia and North America.
- *Gnometa* spp. – recorded defoliating *Pinus*, *Cupressus* and *Juniperus* in East Africa.
- *Malacosoma* spp. (8–10 spp.) (Tent Caterpillars) – larvae polyphagous on broad-leaved forest and fruit trees; throughout Europe, Asia and North America.
- *Nadiasa* spp. – on a wide range of trees in East Africa, including *Sapium*, *Pinus*, *Maesopsis* and *Chlorophora*.
- *Pachypasa* spp. – larvae defoliate *Acacia*, *Pinus*, *Eucalyptus*, *Celtis*, etc., in East and South Africa.

(q) Family Saturniidae (Emperor Moths; Giant Silkmoths)

A large family of large forest moths best developed in the tropics. In wooded countries such as central Uganda, these large moths (six- to eight-inch wingspan) can be very conspicuous around lights at night. In the morning they can be found sitting on nearby bushes and trees. The larvae of many species produce silk for the construction of the pupal cocoon: pupation of some of the more conspicuous and better known species takes place in a silken cocoon attached to a twig or leaves, although many forest species pupate in leaf litter or in the soil. Isolated trees are sometimes defoliated by these huge caterpillars (they are solitary but there may be a couple of dozen scattered about the foliage of a small tree). They are not regarded as serious forest defoliators but a few species are viewed as potentially dangerous to plantations.

The main species frequently encountered include:

- *Actias* spp. (Moon Moths) – recorded on a wide range of tree hosts, including *Betula*, *Moringa*, camphor, tallow and liquidambar; India, South East Asia and China (Fig. 5.195).
- *Anisota* spp. (4+ spp.) (Oakworms, etc.) – on oaks and a wide range of forest trees in the USA and Canada.

Fig. 5.289 Giant Emperor Moth (*Bunaea alcinoe* – Lep.; Saturniidae); wingspan 165 mm; Alemaya, Ethiopia.

- *Antheraea* spp. (Oak Silkworms, etc.) – destructive to the foliage of many trees, including oaks and walnut in India, China and Japan; also *Eucalyptus* in Australia.
- *Attacus* spp. (6 spp.) (Atlas Moths) – these magnificent moths have huge larvae covered with white wax. They have been recorded on many different woody hosts including camphor, cinchona and walnut; pantropical (Fig. 5.196).
- *Bunaea* spp. – on *Sapium* and other trees; East Africa (Fig. 5.289).
- *Caligula* spp. – on a wide range of trees; North India through South East Asia to Indonesia.
- *Epiphora* spp. – *Maesopsis* is defoliated and other trees attacked; East Africa.
- *Eriogyna pyretorum* (Giant Silkmoth) – in South East Asia the larvae eat foliage of camphor trees (Fig. 5.290).
- *Imbrasia* spp. – on *Acacia*, tamarind, *Maesopsis*, etc., sometimes causing defoliation; East and West Africa.
- *Lobobunea* spp. – on *Acacia*, *Eucalyptus*, *Celtis*, etc.; East Africa.
- *Nudaurelia* spp. (Silkworms, etc.) – on tung, *Ficus*, *Acacia*, *Eucalyptus*, *Chlorophor*, *Pinus*, *Schinus* and other trees; East and South Africa.

(r) Family Sphingidae (Hawk Moths; Hornworms) (1000 spp.)

Another group of large, conspicuous, nocturnal moths with large and spectacular caterpillars. Mostly found on shrubs and herbaceous plants; a few occur on trees but are most damaging to nursery stock and saplings. Pupation takes place in an earthen cell in the soil. The larvae are solitary, but a bush or small tree (*Gardenia*, for example) can be defoliated by a couple of dozen caterpillars. The group includes:

- *Acherontia* and *Hyles* – polyphagous larvae that occasionally use small trees as hosts, worldwide.
- *Hyloicus pinastri* (Pine Hawk Moth) – larvae eat pine needles; found throughout Europe and temperate Asia (Fig. 5.291).
- *Laothoe populi* (Poplar Hawk Moth) – on *Populus* and *Salix*, and occasionally on ash and birch; western Palaearctic.

(a)

(b)

Fig. 5.290 (a) Giant Silkmoth (*Eriogyna pyretorum* – Lep.; Saturniidae), wingspan 110 mm; and (b) caterpillar on foliage of camphor; Hong Kong.

Fig. 5.291 Pine Hawk Moth (*Hyloicus pinastri* – Lep.; Sphingidae); wingspan 85 mm; Cambridge, UK.

Fig. 5.292 Eyed Hawk Moth (*Smerinthus ocellata* – Lep.; Sphingidae); wingspan 80 mm; Cambridge, UK.

- *Mimas tiliae* (Lime Hawk Moth) – a polyphagous species found on lime, alder, ash, birch, elm, oak and walnut; Europe and Asia.
- *Smerinthus* spp. (Fruit Hornworms, etc.) – on a wide range of fruit and forest trees, especially *Salix*; Holarctic (Fig. 5.292).
- *Sphinx* spp. – on many fruit trees and some forest trees including ash; Europe, Asia and North America.

(s) Family Notodontidae (Prominents, etc.; Processionary Caterpillars)

Stout-bodied nocturnal moths of moderate size, widely distributed but scarce in Australasia. Larvae may be solitary, gregarious or highly gregarious and processionary, and most feed on leaves of shrubs and trees at night. On forest trees (even in plantations) they are seldom of importance, but can be damaging to isolated trees, specimen trees, ornamentals and nursery stock.

Some species of note include:

- *Cerura* spp. (Puss Moths, etc.) – on *Populus* in East Africa and *Salix* and birch in Europe.
- *Datana* spp. (Walnut Caterpillar, etc.) – on walnut, oak, elm, chestnut, beech, lime, etc., and fruit trees; USA and Canada.
- *Desmeocraera* spp. – on *Eucalyptus*, *Acacia*, guava, etc.; East and West Africa.
- *Phalera* spp. (Bufftip Moths, etc.) – *P. bucephala* is a common gregarious pest of broad-leaved trees and shrubs in the UK and continental Europe, especially on beech, birch, chestnut, elm, lime, hazel, oak and *Salix*. Other species occur in India and East Asia.
- *Thaumetopoea* spp. (processionary caterpillars) – gregarious pests of *Pinus* in the Mediterranean region and other trees in Africa.

(t) Family Arctiidae (Tiger/Ermine Moths; Woolly Bears) (10 000 spp.)

A large and widespread group of small to medium-sized moths of distinctive appearance. Most larvae are very bristly with long urticating setae. The majority of the group feed on low-lying herbaceous plants, but some use shrubs and trees as hosts – nursery stock and saplings are most likely to be attacked. The few forest pest species include:

- *Amsacta* spp. (Woolly Bears) – some shrubs

and trees damaged; India and South East Asia to Japan.
- *Arctia* spp. (Tiger Moths) – in temperate regions of Europe and Asia several species are common. They sometimes eat the leaves of young deciduous trees (oak, alder, birch, poplar, *Salix*, hazel, etc.).
- *Diacrisia* spp. (Tiger Moths/Buffs) – larvae are polyphagous and have been recorded on trees (*Sapium*, *Erythrina*, etc.); Africa and South Asia.
- *Hyphantria cunea* (Fall Webworm) – a major pest of North America which spread to continental Europe in 1940; attacks deciduous trees. In the USA Fall Webworm is found on at least 88 species of tree; in Europe 230 host plants (including herbs) and in Japan 317 plant species.

(u) Family Noctuidae (=Agrotidae) (25 000 spp.)

A large family of medium-sized nocturnal moths, probably the most important as pests of cultivated plants, although less important in forestry. Seedlings and saplings are susceptible to cutworms in the soil which may eat roots or cut soft stems. A few caterpillars will eat foliage of trees. Cutworm pests include the following:

- *Agrotis* spp. (10–12 spp.) (Cutworms) – worldwide but most abundant in the Holarctic region; several serious pest species (Fig. 5.293).
- *Euxoa* spp. (14+ spp.) – a large genus, worldwide in cooler regions; most of the cutworms occur in the USA and Canada.
- *Noctua pronuba* (Large Yellow Underwing) – polyphagous; Palaearctic.
- *Xestia* spp. (Spotted Cutworms) – many polyphagous; worldwide.

The following species will attack foliage:

- *Acronicta* spp. (6+ spp.) (Dagger Moths) – a large Holarctic genus with many pest

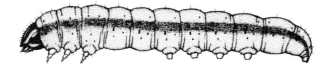

Larva

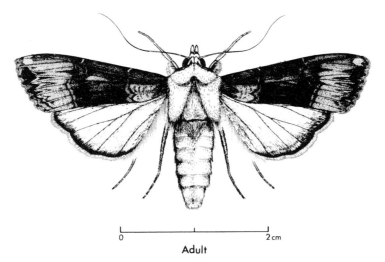

Adult

Fig. 5.293 Black (Greasy) Cutworm (*Agrotis ipsilon* – Lep.; Noctuidae); Cambridge (MAFF), UK.

species attacking many deciduous trees (but not oaks).

- *Carea* spp. – larvae found on *Eugenia* and *Eucalyptus*; South East Asia to China.
- *Cosmia trapezina* (Dun-bar Moth) – on oak, ash, birch, elm, poplar, *Acer*, *Salix* and *Sorbus*, etc.; Europe.
- *Lithophane* spp. (Green Fruitworm, etc.) – one species is found in the Mediterranean on juniper and *Cupressus*; another in the USA on a wide range of deciduous trees and shrubs.
- *Mamestra* spp. (Cabbage Moths, etc.) – young trees are attacked (oak, birch, larch, *Salix*, *Prunus*, etc.); in Europe, Asia and North America.
- *Melanchra* spp. – these occasionally attack trees (such as birch, larch, *Salix* and alder); Europe, Asia and North America.

- *Orthosia* spp. (Drab Moths, etc.) – on temperate fruit and forest trees (oak, elm, *Salix*, beech, birch, poplar, etc.); Europe and Asia.
- *Scoliopterys libatrix* (Herald Moth) – a common pest on *Populus* and *Salix* throughout Europe; young trees can be defoliated.

(v) Family Lymantriidae (Tussock Moths) (2500 spp.)

Medium-sized moths, some females are flightless; larvae very bristly, some with urticating bristles. Most larvae are polyphagous and feed on the foliage of woody hosts; some are pests of forest trees. Larvae may be gregarious and use silken tents for shelter when not feeding.

The more important pests of forestry include:

- *Calliteara pudibunda* (Pale Tussock Moth) – widespread in the UK on many trees and shrubs (as well as hop) including beech, birch, elm, poplar, hazel, oak, willow and walnut.
- *Dasychira* spp. – polyphagous on a wide range of plants in the Old World; 400 spp. have been recorded.
- *Euproctis* spp. – 600 species in the Old World, many of which are polyphagous. Important pests include Yellow-tail and Brown-tail Moths on many different deciduous trees and shrubs; some have now been introduced into North America.
- *Leucoma salicis* (White Satin Moth) – on poplars, willows and also birch in Europe and Asia, now introduced into North America.
- *Lymantria* – 150 species in the Old World, mostly on forest trees. *L. dispar* (the Gypsy Moth) is a famous forestry and fruit tree pest: native to North East Asia, it spread to Europe where it caused much damage; in 1870 it was taken to the USA where it escaped and subsequently became the major pest of deciduous trees. It prefers alder, apple, hawthorn, oak, poplar and willow, but will also attack elm, *Acer*, etc. and some conifers. *L. monacha* (the Nun Moth) is also found on deciduous forest trees in Europe (oak, etc.) but prefers conifers (firs, *Pinus*, spruce). In 1994 it caused concern in Poland for thousands of hectares of pine forest have been defoliated.
- *Orgyia* spp. (60+ spp.) – Holarctic genus of polyphagous pests, with some on both deciduous trees and conifers (Fig. 5.294).

5.5.10 Order Hymenoptera (Sawflies, Ants, Bees, Wasps) (100 000 spp.)

A very large group of what are now regarded as being the most advanced insects. There is great specialization and diversity within the order, both in anatomy and in lifestyles. Most of the parasitic and predatory forms are only indirectly of importance to forestry as natural enemies of the phytophagous pest species. However, in a few cases this natural (biological) control is very important, for major defoliating pests are being kept in check.

The main pests of forestry are the sawflies, which reach their greatest level of development in northern temperate regions. They occur in both deciduous and coniferous forests, and the gregarious species can defoliate whole swaths of forest and plantation. Not only is the taiga conifer forest inhabited by sawflies but some species are even sub-arctic, feeding on conifer needles.

Suborder Symphyta (Sawflies)

The main ways in which sawflies are forest pests are as follows:

1. Leaf lamina-eaters
   – solitary; most Diprionidae and Tenthredinidae.
   – gregarious; some Diprionidae, *Hemichroa* and *Croesus*.
2. Leaf-skeletonizers
   – *Caliroa* and *Endelomyia* (Caliroini).
3. Leaf-rollers
   – Pamphiliidae.
4. Leaf-miners
   – *Fenusa*, *Metallus*, *Schizocerella*, etc.
5. Leaf gall-makers
   – *Pontania* on *Salix* leaves.
6. Shoot- and stem-borers
   – some Cephidae (*Janus*); *Ardis*.
7. Fruit- or flower-eaters
   – *Hoplocampa* (Tenthredinidae); Xyelidae.
8. Wood-borers
   – deciduous trees – *Tremex* (Siricidae).
   – evergreens – many Siricidae.

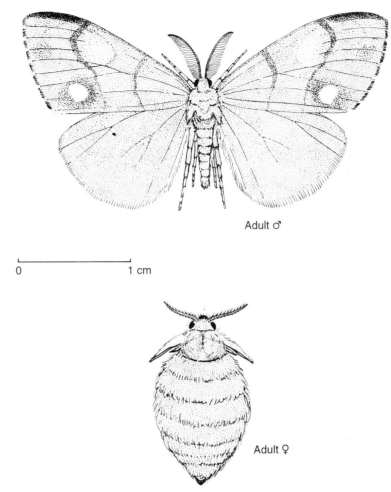

Adult ♂

0   1 cm

Adult ♀

Fig. 5.294 Vapourer Moth, male and female (*Orgyia antiqua* – Lep.; Lymantriidae); Cambridge, UK.

The main pest groups in respect of forestry include:

(a) Family Pamphiliidae (Leaf-rolling/ Web-spinning Sawflies)

Larvae live gregariously in silken webs or in rolled leaves. Pests include:

- *Acantholyda* spp. – on *Pinus*, *Picea*, *Abies* and *Larix*; Asia, North America.
- *Pamphilus* spp. – larvae in rolled leaves of birch, oak, *Salix* and aspen; Europe, Asia and North America.

(b) Family Siricidae (Wood Wasps; Horntails)

These large conspicuous insects have larvae that tunnel in tree trunks – sickly trees are preferred as hosts and serious damage can result. Most are found in northern temperate forests and many attack conifers, but others are found in deciduous trees. They include:

- *Sirex* spp. (4+ spp.) (Blue Wood Wasps) – larvae bore in *Pinus*, spruce and other conifers; Europe, Asia and North America. *S. noctilio* is now established in Australia

(a)   (b)

Fig. 5.295 Giant Wood Wasp; (a) adult female and (b) adult male of *Urocerus gigas*; Cambridge, UK.

and Tasmania and causes serious damage to *Pinus* and other important conifers.

- *Tremex* spp. – larvae bore in a wide range of deciduous trees; Europe, Asia and North America.
- *Urocerus* spp. (Giant Wood Wasps; Giant Horntails) – larvae bore in *Pinus* and firs; Europe, Asia, the USA and Canada (Fig. 5.295).

(c) Family Cephidae (Stem Sawflies)

Most bore inside stems of Gramineae (*Cephus*, etc.). Some *Janus* spp. bore in stems of *Salix* and some other trees in Asia and the USA.

(d) Family Argidae (Rose/Birch Sawflies, etc.) (800 spp.)

Many of these sawflies have distinctively spotted larvae. They are solitary leaf-eaters. They include *Arge* spp. (Birch Sawflies, etc.), the larvae of which eat leaves of birch and *Salix* in Europe, Asia and North America.

(e) Family Cimbicidae

Stout-bodied, largish sawflies with distinctly clubbed antennae. The larvae are solitary and eat the leaf lamina; common but usually minor pests. The group includes *Cimbex* spp. (Birch Sawflies), the larvae of which feed on birch, *Salix*, elm, poplar, alder, lime, etc. in Europe, Asia and North America.

(f) Family Diprionidae (Conifer Sawflies)

Stout-bodied sawflies; females have serrate antennae and males pectinate. Larvae feed gregariously on needles and foliage of conifers in northern temperate regions. Several serious pests are European, but have been accidentally introduced into North America, where spectacular population outbreaks have resulted in devastating defoliation of vast tracts of coniferous forest. The main forest pests are:

- *Diprion* spp. (Pine Sawflies) – larvae feed entirely on *Pinus* needles and shoots; Europe, Asia and North America.
- *Gilpinia* spp. (Spruce Sawflies) – mostly

on spruces; Europe, Asia and North America.

- *Monoctenus* spp. (Cupressus Sawflies) – larvae feed on *Cupressus*; Europe, Asia and North America.
- *Neodiprion* spp. (15+ spp.) (Pine Sawflies, etc.) – on *Pinus*, *Picea*, etc.; Europe, Asia and North America.

(g) Family Pergidae

A large group with diverse habits but best developed in Australia and South America. A few species are pests of trees – the larvae feed gregariously on the foliage. They include:

- *Acordulecera* spp. – larvae on leaves of oaks and hickory; USA and Central America.
- *Perga* spp. (Eucalyptus Sawflies) – larvae eat leaves of *Eucalyptus*, sometimes defoliating; Australasia.
- *Phylacteophaga* spp. (Leaf-miners) – larvae make blotch leaf mines on *Eucalyptus* and *Tristania*; Australasia.

(h) Family Tenthredinidae (typical Sawflies) (4000 spp.)

This is the main family of sawflies and there is great diversity shown in terms of habit and lifestyle, although the great majority are found in cooler temperate regions. The female uses her saw-like ovipositor to deposit eggs into the plant tissues (usually leaves). Some of the pests of forest trees include:

- *Allantus* spp. (Cherry/Rose Sawflies, etc.) – larvae eat leaves and sometimes tunnel into the pith of shoots; some species on *Rosa* and others on *Quercus*; Europe, Japan and the USA.
- *Apethymus* spp. – some species on oak leaves in Europe; another on chestnut in Japan.
- *Caliroa* spp. (Slug Sawflies) – larvae skeletonize leaves of oak, lime, beech, birch, *Salix* and Rosaceae; Europe, Asia and the USA.

- *Croesus* spp. (Hazel Sawflies, etc.) – gregarious larvae eat leaves of hazel, birch, alder, *Salix*, poplar and rowan; Europe, Asia and the USA (Fig. 5.296).
- *Dineura* spp. – solitary larvae on leaves of birch, rowan and hawthorn; Europe.
- *Fenusa* spp. (Birch/Elm Leaf-mining Sawflies) – larvae mine leaves of birch, elm and alder; Europe, Asia and the USA.
- *Hemichroa* spp. (Alder/Camphor Sawflies) – gregarious larvae eat leaves of camphor in China and alder in Europe and North America.
- *Heterarthrus* spp. (Alder/Acer Leaf-miners, etc.) – larvae mine leaves of alder, *Salix*, sycamore and field maple; Europe.
- *Nematus* spp. (40+ spp.) (Poplar Sawflies, etc.) – gregarious larvae on poplars, birch, *Salix* and alder; Europe and the USA.
- *Periclista* spp. – larvae (spiny) eat leaves of oak; Europe.
- *Pontania* spp. (Willow Gall Sawflies) – red galls on leaves of *Salix*; Europe and the USA (Fig. 5.297).
- *Pristiphora* spp. (Birch/Larch Sawflies, etc.) – larvae eat leaves of both deciduous trees and conifers; Europe, the USA and Canada.

Suborder Apocrita

These are the parasitic and stinging forms; the majority of species are placed here. The Parasitica (Ichneumonoidea and Chalcidoidea) are very important as parasites controlling the pest populations of insects. However, only the few phytophagous forest pest species will be mentioned here.

(a) Family Cynipidae (Gall Wasps) (2000 spp.)

A very interesting family which have very complex life cycles. Most make galls on leaves and twigs of oaks (*Quercus* spp.), other Fagaceae and *Rosa*, and a few on herbaceous Compositae. Alternation of generations and distinctive polymorphism are shown by many species. Some of the more

Fig. 5.296 Willow/Hazel Sawfly (*Croesus septentrionalis* – Hym.; Tenthredinidae); (a) gregarious larvae eat *Salix* leaves; Gibraltar Point, UK.; (b) young larvae skeletonize leaves.

conspicuous and important species to be found on oak trees include:

- *Andricus fecundator* (Artichoke Gall Wasp) – the gall is a swollen shoot/bud; Europe and Asia.

- *Andricus kollari* (Marble Gall Wasp) – spherical galls on twigs of *Q. robur* for overwintering agamic generation; Europe and Asia (Fig. 5.298).

- *Andricus quercuscalis* (Acorn Gall) – recently introduced into the UK from North

Fig. 5.297 Willow Bean-gall Sawfly (*Pontania proxima* – Hym.; Tenthredinidae) on *Salix*; Gibraltar Point, UK.

America; a major pest for it destroys the acorn (Fig. 5.299).

- *Biorrhiza pallida* (Oak Apple) – the gall is a swollen shoot; Europe.
- *Cynips* spp. – many different galls on leaves of *Quercus* spp.; Europe.
- *Neuroterus* spp. (Spangle Galls, etc.) – the asexual generation makes small round galls on the undersides of oak leaves; Europe, Asia and North America (Fig. 5.300).

In North America some of the commonest and most distinctive species include *Dryocosmus*, *Callirhytis*, *Amphibolips* and *Andricus*. These produce leaf and twig galls, some of which are similar to those found in the UK and others which are quite different.

The genus *Quercus* is, of course, Holarctic, and represented by a number of species on

Fig. 5.298 Marble Gall on oak (*Andricus kollari* – Hym.; Cynipidae) showing adult emergence hole; Alford, UK.

Fig. 5.299 Acorn Gall (*Andricus quercuscalis* – Hym.; Cynipidae); Alford, UK.

Fig. 5.300 Spangle Galls on underside of oak leaf (*Neuroterus quercusbaccarum* – Hym.; Cynipidae); Alford, UK.

both continents. These are closely related forms such as *Lithocarpus* in Asia which also have galls, so the worldwide Cynipid/oak tree complex of galls is very extensive. Most species are, however, of little economic importance, even though they are biologically fascinating.

(b) Family Agaonidae (Fig Wasps) (1500 spp.)

An unusual group that live in a state of obligatory mutualism with *Ficus* trees (Moraceae) and are responsible for flower pollination in the syconia (figs). Without these insects there would be no sexual reproduction of *Ficus* and the tropical forests would be the poorer as many species could not exist. An introductory account of these insects is given in Hill (1967).

(c) Family Torymidae (Seed Chalcids, etc.) (1500 spp.)

A diverse group, many of which are phytophagous and feed on the seeds of many plants. The genus *Megastigmus* (Seed Chalcids) has some 12 species that feed on the seeds of conifer cones and can be very damaging to seed crops – up to 15% of seed crops may be destroyed. They are found in Europe, Asia, the USA and Canada.

(d) Family Formicidae (Ants) (15 000 spp.)

Tropical forests abound with ants and there are some in temperate forests. Much of the time it is not clear whether they are harmful or beneficial, and in point of fact some come into both categories. In many forests ants are an important element of the insect fauna. The direct pests are the phytophagous sap-feeders that will bite shoots and seedlings to suck the sap. Some of these will have nests in the ground and climb the tree to reach the shoots and soft tissues; others will have a carton nest in the tree foliage (Fig. 5.301), and their feeding can kill the shoots – the genus

Crematogaster (1000 spp.) makes such nests. Many of these species are indirect pests in that they 'farm' aphids and Coccoidea on the tree foliage for the honey-dew they excrete. In Ethiopia, ants of the genus *Dorylus* (sp. nr *brevinodosus*) often ring-bark young trees (up to two to three inches diameter) at ground level to drink the sap and the trees usually die. The Leaf-cutting Ants of the tropical New World (Attini) are forest pests directly in that they cut pieces of leaf lamina in order to make fungus gardens on which they feed. They are found throughout Central and most of South America, and in the southern states of the USA. There are some 18 or more species of *Atta* and many species of *Acromyrmex*.

Some species are arboreal nesters in trees, palms and shrubs, and they may be aggressive species that attack forest-workers. Other species have a reputation for being aggressive but in fact are not very fierce at all; however, their presence in the foliage may be a serious deterrent to native labourers.

(e) Family Vespidae (Social Wasps) (800 spp.)

As with the aggressive ants, the nuisance value of social wasps with their arboreal nests is their readiness to sting human interlopers. In established forests in the tropics, the wasps' nests are usually high in the canopy and would only be disturbed by felling. With smaller trees, however, the nests can be quite low and easily disturbed accidentally. Nests of *Vespa* can be made at ground level in litter and low herbage and some species have underground nests. The two main genera worldwide are:

- *Polistes* (Paper Wasps) – many species, pantropical; build a single comb hanging from a petiole in foliage; most colonies number less than a hundred.
- *Vespa/Vespula* – tropical and temperate wasps; nests either hang from tree foliage or in cavities; most colonies contain up to

Fig. 5.301 Carton nest of Black Tree Ant (*Polyrachis dives* – Hym.; Formicidae); nest diameter 15 cm; Hong Kong.

20 000 individuals. They can be very aggressive in defence of the nest.

(f) Family Megachilidae (Leaf-cutter Bees)

These bees cut circular pieces of leaf material with which to construct their small subterranean nest cells. The genus *Megachile* is worldwide, and its leaf damage is unsightly on nursery stock for sale and on specimen or ornamental trees (Fig. 5.302).

(g) Family Xylocopidae (Carpenter Bees)

These large stout-bodied bees are found throughout the tropics and also in some warmer temperate regions such as the south of France and southern USA. *Xylocopa* species make breeding tunnels up to 10–12 mm diameter and 20–30 cm long in dead wood and building timber and also in thick bamboo stems (Fig. 5.303).

5.5.11 Class Arachnida, Order Acarina (Mites)

The main characteristics of the Arachnida are that the adults have four pairs of walking legs and no wings, and the body is not divided clearly into head, thorax and abdomen. The mouthparts consist of paired chelicerae modified into stylets for rasping the leaf surfaces so that exuding cell sap can be sucked up.

Fig. 5.302 Leaf-cutter Bee damage to Bougainvillea foliage (*Megachile* sp. – Hym.; Megachilidae); Bird Island, Seychelles.

(a) Family Eriophyidae (Gall/Rust/Blister Mites)

These are minute mites, only visible under a hand lens, with reduced legs and an elongate tapering body. Most species induce erinea or galls on the plant host, usually of characteristic shape and colour. (An erinium is a warty outgrowth of dense hair-like structures on the damaged lower epidermis which on the upper surface appears as a pale raised area. The mites live between the hair-like outgrowths.) Some species are known as 'Bud Mites' and they inhabit buds on woody hosts. Generic names are something of a problem and many changes have been made in recent years. The group includes:

- *Aceria* spp. (Bud Mites) – attack a range of woody hosts; worldwide.
- *Aculus* spp. (Elm/Willow Leaf Gall Mites, etc.) – on ash, elm, *Prunus* and *Salix*.
- *Artacris* spp. (Sycamore/Maple Leaf Gall Mites) – small, unilocular red galls made on upper leaf surface; Europe (Fig. 5.304). Similar galls on maple in the USA are attributed to *Vasates* spp. Nine different Gall Mites attack *Acer* in Europe, and about 20 in the USA – mostly different species of mite on the different maple species.
- *Cecidophyopsis* spp. (Big-bud Mites) – on yew, *Ribes* and other woody hosts; Europe.
- *Eriophyes* spp. (Bud/Blister Mites) – a wide range of species on a wide range of tree hosts, some causing erinea (Fig. 5.305), sometimes so many that leaves are totally distorted (Fig. 5.306).
- *Phyllocoptella avellaneae* (Filbert Big-bud Mite) – swollen, unopened buds die; Europe, Asia, Australia and North America (Fig. 5.213).
- *Vasates* spp. (Gall Mites) – on several tree species, especially maples in the USA.

(a)

(b)

Fig. 5.303 (a) Bamboo Carpenter Bee (*Xylocopa iridipennis* – Hym.; Xylocopidae); body length 26 mm; south China; and (b) a bored bamboo stem.

Some Old World species are now placed in other genera.

(b) Family Tetranychidae (Spider Mites)

A large group, worldwide in distribution, on a wide range of hosts. The adults are very small, but can just be seen with the unaided eye, and with long legs look rather spider-like. Their feeding produces a surface scarification of the leaves, which can be killed, and they sometimes make a silken web over the foliage. On mature trees, damage is negligible, but on ornamental trees it can be unsightly; nursery stock can suffer set-back.

The main pests are included in the following genera:

- *Bryobia* – worldwide; most species polyphagous on a wide range of hosts.
- *Eotetranychus* spp. – worldwide, but most pest species are on trees (in small colonies underneath the leaves) in the USA on

Fig. 5.304 Sycamore Leaf Gall Mite (*Artacris macrorhynchus* – Acarina; Eriophyidae); Gibraltar Point, UK.

Fig. 5.305 Gall Mite erinea (*Eriophyes* sp. – Acarina; Eriophyidae) on leaves of *Celtis*; Alemaya, Ethiopia.

Fig. 5.306 Celtis leaves totally deformed by Gall Mite infestation (*Eriophyes* sp. – Acarina; Eriophyidae); Alemaya, Ethiopia.

oaks, chestnuts, pecan, fruit trees, elms and camphor. There are several important pests in Europe on oaks, alder, *Acer*, lime, poplars and *Salix*.

- *Eutetranychus* spp. – several species are pests on trees in the New World and parts of Africa and Asia (Fig. 5.307).
- *Oligonychus* spp. – a large genus, in appearance like *Tetranychus*; worldwide, one group infests forest, nut and fruit trees in both tropical and temperate regions; common tree pests in the USA, where many conifers are infested.
- *Panonychus* spp. (Fruit Tree Red Spider Mite, etc.) – a small genus. Widespread and common; several species are important tree pests, especially *P. ulmi* which is polyphagous on deciduous trees (Fig. 5.308).
- *Tetranychus* spp. (Carmine/Two-spotted Spider Mites, etc.) – a large genus with many pest species worldwide on a wide range of hosts; some are host-specific but many are polyphagous; infested plants usually webbed with silk. Conifers are not usually attacked.

(c) Family Tenuipalpidae (False Spider Mites)

A small group but with a few important pests. They resemble Spider Mites but do not produce silk. The main pests are *Brevipalpus* spp. (Scarlet Mites, etc.) which are polyphagous on a wide range of cultivated plants and trees. They are cosmopolitan throughout the warmer regions of the world (Fig. 5.309).

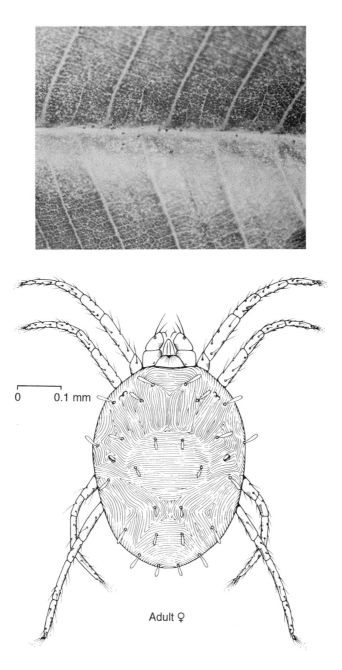

Fig. 5.307 Oriental Mite (*Eutetranychus orientalis* – Acarina, Tetranychidae) infesting Franjipani leaf; Hong Kong.

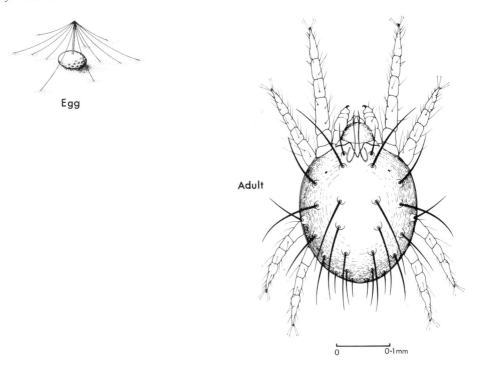

Egg

Adult

0   0·1mm

Fig. 5.308 Citrus Red Mite (*Panonychus citri* – Acarina; Tetranychidae); south China.

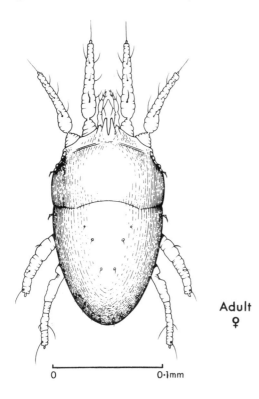

Adult
♀

0   0·1mm

Fig. 5.309 Red Crevice Mite (*Brevipalpus phoenicis* – Acarina; Tenuipalpidae).

6

INSECT PEST CONTROL

There are many different methods of insect pest control available to applied entomologists, and in general it is clear that the basic orientation is usually towards the host population as a whole rather than individuals. However, with some humans and some livestock, as well as some expensive horticultural plants, the welfare of individuals is of concern. Of course, even in agriculture there are occasions when a single insect pest can have serious consequences, especially in the case of quarantine inspection or grading produce quality.

When deciding which method of control to use, the main factors to consider can be grouped under the following headings.

1. Expected damage – both the nature and extent of damage. Some pests can be lethal to the hosts and others can be quite debilitating, but often the damage is really inconsequential – the insects may be only a nuisance, the blood loss trivial or the plant damage only cosmetic. In general, slight damage is best tolerated and accepted; we should not over-react to the presence of a few insect pests.

2. Degree of risk – some organisms in some locations are at high risk, because of 'resident' large pest populations, and the hosts can expect to suffer attack. In these situations, preventative measures (sometimes called insurance measures)

are usually justified. These range from malarial prophylactics and mosquito nets, to cattle dipping, seed dressings and soil sterilization.

3. Nature of the pest complex – usually the host organism will be subjected to attack by several different pests which will be interacting with the host and each other in the form of a pest complex. An extreme case would be the introduction of a new agricultural crop into an area; it will be attacked by a wide range of pests – mammals, birds, insects, mites, nematodes, molluscs, viruses, fungi, bacteria and weeds. Clearly, pest control measures need to be orchestrated against the whole pest complex rather than just a single major pest at a time. This is one of the reasons for the ever-increasing use of the integrated pest management (IPM) approach, see Chapter 2. With many agricultural crops, for example, there may be four to eight major species in the insect pest complex and ideally they require simultaneous control

4. Economic factors – the main aspect of concern is the value of the expected or actual damage as opposed to the cost of the control measures. Basically, the cost of the control measures must be less than the value of the damage. Losses can be either direct or indirect, involving death of the organism or destruction of

the living produce (fruit, grain, timber, etc.); alternatively in terms of international trade contamination, loss of reputation and other indirect losses can be serious. The three main costs are expected to be in terms of manpower, the pest control products required and any other specialized equipment or facilities which may be needed.

5. Biological/ecological factors – the first source of concern is usually the extent of the existing level of natural control. All insect species are preyed upon by other organisms and parasitized by other insects and micro-organisms. With any planned pest control project it is vitally important that this natural control be undisturbed. The second source of concern is usually the expected extent of environmental contamination or destruction. Some pest control measures involve deliberate vegetation destruction; others use heavy equipment that can cause extensive accidental vegetation destruction and lead to erosion. Many control projects involve the use of pesticides or other poisons or drugs, all of which will affect more than just the target organisms so that local food webs can be affected. Chemical poisons can also end up in the local water systems and may affect a wide range of other living organisms. In dry agricultural areas, the availability of water to make solutions of insecticides can be important – in such situations it is pointless to think of using high-volume sprays.

Before any decision is made as to the most appropriate method of pest control, it is assumed that the pests are correctly identified and that their general biology is known. Over the years, a great deal of effort and money have been wasted on inappropriate pest control projects.

There is now ever-increasing pressure to use the IPM approach to pest control for a range of now quite obvious reasons (see Chapter 2 and p. 355). It is, however, more useful to regard the different basic methods separately in this chapter, starting with the most important, convenient or desirable methods. It should be remembered that the IPM approach will usually employ a mixture of all or most of the other methods considered.

6.1 LEGISLATION

Some insect pests are so potentially damaging that they are regarded as international pests. The organizations most involved are the World Health Organization (WHO) for medical pests, and the Food and Agriculture Organization of the United Nations (FAO) for agricultural pests. There are also regional organizations in most parts of the world, such as the European Plant Protection Organization (EPPO) in Paris. International regulations are expressed through a series of quarantine laws. The FAO-designated phytosanitary regions of the world are shown in Fig. 6.1.

Quarantine legislation applies both to pests of international importance and also to pest situations within a country. The most serious international pests are clearly micro-organisms such as viruses and bacteria, which cause diseases in humans and livestock. However, a number of agricultural pests can be rated as internationally serious. Historically, the introduction of exotic pests has been a major concern. The major world crops each originated in a fairly restricted area, such as potato in the highlands of South America, maize in Central America, rubber in Brazil, oil palm in West Africa, rice in South East Asia and coffee in East Africa. These highly desirable crops were taken around the world and established everywhere where conditions were suitable. Very often in some of the new locations the crop prospered and yields were often higher (sometimes much higher) than in the in-

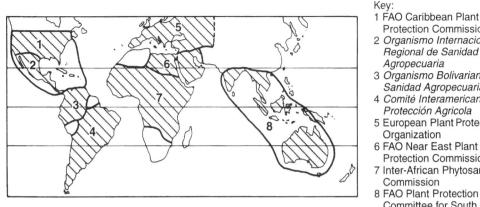

Key:
1 FAO Caribbean Plant
  Protection Commission
2 *Organismo Internacional
  Regional de Sanidad
  Agropecuaria*
3 *Organismo Bolivariano de
  Sanidad Agropecuaria*
4 *Comité Interamericano de
  Protección Agricola*
5 European Plant Protection
  Organization
6 FAO Near East Plant
  Protection Commission
7 Inter-African Phytosanitary
  Commission
8 FAO Plant Protection
  Committee for South East
  Asian and Pacific Region

Fig. 6.1 The FAO phytosanitary regions of the world (from Hill and Waller, 1982; after Mathys, 1971).

digenous country. This was often due to the absence of native pests and diseases in the new location. Over the years, however, many of the native pests were also introduced, mostly by accident. For example, the success of coffee in Brazil was due to the absence of Coffee Rust, Coffee Berry-borer and Antestia Bugs (all East African pests), but this initial advantage has now been lost as both Rust and Berry Borer have now become established in South America.

A few introductions of noxious organisms have been deliberate, such as rabbit and *Opuntia* in Australia, the African Honey Bee into South America and Gypsy Moth (from Europe) into North America; there are many more examples! But most have been accidental through carelessness and ignorance. The importance of such accidental introductions was quickly apparent, and in 1953 the FAO established a Plant Protection Convention which drew up a unified system for plant import and export. This involved the International Phytosanitary Certificate (Fig. 6.2) – nowadays certain plant materials can be taken out of one country and imported into another provided they are accompanied by completed certificates to declare they have

been inspected and declared pest-free. However, if a high risk from a particular pest is involved it is usual for that crop to be prohibited. Thus citrus fruit cannot be imported into some tropical countries because of the danger of accidental importation of either San José Scale or Medfly. Postal services are now a cause for concern. Recent outbreaks of Medfly in California have been linked to postal packages of fruits from Hawaii. Conversely, over recent years many important insect pests have reached Hawaii from mainland USA via the post. In point of fact, some 90% of the major pests in the Pacific region are exotic introductions. The important insect pest species for a particular country varies according to climate and local preferences – thus maize pests are especially important in Africa, citrus pests throughout the subtropics, potato pests (such as Colorado Beetle) in the UK, and so on.

Livestock are treated similarly in many countries, and the recent introduction of New World Screw-worm into North Africa via infested goats was a reflection of local quarantine failure.

The objective of quarantine legislation is to prevent serious exotic pests from becom-

---

MODEL PHYTOSANITARY CERTIFICATE

Plant Protection Service   No.: ....

of: ....

To: Plant Protection Service(s)

of: ....

This is to certify that the plants or plant products described below were inspected in accordance with the requirements of the importing country and found free of quarantine pests and substantially free from other injurious pests; and that they are considered to conform with the current phytosanitary regulations of the importing country.

Identification and Description of Consignment

    Name and address of exporters: ....

    Name and address of consignee: ....

    Number and description of packages: ....

    Quantity and name of produce: ....

    Distinguishing marks: ....

    Place of origin: ....

    Botanical name: ....

    Means of conveyance (if known): ....

    Port of entry (if known): ....

Disinfestation or Disinfection Treatment

    Date: .... Kind of treatment: ....

    Chemical (Active Ingredient): .... Duration and temperature: ....

    Concentration: .... Additional information: ....

Additional Declaration:

(Stamp of the Service)

Place of Issue: .... Date: ....

....
(Name of Inspector)

....
(Signature)

No financial liability with respect to this certificate shall attach to (Name of Plant Protection Service)...................or to any of its officers or representatives.

Fig. 6.2 International Phytosanitary Certificate (after Hill and Waller, 1982.

ing established in a country. For pests that become successfully established there are other laws and legal rules to deal with their collection and destruction. Thus in the UK the Colorado Beetle Order (1933) stipulates the procedures to be taken by both the public and the Ministry of Agriculture, Fisheries and Food (MAFF) on the discovery of a Colorado Beetle infestation. Similarly, in both Florida and California Medfly introductions are combated by State agricultural departments by both aerial and ground spraying of insecticides. In a recent infestation in California, the State legislature was reluctant to initiate urban spraying; this hesitation prompted Federal action and the urban aerial spraying programme was carried out successfully. The California fruit industry has an annual value of some US$1400 million and so the financial importance of insect pests such as Medfly and San José Scale cannot be overestimated. The enforcement of pest legislation is vital.

6.2 PHYSICAL METHODS

These are desirable methods, for they can be effective and involve no use of chemical poisons or drugs. However, some methods rely on a high level of labour availability, and so their use is often limited.

6.2.1 Mechanical methods

These are not all too readily categorized. Hand-picking of pests was probably the earliest method of pest control and it can still be used profitably for some pests (caterpillars on young fruit trees, ticks on livestock and pets, some grasshoppers and locusts). Sticky bands are used on fruit trees (against ants and flightless female moths), and fly-papers are re-appearing in food shops in Europe after many years of absence. Valuable fruits, especially in gardens and on smallholdings, are often bagged to deter Fruit Flies and Fruit Moths.

6.2.2 Shelter

This can be provided in the form of mesh netting. Mosquito nets are a vital form of physical protection against nocturnal predatory feeding in the tropics, and protective clothing is worn against both ticks in long grass and biting flies in the Arctic.

Protected cultivation of horticultural crops has been practised in Europe for centuries, but in the last couple of decades has become very widespread with more sophisticated microclimate control. The extensive areas of glasshouses that can now be seen are used mainly for climate control, but insect pest avoidance is an important issue (Fig. 6.3). What is also becoming equally widespread is the use of polythene tunnels as a form of protective cultivation (Fig. 6.4). In both the tropics and temperate regions the use of net enclosures for protection against insect pests is common (Fig. 6.5). The netting is nowadays made of plastic, and with fine mesh even aphids can be excluded. For expensive horticultural crops (flowers and some vegetables) net enclosures are effective and inexpensive, and they also provide protection for seed-beds for agricultural crops and tree nurseries.

6.2.3 Physical factors

Physical factors can be very effective for insect pest destruction and control. Lethal temperatures, both high and low, are widely used. In the tropics cool storage of grain in insulated stores will curtail insect infestations. In Hong Kong, beetle-infested high-value animal feed supplements were put into cold storage at $-20°C$ for a week to kill the beetles. In the same store were chests containing expensive fur coats that stay there for ten months of each year to protect them from moth and beetle pests. Grain in bulk stores is ideally kept cool – this does not kill the pests but will slow their metabolism and rate of development, thus reducing the level of damage to the grain (Fig. 5.33).

Fig. 6.3 A modern glasshouse occupying several hectares; near Boston, UK.

Fig. 6.4 Polythene tunnels used as greenhouses; Wainfleet, UK.

Generally tropical insects are best killed by cold and temperate insects by heat. Hot water treatment (dipping) is used on bulbs to kill nematodes, mites, fly larvae and some pathogens, and the heating of cotton seed kills the larvae of Pink Bollworm. Timber insect pests can be killed by kiln treatment – the heat is used both to dry the wood and kill any insects inside.

Drying is important for the preservation

Fig. 6.5 Net-house for insect-free seed-beds; Wainfleet, UK.

(protection) of a wide range of produce, for not only does it make the produce less attractive to insects, it slows down their rate of development and reduces damage by micro-organisms. Drying is one of the main methods of food preservation and is of great economic importance.

The storage of foodstuffs and expensive vulnerable items (such as fur coats) under hermetic conditions is very effective as a means of protection. Grain storage in sealed silos can be made even more effective by the introduction of carbon dioxide to exclude the air: the pests die of asphyxiation.

6.2.4 Electromagnetic energy

This can be used in several ways as a means of insect pest control. Long-wavelength radiation can be used on timber infested by beetles: the living biological material (i.e. live beetle larvae) is selectively heated and dies.

The ionizing radiations (X-rays, γ-rays) are lethal at high levels but low doses are sterilizing. This phenomenon is being widely used in the production of sterile male flies and moths under the Sterile Insect Release Method (SIRM) projects. Screw-worm Flies, Fruit Flies and Pink Bollworm moths have all been successfully controlled using the sterile male technique. A recent development is the use of ionizing radiation for food preservation – with the advent of supermarket retailing worldwide it has become necessary for perishable foodstuffs, especially prepared foods, to have a longer shelf-life. Early research has been very promising and it would appear that this is likely to become a widespread practice, despite some initial public apprehension.

Colour appears to be important to some insects. It has long been known that yellow is attractive to aphids and thrips – presumably because of the colour of the plants they use as food sources. A number of species of flies and some moths are also attracted to the colour, but it is not always clear why. This fact is used in the construction of various insect traps, with those made of yellow material consistently catching more insects. Conversely, blue colours and reflective materials have a repelling effect on aphids and some other insect pests. It has recently been discovered in Africa that Tsetse

Fig. 6.6 Light trap, using a mercury-vapour lamp, emitting ultra-violet light; south China.

Flies are attracted to the colour blue, which was quite unexpected, and this helps considerably in field trapping the flies. It was accidentally discovered that blister beetles are also attracted to blue, and blue traps in the field will catch large numbers of beetles.

Ultra-violet radiation is attractive to many different insect groups, and mercury-vapour lights at night can trap vast numbers of insects. This can be a very valuable pest survey tool, and in some species so many insects are trapped that light traps can appreciably lower the pest population locally. Moths, planthoppers, leafhoppers, some other bugs and some groups of beetles are the groups most attracted to ultra-violet radiation (Fig. 6.6).

6.3 CULTURAL CONTROL

This refers to regular (farm) operations that do not require the use of specialized crop protection equipment or skills designed to control insect pests or to minimize their damage. It is generally more applicable to agriculture than other areas of human endeavour, but not entirely. These methods do not usually give a high level of pest control, but typically involve minimal extra labour and cost. These were generally among the earliest methods of pest control to be used, but with the advent of the organochlorine insecticides (DDT, etc.) in the 1940s their use fell into abeyance, because the level of control they offered was too low. More recently, the general disillusionment with chemical poisons and the keen interest in the IPM approach to pest control, has led to a revival of interest in the use of cultural methods of control as part of the management programme.

6.3.1 Optimal growing conditions

Optimal conditions for all organisms are vitally important. A healthy plant or animal has considerable natural tolerance to both pests and diseases, both physically and physiologically. Conversely, a host that is in poor health, injured or stressed is predisposed to pest attack and this susceptibility results in an exaggerated level of damage. This is a major reason why cultivation of marginal land is often unsuccessful. Similarly, rearing of livestock under unfavourable conditions is seldom successful. The importation of Friesian cattle into tropical Africa was usually disastrous until suitable hybrids were bred. When Hong Kong was first settled as a British colony, few of the young expatriates lived to see 30 years of age, for conditions were just not suitable for their lifestyles; wearing thick, woollen clothing and eating heavy stodgy meals were allied with malaria and plague in a lethal combination.

Avoidance of an area where a plant or animal would be at high risk from a particular pest is an easy method to use. It is most effective against soil fungi and nematodes, and certain parasites. Sometimes the decision for avoidance is hardly voluntary – the historical reason why large areas of tropical Africa have not been colonized and over-run with cattle herds is because of the presence of the Tsetse. They abound in certain restricted areas referred to as 'fly-belts' where neither humans nor cattle (or horses) can really compete with them for living space. The local game animals have, of course, had thousands of years in which to adapt to the flies and the trypanosomes they carry.

Often it is in fact difficult to use avoidance as a technique against insects, because of their mobility and range. However, against soil pests in a particular field it can be used easily, and the use of a non-susceptible host variety could be regarded as a form of 'avoidance'.

6.3.2 Time of sowing

This is an important factor in agriculture for the avoidance of either the egg-laying period of a pest or to pass the vulnerable stage of plant growth before the insect pest numbers reach pest proportions. Mostly, sowing early is the technique, but it can be sowing late.

The recent increase in the practice of sowing autumn cereals and beans in the UK has reduced the danger from aphid attack very greatly. When the aphids become abundant in the spring the host plants have long since passed their vulnerable stage of growth. Agriculturally, many plants are vulnerable to a particular pest only at a particular time or stage of growth; at other times inflicted damage is usually slight and relatively unimportant. The vulnerable times for most plants and animals are: when they are young and becoming established; at the time of breeding (for then much body energy is diverted to the gonads and the reproductive system); and as the organism approaches old age. With a plant the sown seed is vulnerable, the seedling especially so as it is small and delicate; a transplant is always vulnerable during the period of root establishment; and some plants are also vulnerable at flowering and fruit production.

6.3.3 Time of harvest

This applies to certain crop pests; maize and beans suffer from weevil and bruchid attack in the field, and this can be minimized by prompt harvesting. In Africa it is clear that many maize and bean crops are not harvested as soon as they are ripe; they are often left in the field for weeks and are heavily infested at the time of harvest.

6.3.4 Deep planting/sowing

Deep planting can sometimes reduce the level of insect pest attack. Most seeds have a natural preferred depth of sowing, and some require a surprisingly shallow seed-bed – the Ethiopian 'tef' holds the record as it cannot tolerate any covering of soil and has to lie on the surface! Many seeds accidentally sown too shallowly are usually eaten by insect pests – but mostly by birds and rodents. Tuber crops generally suffer less pest damage if they are deeper in the soil – tubers

Fig. 6.7 Seagulls following a plough; Friskney, UK.

and root crops exposed on the soil surface are liable to be attacked by both insect and vertebrate pests, so the traditional 'earthing-up' is sensible practice.

6.3.5 Deep ploughing

This is a useful agricultural practice to bring soil pests to the surface, as well as to loosen and aerate the soil. Most soil pests live in the top 10 cm and the great majority within the top 20 cm of the soil, and deep ploughing exposes them to predators and to dehydration by the sun. In temperate regions the usual bird predators are crows and gulls, and in the tropics egrets, and they can be seen following ploughs (Fig. 6.7), sometimes in their hundreds.

6.3.6 Close season

This is more applicable to the tropics where insect breeding is more or less continuous, rather than the temperate regions where the cold winter period is a natural close season. For a number of pests, if they can be denied a host for a specified period of time then they will suffer a major population decline. With some medical and veterinary pests, if the host is denied access to the habitat for a critical period of time, then the pest population can be severely depleted.

The use of fallow fields was an ancient tradition to allow the land to recover after a period of nutrient depletion caused by growing crops. There were no artificial fertilizers available in those days; compost and animal dung were the main sources of nutrients, as was human 'night-soil'. Typically, every third, fourth or fifth (but usually fourth) year was the fallow year for most lands. Most pest organisms suffer population depletion during a period of fallow as the host organism is not available.

The new European/US practice of agricultural 'set-aside' is a form of fallow, but under this scheme areas of land (usually marginal land) may be left for several successive years.

6.3.7 Crop rotation

Crop rotation was traditionally linked with a period of fallow. Over a typical four-year rotation one year was fallow and each of the other three was used for the cultivation of a crop unrelated to the others. Thus there could be wheat, followed by potato and then a *Brassica* crop. Continuous cultivation of a single crop will lead to mineral depletion and will encourage pest and disease build-up. But of course plantation and orchard cultivation is very long term, and there is no possibility of rotation, similarly with forestry. A major problem with this is that some crops require special growing conditions and so effective rotation may not be feasible. In the fens of East Anglia, the peat soils are ideal for growing carrots, parsnips and celery, but this leads to a build-up of Carrot Fly, and here also grow the wild Umbelliferae that are alternative (wild) hosts. The end result is an over-dependence upon chemical pesticides to protect the crops that grow so well in the peat soils. This principle can also apply to livestock, but more usually against liver flukes, although ticks and some other parasites can be avoided by a rotation of livestock types. Generally, rotations are effective against monophagous and oligophagous pest species, but not against those with strong powers of dispersal or migratory pests.

6.3.8 Secondary hosts

These are sometimes called wild or reservoir hosts, and can be of great importance in helping to maintain a pest population between crops or host availability. Most insect pests are not monophagous and so will live on other hosts. Some parasites are extremely host-specific, and these generally respond to cultural control methods. In agriculture, many insect pests will feed on weeds belonging to the same family as the crop, and it is difficult, for example, to control grass weeds around rice paddy fields; similarly, Chenopod weeds in a sugar beet field are very difficult to control. Control of secondary hosts is often an important part of most IPM projects.

6.3.9 Weeds

Weeds are of vital importance to agriculture and horticulture partly because of direct competition with crop plants, but also because they act as reservoirs of disease organisms and as alternative hosts for insect pest. Some major crop pests actually prefer to lay their eggs on weeds rather than on the crop plants.

6.3.10 Intercropping and the use of trap crops

These refer to the practice of mixing cultivation of plants. Trap crops are areas sown with a plant which is actually preferred by a particular pest. The pest is lured on to these other plants where it can either be ignored or else more easily controlled. For example, in north Thailand maize crops are attacked by Bombay Locust, but if interplanted with a low-growing legume, the grasshoppers will mostly be found in the ground cover where they can be caught and eaten by ducks. With careful intercropping there can be a significant reduction in the insect pests on both crops. At the smallholder level this approach to pest control is very strongly recommended at the present time. It can also apply to other insect pest situations: for example, if a rented house or flat is flea-infested then bring a cat into the place and the fleas will feed on it rather than on you.

6.3.11 Crop sanitation

This is a general term covering several different aspects of crop cultivation. It applies especially to forestry. The term is mostly used by plant pathologists rather than entomologists, and refers to the removal of sources (foci) of diseases. The removal and destruction (roguing) of diseased, badly damaged or heavily infested plants is of great importance in the curtailing of a pest population. The same practice applies to livestock and humans with infectious diseases or parasites, when individuals or groups are kept in isolation until the parasite is killed. Fifty years ago every town in the UK had its own isolation hospital for this purpose, and all quarantine centres have isolation facilities. Removal of rubbish on and in plantations, farms and forests is of importance in that some serious insect pests, such as beetles, breed in such rotting rubbish. Fallen fruits often contain pupating Fruit Flies and caterpillars and should always be collected and destroyed. Crop residues should similarly be removed and destroyed – cereal stalk-borers often pupate in the lower parts of stems of maize and sorghum. Tree stumps should be left as low (small) as possible, and cut branches should be severed as close to the trunk as possible – pests and diseases gain access to healthy trees via pruning scars and dead branch stumps. The collection and removal of domestic garbage and sewage are vitally important to keep down populations of urban flies and various human parasites. The recommended method of destruction of all diseased/infected plant and animal material is by burning – other measures such as burying may not kill the pest organisms.

6.4 BREEDING/GENETIC METHODS

Host resistance to pest attack is shown by all groups of plants and animals to the whole range of pest organisms including pathogens. There are always some individuals in a population that either harbour far fewer pests than the others or else show fewer signs of damage. These individuals usually represent a different genetic variety

from the remainder of the population, and they are said to show **resistance** to the pest. Conversely, some individuals will show more severe signs of damage and will carry larger numbers of pests, and these are termed **susceptible**. Usually these plants are killed by the pests and so will not breed and pass on their disadvantageous genes.

The main use of resistant varieties of crop plants in agriculture was originally against plant diseases (Russell, 1978) and then plant-parasitic nematodes (eelworms), and in general has been very successful. Now that the problems connected with widespread insecticide use are very apparent, there is renewed interest in the breeding of plants that show resistance to important pests and diseases. As will be mentioned later many interesting aspects of genetic manipulation are now being researched and developed.

Varietal resistance to insect pests was promoted by Painter (1951) and later enlarged by Russell (1978), and the present interpretation is that there are four types of resistance:

1. Pest avoidance.
2. Non-preference
   (=non-acceptance; =antixenosis).
3. Antibiosis.
4. Tolerance.

6.4.1 Pest avoidance

This is when the host plant escapes infestation by not being at a susceptible stage when the insect pest population is at its peak. Some apple varieties escape infestation by several insect pests in the spring by having buds that do not open until after the main emergence period of the pests – thus the final amount of damage done is reduced.

6.4.2 Non-preference

This is shown when insect pests refuse to colonize a particular host, or else go reluctantly to the host but will only eat a small amount and will oviposit fewer eggs than usual. The insect reactions in these cases are easily observed but the details (i.e. behaviour, physiology, chemistry, etc.) of the host resistance are usually not known as yet.

6.4.3 Antibiosis

Antibiosis is one of the most common and widespread forms of plant resistance, occurring in both a physical and physiological or biochemical form. The physical aspects include anatomical features such as a thickened cuticle, hairy stems and leaves, a thickened stem and reduced pith in cereals; or compactness of cereal panicle (especially sorghum); and tightness of leaf sheath and husk of maize. Graminaceous stems can also be hardened by the extra development of more silica deposits in the stem tissues and also in leaves.

Biochemical aspects include the addition of chemicals that will deter feeding (such as alkaloids) or tannins that make the host indigestible. Sometimes the plant variety is deficient in certain chemicals that are needed as feeding stimulants. It is interesting to note that current opinion inclines to the view that most of the mechanisms used by plants to repel or deter phytophagous grazers and browsers were actually developed initially against the phytophagous reptiles and then against the earliest mammals.

6.4.4 Tolerance

This is the term used when the host (plant) supports a sizable pest population but suffers little actual damage. This is in part characteristic of healthy vigorous hosts, growing under optimal conditions that are able to compensate for damage caused or infestation by pests. Sometimes it is a matter of biochemistry – some people, especially in

middle age, do not react to insect bites, and mosquitoes, horse flies and the like can feed on arms and legs without any feeling of discomfort on the part of the host. Most animals can in fact spare small quantities of blood without any ill-effects, and all plants produce more foliage and fruits than they need – it appears that they are 'anticipating' being grazed by phytophagous animals.

Many crop plants have more useful genetic characters than most agriculturalists realize, and there can often be considerable scope for the development of selected varieties for special purposes. Clearly, pest resistance is one of a number of desirable properties. Sometimes the only people who really appreciate the qualities of a crop are the indigenous peoples of the centre of evolution of that species. For example, as reported in *New Scientist* (13.1.90), the South American Indians grow cassava as their staple carbohydrate. In Brazil, 22 varieties are recognized, in Peru 50, but in the Rio Negro 140 varieties. The qualities that the local farmers select for are early harvest (after six months) and late harvest (12 months) – most require nine months. Some are better for flour (farina flour), some for tapioca and others for gum. Some varieties grow better in different soils. Some give a better yield; some keep better in storage; and some have leaves more suitable as vegetables. Many crop plants and some animal species quite possibly have a number of useful qualities that have not yet been exploited.

Genetic manipulation/engineering is the latest weapon in the armoury of the applied entomologist in the quest to kill insect pests, or at least to control the damage they inflict. At the present time more than 60 plant species have been genetically engineered, and many disease and insect-resistant genes have been identified. A gene encoding an insect toxin has been found in the bacterium *Bacillus thuringiensis* and this has been incorporated into cotton plants (by Monsanto) to give resistance to cotton bollworms, and into maize (by Ciba Seeds) to give resistance to the European Corn-borer (*New Scientist*; 7.1.95). Both of these transgenic crops (seeds) are at present pending approval by both the Environmental Protection Agency (EPA) and the United States Department of Agriculture (USDA), and seeds could be available in 1996. In the UK a recent experiment at Oxford was given approval and a baculovirus with an added gene to produce scorpion venom was sprayed in an aqueous solution over cabbage plots to kill Cabbage Looper caterpillars. Whether this technique would be effective and safe for commercial use is not yet clear.

A common soil bacterium (*Agrobacterium tumefaciens*) has been given a gene from peas, known as the pea lectin gene, and has been incorporated into Desirée potatoes in Norfolk. In theory, the lectin gene interferes with insect digestion and will debilitate insect pests on the potatoes. Most plants produce lectins, but some are more effective at resisting insect attack than others. The snowdrop produces a protein very effective against aphids. This gene has been isolated and has been transferred into tobacco plants and work is in progress with potato and rape.

The whole process of manufacturing and releasing biopesticides (genetically modified organisms) to kill insect pests is giving rise to concern from organizations such as the EPA, MAFF and USDA, and a number of scientists have expressed their disquiet. Only time will show how safe the process is.

6.5 BIOLOGICAL CONTROL (BIOCONTROL)

This can be viewed either very narrowly or broadly, but the use of resistant varieties and other genetic aspects are best dealt with separately because of their importance. The topics remaining still present a considerable diversity and in other texts may be grouped differently according to the basic emphasis (i.e. behaviour, genetic, biochemical, interference, etc.).

6.5.1 Natural control

Natural control is the existing population control being exerted by the naturally occurring predators and parasites in a local agroecosystem. Typically it is not very apparent, and in the past has often been overlooked. However, after careless use of broad-spectrum and persistent insecticides such as DDT and dieldrin, which killed the natural enemies, there followed new and more severe pest outbreaks. It was then realized that the local natural enemies had in fact been keeping the pest population suppressed. Work in South East Asia showed that rice stalk-borers usually suffered an egg predation level of 90–95%, and this was followed by larval and pupal parasitism. It does appear that in many cases common crop pests are subjected to an 80–90% loss of population by the local natural enemies, and it is vitally important that this predation is undisturbed by whatever control programme is chosen.

Natural predators of importance include birds, spiders and several groups of insects (some bugs, Hover Fly larvae, wasps, ants and many beetles), as well as predatory mites. Insectivorous mammals, reptiles and amphibia, and many other insects such as Dragonflies and Praying Mantises, are only of importance in that they form part of the natural enemy complex.

The main groups of insect parasites are Tachinidae (Diptera), Ichneumonidae, Braconidae and Chalcidoidea (Hymenoptera; Parasitica), as well as a few entomophagous nematodes.

Pathogens, sometimes referred to as 'micro-organisms', include fungi, bacteria and viruses that use insects as host organisms and usually kill them. A number of species are being widely used in IPM projects with great success as biological insecticides.

6.5.2 Biological control

Sensu stricta is the deliberate supplementation of existing natural control by the introduction of selected predators or parasites (or pathogens) into the pest/crop ecosystem. The measure is designed to reduce the pest population to a level at which damage would not be serious. Much of the early work was started at the turn of the century by the staff at the University of California (Department of Biological Control), especially for the control of scale insects (Coccoidea) on *Citrus*, for when *Citrus* and peach were brought from southern China the scale insects were imported along with the planting stock. During the last century there have been many successful biocontrol projects involving insect pests and phytophagous insects controlling weed species (*Opuntia*, etc.), but also a number of failures. Today most biocontrol projects tend to be part of IPM programmes rather than projects on their own. Generally biocontrol is only successful on long-term crops such as fruit and plantation crops because of the time factor – populations of predators and parasites always lag behind in time after the prey/host, and so are seldom effective in controlling pests on annual crops. As mentioned previously, ploughing exposes soil-dwelling insects to predation by plough-following crows, seagulls and egrets. Bird predation has been used as a means of insect pest control very successfully in several different situations. In East Africa chickens are often used in smallholder cotton fields to eat Cotton Stainers, etc., and ducks in northern Thailand eat nymphs and adults of Bombay Locust in the maize fields (particularly if the rows are interplanted with a low-growing legume). The Chinese have a long tradition of using natural predators against crop pests and they use young ducks in the rice fields with notable success. This has led to a recent recommendation in North America to use ducks in pig-pens and calf-sheds to control flies. In the Mahé fish market (Seychelles) many Little Egrets can be seen stalking between the stalls, encouraged by vendors and ignored by shoppers, where they feed entirely on the huge fat bluebottles attracted to the mounds of fish. Many texts now give a synopsis of the classical cases of biocontrol, both successful and unsuccessful. These include Huffaker, 1971; DeBach, 1974; Delucchi, 1976; Greathead, 1971 and Hill, 1987.

6.5.3 Insect sterilization (=autocide)

This is the technique whereby males are sterilized without affecting their sexual behaviour. A characteristic of most female insects is that they only mate once, using their spermatheca for sperm storage, and they cannot distinguish between fertile and sterile males. This technique is commonly referred to by several different titles, such as the 'sterile male technique' (SMT), 'sterile insect technique' (SIT) and 'sterile insect release method' (SIRM). Basically, a large number of male insects are sterilized, usually by exposure to X-rays or γ-rays (but chemosterilization has also been used), and then released into the environment where they outnumber normal males. Most females therefore mate with the sterile males and produce no offspring. This method has proved successful against Fruit Flies and also Screw-worm flies which attack livestock. There are now several facilities available in Europe, Mexico and the USA where insect irradiation is an important adjunct to a nuclear reactor, and large numbers of insects can be treated and exported by air transport to distant destinations. The recent accidental introduction of New World Screw-worm into North Africa was very successfully controlled in a couple of years by an extensive (and expensive) control programme based on the release of sterile male flies. The technique works best on isolated pest populations, such as on islands (Curaçao, Hawaii, etc.) but also works on discontinuous populations such as in southern Texas and North Africa.

6.5.4 Semiochemicals

Semiochemicals are defined as chemicals produced by one organism that incite a response by another organism. Several different types of chemical are involved in insects' lives that can be used in population control.

The most commonly encountered chemicals in this context are probably the **pheromones** which were once classed as ecto-hormones – these are glandular secretions used for communication between individuals within the same species. There are three main types, firstly **sexual pheromones** that are usually (but not always) secreted by female insects to attract males for the purpose of mating; the most spectacular results are seen in Lepidoptera, and some Emperor Moth (Saturniidae) males are recorded flying five miles upwind to a receptive female. Some timber beetles (Scolytidae) have aggregation pheromones, a type of **social pheromone**, that cause the collection of a large population of beetles on a sickly tree that is suitable for colonization. Social wasps secrete an attack pheromone if the nest is disturbed, and the result is a stinging frenzy which can be very painful for the intruder. A few insects have **dispersal pheromones**, which are chemicals that induce a state of alarm in the recipients and cause the population to disperse rapidly over the host plant. It is now known that some aphids produce an alarm pheromone from their siphunculi. This has been mimicked by a potato variety that produces a chemical from secretory hairs on the leaves that acts as an alarm pheromone for aphids.

The main use of sexual pheromones is for pest population monitoring. Small paper or plastic shelters, often coloured yellow, are hung in tree foliage in orchards or on stakes in field crops (Fig. 6.8); inside there is a sticky surface and a tiny vial of pheromone as the attractant. Male moths, Fruit Flies and some other insects, are attracted by the odour and enter the trap where they stick to the adhesive.

Fig. 6.8 Pheromone trap (yellow) positioned in fruit orchard under apple trees; Skegness, UK.

The list of pheromones now available commercially is quite extensive. There are also some chemicals called 'attractants' which are not related to pheromones but serve the same function for trapping. If enough traps are used, there could be sufficient male insects caught to seriously deplete the population; this technique is termed 'trapping-out' or 'mass trapping'. If the area is flooded with sex pheromone, then mating becomes disrupted as the males are unable to locate the virgin females – this is called 'disruption technique' or 'mating disruption'.

Aggregation pheromones are used in forestry, sometimes as bait in traps to collect large numbers of bark beetles (Scolytidae), and also to attract the adult beetles on to healthy, resistant trees which will kill them in their bark tunnels by exuding copious sap.

6.6 CHEMICAL CONTROL

The chemical poisons that kill insects are known as insecticides, with acaricides ('miticides'!) killing mites and ticks (Acarina). Before reviewing the main types of insecticides, it is useful to look at the different modes of action. It must be emphasized that chemical control is quick in action (minutes, hours or a few days), but it is essentially repetitive. Commercial development of chemical pesticides started mainly with the need to protect cotton crops in the tropics (especially Sudan and Egypt) which have always suffered from a wide range of pests. The need to spray chemical poisons has persisted, and to date (1992) in both Uganda and Ethiopia the practice in cotton protection requires 15 insecticide sprays at one-week intervals during the growing season. Modes of action for insecticides include the following:

1. Repellents – these are designed to keep the insects away from the host, and are usually employed against mosquitoes and other biting flies that are medical pests. On plants, the pest might alight but will not feed (maybe just a test probe) and soon leaves.

2. Antifeedants – certain chemicals block part of the feeding response that most phytophagous insects show. The overall response is often complicated and it can be disrupted at several points in different ways.

3. Fumigants – this is the first group of the chemical poisons listed to enter the insect body and kill the host (pest). These are volatile chemicals that vaporize, the toxic gas killing the pest within enclosed containers or the soil.

4. Smokes – the insecticidal powder is finely divided and mixed with a combustible material. It is dispersed as 'smoke'; this technique is obviously only of use in enclosed spaces (stores, buildings, greenhouses, etc.). A type of 'smoke' is produced by the burning of mosquito coils which release pyrethrum into the air inside bedrooms and buildings. Some anti-mosquito smokes are repellent and others toxic, so there is some overlap between these action types.

5. Stomach poisons – these were the earliest forms of insecticide given as inorganic salts (arsenic, lead, etc.) that had to be ingested to be effective. Against insects they are usually only effective for phytophagous forms and scavengers; either sprayed on to foliage (to be eaten) or mixed with a bait to encourage ingestion.

6. Contact poisons – these are typically absorbed directly through the cuticle of the insect. There are two basic types:
(a) Ephemeral – short-lived; on plants usually as a foliar application; also used for flying medical pests (mosquitoes, etc.) with rapid 'knock-down' effect.
(b) Residual – persistent (i.e. long-lived); usually soil or foliage application; some medical and veterinary pests are killed by

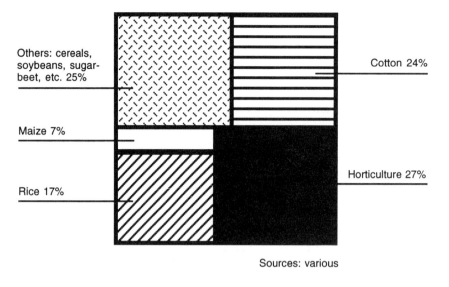

Others: cereals, soybeans, sugar-beet, etc. 25%

Cotton 24%

Maize 7%

Horticulture 27%

Rice 17%

Sources: various

Annual global losses in crop production due to pests have been estimated at US$ 300 thousand million. The annual cost of pesticides used in agriculture worldwide is US$ 20 thousand million. Approximate expenditure on pest control projects by international development agencies has been in the region of US$ 150 million annually.

Fig. 6.9 World insecticide market by crop (1992) (from *Bull. Pest Management*, **2** (1994), p. 10).

spraying their resting places (houses, tree trunks, etc.)

7. Systemic poisons – against plant pests, these chemicals are watered into the soil or sprayed over the foliage; they are absorbed and translocated by the plant to all parts. They are used mostly against sap-sucking insects and may also be formulated as granules for slow release or for safety. On trees they can be applied directly into the phloem system by injection. They are now used widely with livestock against a range of ectoparasitic insects, mites and ticks, either injected, ingested or applied to the skin.

As already mentioned, the original main stimulus for the commercial development of insecticides was to protect very valuable cotton crops. It is interesting to note that cotton is still the most insecticide-intensive crop worldwide (Fig. 6.9).

For the last half-century the main weapon in the human arsenal against insect pests has been chemical insecticides, for they have been effective (high kill, rapid acting, predictable in action and usually not too expensive (sometimes even quite cheap) to buy). However, we have become overly dependent on their use, and now that their overall usefulness is declining we are having to rethink our pest control strategies. Part of the problem is that very few new compounds are coming into use; this is partly because we are running out of candidate chemicals and also because of the costs involved with the new development procedures. Over the last 20–30 years there have been great changes in the procedure required for the commercial development and official registration of a new pesticide. This has largely been because of the 1986 'Control of Pesticides Regulations' in the UK and similar legislation in the USA. The official concerns are basically environmental risk (safety) balanced against effectiveness, and the commercial concerns are production problems and cost. The new official requirements are very stringent, and all UK and US agrochemical companies are having to retest and re-register all their existing products within a certain time period.

Between 1960 and around 1980, the time required for development of a new pesticide increased from three to five years to ten years or more; the cost from US$1 million to 10–20 million; and the number of chemicals tested from 2000 to 10 000. The end result is that by 1980 only one or two new insecticides were being produced each year.

6.6.1 Pest resistance

Resistance to insecticides is a major factor to be considered in all pest control management programmes. All synthetic chemical insecticides induce resistance in the target insects. Typically any chemical poison will kill 80, 90 or even 98% of the target insects – the survivors are genetic variants that show natural resistance to that poison. Even the most spectacular kills are not total – 99.5% has probably been the highest recorded. After several generations it is usual for the population to regain its former size, but many of the individuals are genetically changed, and a large proportion will now show resistance to the poison. One of the main features of the organochlorine insecticides was their spectacular kill rate – 98 and 99% was not uncommon, but nowadays most of them are useless in most locations because of the high rate of resistance that has been developed against them by most of the major pests. It was generally thought that resistance to natural insecticides (that is, those of biological origin) would not occur, and initially this did seem to be the case; however, it is now clear that some resistance to these products is developing. A general conclusion is that some 10–15 generations of insect pests

are required for the manifestation of resistance. At the present time more than 500 species of insect show resistance to one or more chemicals, and a few serious pests show resistance to almost all of the poisons available.

6.6.2 Insecticides

The classification of the substances that are used to control (i.e. kill) insect pests into convenient categories is not easy: a few groups of chemicals fall very clearly into definite categories but with others such designation presents problems. The result is that in most texts several different approaches are used, but the main problem is one of definition. The approach used here is a mixture of function, chemical structure and origin.

(a) Repellents

Repellents are only used occasionally on crops – one major problem is that the repelled pests will move off on to another host so another crop may be damaged. Various plant extracts are known to possess repellent properties and some are being used successfully to protect stored grains and foodstuffs in the tropics; neem, chilli peppers and tobacco are some of the plants used, and this is a field currently being seriously researched in several tropical countries. With mosquitoes and other biting flies, various toluamides are being used quite effectively as deterrents, both applied to exposed skin and used to impregnate clothing; 'deet' (diethyl-toluamide) is one of the most widely used, as is ethyl hexanediol and hexahydro-1H-azepine. Mites and ticks are repelled using dimethylphthalate, dibutyl-phthalate and benzyl benzoate, as well as several mixtures of toluamides including 'deet'. Most of the research in this field is being done in North America and the USSR.

(b) Antifeedants

With many phytophagous insects, research has shown that the feeding process is quite complicated. There are specific responses, first at a distance from the crop (host), then at close range and finally after the insect alights on the foliage. The foliage is then tasted and tested before an exploratory bite or probe is made for further tasting. The critical factors during this process are mostly olfactory, but sight can also be important, and a number of different chemicals can be involved. If any part of the olfactory sequence is broken, then the insect refrains from feeding. Such interference has been the basis of several successful pest control projects. Neem is one of the plants to produce chemicals that act as antifeedants.

(c) Inorganic salts

These were probably the earliest used compounds for pest control; most are stomach poisons and have to be ingested to be effective. They are mostly naturally occurring compounds and most used were salts of heavy metals. The most important were probably copper acetoarsenite, lead arsenate, cryolite (sodium aluminofluoride), mercurous chloride, sodium fluoride, cyhexatin and sulphur. Some are still being used at the present time.

(d) Organic oils

Organic oils were also among the earliest used chemicals, with tar oils and kerosene (T.V.O.) being the first to be adopted. As a group these are the hydrocarbon oils, naturally occurring in some regions but now distilled from coal tars and crude mineral oils. They are mostly very phytotoxic and make good herbicides and fungicides; they are also very effective for wood preservation (creosote, etc.). The heaviest components are the tar oils, used as winter washes on fruit trees. Medium weight fractions include kerosene, and the lighter parts are the petroleum oils. Very refined oils are now produced as white oils and some have such a low level of phytotoxicity that they can be used as carrier fluid for foliar sprays.

Some of these oils are widely used on water bodies to control mosquito larvae by suffocation. Creosote applied to timber will kill all insects and fungi.

(e) Fumigants and sterilants

These are mostly volatile liquids that vaporize under ambient conditions, releasing gases toxic to insects and other pest organisms. The gases are mostly used in food stores and other enclosed spaces. Some of the volatile liquids can be injected into soil and used for soil sterilization to kill soil pests. An important aspect of plant quarantine procedures is the fumigation of produce for export or import. Some solids are used for this purpose – camphor and dichlorobenzene crystals vaporize and, in a small enclosed space (i.e. insect cabinet, clothes cupboard), can be effective fumigants. A few chemical powders react with atmospheric moisture to release a toxic gas – aluminium phosphide ('Phostoxin') tablets release phosphine on exposure to air; this is a very effective fumigant and is widely used in the tropics for grain protection. Other useful compounds include carbon disulphide, ethylene dibromide, formaldehyde and methyl bromide. However, some of these chemicals have been shown to be carcinogenic and their use is now curtailed.

(f) Chlorinated hydrocarbons (=organochlorines)

These were the first of the 'modern' insecticides, nowadays sometimes referred to as 'second-generation insecticides' whose development started with DDT in 1940. They are characterized by being long-lived (persistent), with a broad spectrum of action, as both contact and stomach poisons. The main chemicals used were aldrin, dieldrin, DDT, endosulfan, γHCH and a few others. They regularly gave control levels of 98% or more, but time showed that they were equally devastating on the populations of predators and parasites. It was thought in the early 1950s that these chemicals heralded a new era in agriculture and preventative medicine, and that our insect pest problems were virtually over. But the destruction of natural enemies, including wild birds, environmental contamination and the development of resistance, led to the realization that these chemicals were not the universal panacea after all. For example, malaria is rapidly becoming just as widespread a pest in the tropics as it was before the advent of DDT. The final blow to the organochlorines is that most are now banned both internationally and nationally by most countries.

(g) Phenols

Phenols are a small group of chemicals, sometimes called the nitro phenols, used mainly as herbicides and fungicides. A few are effective against mites. The dinitrophenols have very high mammalian toxicity which does restrict their use. Binapacryl, DNOC and pentachlorophenol are still used occasionally.

(h) Organophosphorous compounds

These were first developed during World War II as nerve poisons (gases) and they have a high vertebrate toxicity. They were, and are, however, very effective insecticides. Most are short-lived and have contact and systemic-type actions. Some 50 plus compounds have been used, but some (such as parathion) have been withdrawn for general use because of toxicity dangers. Action appears to be via inhibition of acetylcholinesterase – hence nerve poisons. Some of the newer compounds are more persistent,

and several are very useful and widely used. Diazinon, dimethoate, fenitrothion, malathion, parathion, pirimiphos-methyl and tetrachlorvinphos are some of the best-known examples.

(i) Carbamates

Carbamates (and related compounds) were developed as anti-acetylcholinesterases (i.e. nerve poisons) as an alternative to the organophosphates with their lower vertebrate toxicity. (It should be pointed out that over its period of use the organophosphate parathion has probably killed several thousand spray operators, farmers and bystanders worldwide through its excessive toxicity, including dermal toxicity.) About a dozen carbamates are used as insecticides at present (although more are fungicides and herbicides), the main ones being aldicarb, carbaryl, carbofuran, methiocarb, pirimicarb and propoxur. Some of the most recently developed chemicals are less related to the original carbamates.

(j) Synthetic pyrethroids

These are being developed as alternatives to the naturally occurring pyrethrum whose main drawback is its short life as an effective chemical. The modern chemical analogues are, however, more and more removed from the original pyrethrins. They do have the same spectacular 'knock-down' effect and the broad action, but they are more stable and far more persistent. They also have a higher mammalian toxicity. These chemicals include cypermethrin, deltamethrin, resmethrin and fenvalerate, and the more recently developed bioallethrin, permethrin and tetramethrin.

6.6.3 Bioinsecticides

It is debatable whether this category should be presented as a separate group, but for teaching purposes there are some advantages in this approach. It was originally thought that these 'natural' insecticides would not induce resistance in the target pests, but it is now known that this idea is not true: some resistance has been established, although less than that shown with the synthetic chemicals. The main sources of these biological insecticides are shown below – the first three categories are often combined as 'microbial insecticides'.

(a) Fungi

Some species of fungi produce natural antibiotics that have been used against plant diseases for a long time. It has been recently shown that entomophagous fungi can be used for pest control. Greenhouse pests in Europe (aphids and whitefly) can be controlled using the fungus *Verticilium lecanii*, and it does appear that some field crops can be protected using fungal pest control. Some 750 species of entomopathogenics are known, but at the present time only a few are being used for pest control. Species of *Entomophaga* are being used to kill grasshoppers in pastures in both the USA and Australia with considerable success, and fungi are now being used for locust control in Africa and Asia. It is likely that the use of fungi for insect pest control is very much in its infancy and the future may see great development in this field. One major problem, however, is that fungi need high humidity for survival and reproduction, which may prevent their exploitation on field crops.

(b) Bacteria

At least four species of *Bacillus* are entomophagous. *B. thuringiensis* (Bth) has long been known to be a killing agent against caterpillars, but several distinct varieties are now know to be effective against some Diptera and some Coleoptera. This bacterium is being further exploited in that it is being introduced into different plants so that the plant tissues contain the insecticidal toxin. It should perhaps be stressed that Bth does not harm other insects, spiders or vertebrates – it is very host-specific. Research on entomophagous bacteria has now become intensive and more than 50 different toxins have been isolated, effective against a wide range of insect and mite pests. During the 1980s sales of Bth toxins increased four-fold to a present total value of US$100 million. The development of transgenic crop plants with Bth toxins in their tissues is a very active line of research, and in the next few years extensive progress should be made. Sad to say, however, some pest populations now appear to have developed resistance to Bth toxins, but the extent of this problem is not clear at the moment.

(c) Viruses

Viruses belonging to several different groups cause disease in insects, but only those from two families appear to be really suitable. The granulosis viruses (GV) and the nuclear polyhedrosis viruses (NPV) are both Baculoviridae, and the cytoplasmic polyhedrosis viruses (CPV) belong to the family Reoviridae. Most of the entomophagous viruses show very definite host-specificity and a number have been commercially developed for use against particularly serious crop pests – these include Codling Moth GV, Cabbage White Butterfly GV and *Neodiprion* PHV.

(d) Plant extracts

Plant extracts with toxic effects have been known since time immemorial in India and other parts of tropical Asia. The earliest were used as fish poisons, such as the extracts of *Derris* spp. (Rotenone) in South East Asia. In India many have been used to protect rice and other foodstuffs in storage. In the last couple of decades research interest in the use of plant extracts for plant and food protection has escalated, as well as the use of some of these chemicals to kill mosquitoes and other medical and veterinary pests.

More than 2000 species of plants are known to produce chemicals that have insecticidal properties. The most famous is probably Pyrethrum (*Chrysanthemum cinerariaefolium*) which is native to the mountains of Eastern Europe and West Africa and is now grown in the Kenyan Highlands. The crude extract from the flower heads is called 'pyrethrum' (which is also the trade name for retailing) and is a mixture of pyrethrins and cinerins. This is a powerful contact insecticide which causes a rapid paralysis ('knock-down'), but is unstable to sunlight and rapidly hydrolysed. It is harmless to vertebrates and most insect predators. With the development of the very toxic synthetic insecticides the commercial interest in natural pyrethrum declined. At the same time there has been development of synthetic pyrethroids, which proved to be very effective, albeit quite expensive. However, with the escalating problem of pesticide resistance there has been renewed interest in natural pyrethrum. Now the Kenyan industry has a new threat in that American genetic engineers have taken genes from the plants and translocated them into yeast cells. Pyrethrin-producing yeast is now being grown in huge vats under controlled commercial conditions.

In India the multitudinous uses of neem (*Azadirachta indica*) have long been known. In the last two decades the interest has become worldwide, with several international conferences on the subject of using extracts of neem as insecticides. In the poorer regions of the tropics there is great interest in the use of indigenous plant materials as insecticides, and extensive research is now being conducted into these convenient and cheap alternatives to expensive synthetic chemicals.

(e) Insect growth regulators

These are basically juvenile hormones whose presence will prevent the insect from maturing. As insecticides they are quite successful, but tend to operate only at the time of larval metamorphosis. Their use is therefore more appropriate against species that are pests in the adult stage, such as mosquitoes, House Flies and cockroaches.

6.7 IPM

The original concept of integrated control was developed to stress the need for understanding the complicated and basically antagonistic relationship between the use of chemical insecticides and biological control. California was the location for the first of the large citrus orchards at the turn of the century, and it soon became evident that the Coccoidea and some other citrus pests were best controlled by their natural enemies, whereas others needed to be controlled by chemical sprays. The earliest use of the integrated approach was thus the careful application of chemical insecticides so as not to upset the existing natural control while presenting no conflict with the introduction of any other predators or parasites. This was a monumental step towards careful and rational pest control, and led inevitably to the management approach.

The definition of IPM made by Smith and Reynolds (FAO, 1966) in the FAO Symposium on Integrated Pest Control (1965) was that it is a pest management system that, in the context of the associated environment and the population dynamics of the pest species, utilizes all suitable techniques and methods in as compatible a manner as possible while maintaining pest population levels below those causing economic injury.

Insect pest management follows (also IPM) IPM principles but is basically concerned with the management (control) of insect populations. (IPM deals mainly with crops (or cultivated plants) and includes all of the pests, that is mammals, birds, insects, mites, nematodes, weeds and pathogens acting in concert, and needing an overall approach to their collective control.) The present interpretation is the use of a multi-pronged approach with the intention of overall pest population control and the minimal use of chemical insecticides. In recent years, much progress has been made in the development of insect pest management (and IPM) projects for many major crops in many part of the world. One outstanding feature is that on average an adequate level of insect pest control has been achieved with only 50% of the usual application of insecticides, and local ecological disruption has been kept to a minimal level.

A series of publications has been written in recent years on the topic of IPM (and also insect pest management). They started with general works like the FAO symposium papers (FAO, 1966; Rabb and Guthrie, 1970) and these led to papers on IPM projects for specific crops, such as maize (FAO, 1979a) and sorghum (FAO, 1979b). The University of California then started a Statewide Integrated Pest Management Project which involved some 30–50 collaborators per crop: the first books were titled *Integrated Pest Management for Rice* (Pub. No. 3280) (1983); *Pest Management for Cotton* (Pub. No. 3305) (1984); *Pest Management for Potatoes* (Pub. No. 3316) (1986) and *Pest Management for Citrus* (Pub. No. 3303) (1984). Other books include those by Metcalf and Luckmann (1975) and Burn, Coaker and Jepson (1987), where specific crops are concerned, and finally the general text by Dent (1991).

6.8 PEST ERADICATION

The main object of most pest control programmes is to reduce the insect population to a level at which the damage being done is not important. However, occasionally a situation occurs when the insect pest is so damaging, especially in a new location, that it has to be destroyed completely. Some medical and veterinary pests (diseases) come into this category, including plague, yellow fever, dengue and malaria, as well as some pests of plants.

In some countries Screw-worm outbreaks were very successfully controlled by the earliest male sterilization projects. In 1988 the accidental introduction of New World Screw-worm (*Cochliomyia hominovorax*) into North Africa was combated by a concerted international effort under the aegis of the FAO in Rome. At a cost of about US$80 million over a period of four years, the mass release of sterilized male flies is thought to have successfully eradicated this pest population.

In most countries there is legislation to ensure that eradication measures are taken against particular insect pests that pose major problems. People, livestock and crop plants in most regions will be at risk from certain diseases and certain insect pests that are vectors, as well as the insects that are direct pests themselves. In Hong Kong there has been for many decades a weekly survey of the Tropical Rat Flea population on local urban rats in case they should reach plague proportions. In the UK bed bugs, lice and other human ectoparasitic infestations come under the responsibility of the local Public Health Inspectorate. Similarly in Hong Kong, such infestations were the purview of the Urban Services Department, as also was the Malaria Control Unit which sprayed oil on the scattered remote water bodies where mosquitoes bred.

In the UK the Colorado Beetle Order (1933) requires all farmers and growers to report any suspected introduction of Colorado Beetle. If the identification is positive, then MAFF staff are compelled to take immediate action; to collect all adult and larval beetles, to spray the potato foliage with insecticide and also to fumigate the soil in the infested area to kill any pupae that might be present.

Similarly, California and Florida have a multimillion-dollar fruit industry ($14 000 million and $180 million, respectively) which have been defended against a few very serious fruit pests, such as San José Scale, California Red Scale and Medfly. All too frequently, Medfly infestations are recorded and then either State or Federal legislations come into force and mandatory control measures are taken, despite local urban objections to the aerial spraying of malathion.

Eradication programmes work best when the pest population is isolated so that population reinforcement is not possible. Thus the early SIRM project on Curaçao against Screw-worm was greatly facilitated by being held on a small island. Several important forestry pests in North America have also been the subject of eradication programmes. Their damage potential to both coniferous and broad-leaved (deciduous) forests by tree defoliation is very great. Gypsy Moth, Winter Moth, Pine and Larch Sawflies are top of the list of insect pests of forest trees in this region. Extensive aerial spraying with insecticides has been carried out in some areas; in others parasites have been released in large numbers. Some of the serious pests are being kept under a measure of control, especially by the introduced parasites, but no eradication has been achieved, for the area is too large and there are too many pests.

Generally, there is agreement that the eradication of any pest (any species) is a grave biological responsibility and that it should not be done unless really necessary, for the ecological consequences cannot always be predicted.

GLOSSARY

Abiotic Of a non-living nature, such as weather and physical factors (opp. biotic).

Abundance The relative numbers of a population of animals or plants.

Acaricide Material toxic to mites (Acarina).

Acarina An order of the class Arachnida, containing mites and ticks.

Acclimatization Habituation of a species to a different climate or environment.

Adaptation Process by which an organism becomes fitted to its particular environment.

Adaptive radiation Modifications within a group of organisms whereby individual species become adapted (fitted) to different ways of life.

Adjuvant A spray additive to improve either physical or chemical properties.

Aedeagus The male intromittent organ, or penis.

Aerial Inhabiting the air, such as flying insects.

Aeroplankton Anemoplankton; tiny winged insects drifting in air, together with spores, pollen, bacteria, etc.

Aestivation Dormancy (quiescence) during a hot or dry season.

Agamic Non-sexual, parthenogenetic reproduction.

Age class Division of a population into arbitrary age groups for ease of statistical analysis.

Aggregation Grouping of organisms into a cluster, such as a flock, swarm, etc.

Agriculture Cultivation of soil for purposes of crop production, sometimes distinguished as agriculture (field crops) and horticulture.

Agroecology Study of ecology in relation to the practice of agriculture.

Agronomy Scientific land management, concerned with the theory and practice of field crop production and soil management.

Alate Winged insects; possessing wings (opp. apterous).

Allochthonous Exotic; not aboriginal; introduced (opp. autochthonous).

Allopatric Having separate and mutually exclusive areas of geographical distribution (opp. sympatric).

Allotropous Unspecialized; insects that feed on a wide range of flower types.

Alternate host A plant or animal acting as sole host for a certain stage in the life cycle of a pest organism.

Alternative host A plant or animal which acts as one of several hosts to an insect (pest).

Anemophilous Plants adapted for wind pollination.

Annual Living for one year only (cf. biannual, etc.).

Antibiosis The resistance of a plant to insect attack by having, for example, a thick cuticle, hairy leaves or toxic sap.

Antibiotic A substance produced by one organism which kills or inhibits the growth of others.

Antifeedant A chemical that inhibits the feeding of certain insect pests.

Aposematic Warning coloration which serves to deter enemies.

Apterous Wingless; insects without wings (opp. alate).

Arboreal Organisms that inhabit trees, especially a forest canopy; adapted for life in trees; in, on, or among trees.

Arbo-virus An arthropod-borne virus causing disease in humans, livestock and other vertebrates (sometimes used to denote pathogenic viruses endemic to tropical forests).

Arctic Northern; of the North Polar region; animals that live above the Arctic Circle.

Arenicolous Living in sand; psammophilous.

Arista A large bristle, located on the dorsal edge of the apical antennal segment in higher Diptera.

Arrhenotoky Parthenogenetic production of male offspring.

Asymptote The point in the growth of a population at which numerical stability is reached (i.e. equilibrium).

Atrophied Reduced in size; vestigial; rudimentary; part of an organism that has not developed properly or has withered.

Attractant Material with an odour that attracts certain insects; lure.

Autecology Ecological study of a single species of animal or plant, in particular its reaction to the habitat.

Autochthonous Native; aboriginal; indigenous; formed where found (opp. allochthonous).

Autocide Control of an insect pest species by the sterile male technique.

Avoidance Pest control measure that operates by the avoidance of areas of known high risk, such as keeping cattle out of Tsetse fly-belts and not planting a susceptible crop in a field known to contain a major pest.

Bacteria Single-celled microscopic plants that live as saprophytes or parasites.

Bait Foodstuff used for attracting pests (and other animals), usually mixed with a chemical poison.

Beating A method of collecting foliage insects by hitting tree branches over a tray or inverted umbrella.

Belt A geographical zone characterized by certain climatic or biological features; ecological unit characterized by its vegetation; belt transect.

Benthic Living on the bottom of lakes; living on the sea bottom.

Berlése funnel Large funnel with perforated trays, used for dry extraction of insects and small invertebrates from leaf litter or soil.

Biannual Appearing twice in one year.

Biennial Lasting for two years; appearing once in two years; plants that require two years for their life cycle.

Binomial nomenclature The principle that each organism has two names, genus first and then species.

Biocide General biological poison or toxicant.

Biocoenosis Community of organisms inhabiting a given habitat (biotope) and mutually interacting; life assemblage.

Biodiversity The natural diversity of living organisms; the range of types and species of plants and animals in an area.

Biogeography Chorology; geographical distribution of plants (phytogeography) and animals (zoögeography), their origins and adaptations.

Biological control (= biocontrol) Control of pests using biological agents such as predators, insect parasites and some fungi, bacteria and viruses.

Biomass Weight of plant or animal material produced in an area, both as standing biomass and harvested biomass.

Biome Major community of living organisms, usually a climax community, such as tundra, grassland and forest.

Biometrics Statistical study of living organisms and their variations.

Bionomics Ecology; study of organisms in relation to their environment.

Biota The total flora and fauna of a region or area.

Biotic Referring to life or living organisms.

Biotope Place of uniform environmental conditions where organisms survive; the smallest subdivision of the habitat.

Biotype A group of genetically identical individuals.

Birth rate Natality; the population increase factor; number of new individuals (offspring) born into an area over a specific period of time.

Bivoltine Insects with two generations per year (cf. univoltine, multivoltine).

Boreal Northern geographical region, cold or cool, more or less coincident with the taiga and the northern coniferous forests.

Botanical insecticide Insecticidal material produced from plants, such as pyrethrum and neem.

Brachypterous Having short wings that do not cover the abdomen; such insects usually cannot fly (cf. macropterous, etc.).

Bromatia Globular terminal hyphal swellings on fungi cultivated subterraneously in the nest of some termites and ants, used as insect food.

Brood Hatch of young insects produced from eggs; swarm.

Budworm Common name in USA for various tortricid larvae.

Canopy Topmost layer of leaves, twigs, flowers and branches of forest trees and other woody plants.

Carnivore Predator; flesh-eating animals and some insectivorous plants (opp. herbivore).

Carrier Material used as diluent and vehicle for the active ingredients of pesticides; usually in a dust (solvent in a liquid).

Carrying capacity The number of animals that can be accommodated successfully in a particular area or habitat.

Caterpillar Cruciform larva with many legs (polypod) on both thorax and abdomen; larva of Lepidoptera (moths and butterflies).

Caudal Referring to a tail or posterior end of insect body.

Cavernicolous Cave-inhabiting.

Cecidogenous Insect larvae that make and live in galls on plants.

Cecidology Study of plant galls, mostly formed by insect action, but some made by fungi.

Census Statistical enumeration of a population, usually separated into sex and age groups, etc.

Chaetae (= setae) Bristles; thin; elongate outgrowths of body cuticle, some articulated (setae), some not (spines).

Chaetotaxy The arrangement and nomenclature of the bristles on an insect's exoskeleton, in both adults and larvae.

Chemosterilant Chemical which renders an insect sterile without killing it.

Chrysalis Obtect pupa of butterfly or some moths (Lepidoptera).

Circadian rhythm Biological rhythmic activity with a 24-hour periodicity or cycle.

Cladistics A type of classification based on phylogeny and common ancestry.

Classification The establishment of a series of taxa into a defined and systematic hierarchy; taxonomy.

Cleptoparasite (= Kleptoparasite) An organism that steals food from another, usually a female insect stealing stored food from another for the rearing of her own offspring.

Climate The usual succession of states of weather in an area during the year; can be designated as macroclimate (over a large area), microclimate (in a very small area) and mesoclimate (between the two).

Climax vegetation Mature or stabilized stage in a series of plant communities, when dominant species are completely adapted to environmental conditions.

Cline Gradual change in body form over an area; gradient of biotypes; can be gradual or stepped.

Coarctate pupa A pupa enclosed inside a hardened shell formed from the previous larval skin (some Diptera).

Cocoon A protective case, usually made of spun silk, inside which the larva turns into the pupa.

Coleoptera Beetles: the order characterized by having the forewings modified into protective elytra; the largest order of insects.

Collembola Springtails; an order of primitive insects that are wingless and entognathous and grouped in the Apterygota.

Commensal An organism that lives with another and shares food, both species usually benefiting from the association; a form of symbiosis.

Community A distinct unit of vegetation of different types, in a definite area, with associated animal species.

Compatibility The ability of two organisms to live together in a balanced relationship (e.g. host and parasite); also applies to a mixture of chemicals in a pesticide spray.

Compressed Insects with a body flattened from side to side (opp. depressed).

Conservation Preservation of natural habitats; protection of natural resources with a view to sustained yield.

Contagious distribution Condition where, in a given habitat, the individuals of the same species tend to occur close together in groups, separated by areas where they are absent.

Control To reduce pest density, or damage, to a level at which it is insignificant, by many different methods (e.g. cultural control, biocontrol and chemical control).

Convergent evolution When organisms belonging to different groups develop similar characters, often for a specific function (e.g. gills); this does not imply any close phylogenetic relationship.

Co-operation Helpful interaction between animal individuals or species; working together.

Coprophagous Insects that feed on dung.

Coprozoic Insects that live in dung or faeces.

Cosmopolitan Having a worldwide distribution; occurring very widely throughout the major regions of the world, or at least on two or more continents.

Costa A longitudinal wing vein, usually forming the anterior margin (leading edge) of the wing.

Cover Shelter; shade; proportion of ground area covered by plants; proportion of surface area of target plant on which pesticide is successfully deposited.

Crawler The active (dispersive stage) first instar larva of a scale insect (Hem.; Coccoidea).

Cremaster A hooked or spine-like process at the posterior end of an insect pupa (usually Lepidoptera), often used for attachment.

Crepuscular Animals that are active in twilight, pre-dawn and at dusk in the evenings (cf. diurnal, nocturnal).

Crochets Hooked spines at the tips of the prolegs of Lepidopterous larvae (caterpillars) arranged in rows or circles, used for gripping; any small hooked structures.

Crop Plants grown by humans for particular purposes, usually edible plants grown for their fruits or foliage, etc.

Crop rotation The successive growing of different crops on the same area of land.

Crowding Condition of a population when the individuals are packed too closely together.

Cruciform Cross-shaped.

Cryptic Hidden; concealed; may also refer to behaviour when an insect is hiding; also coloration and body shape.

Cryptozoic Fauna hidden under stones, in crevices, under tree bark or in darkness.

Cursorial Adapted for running (cf. saltatorial, etc.).

Day-degrees Expresses the relationship between speed of development and temperature; often employs a base temperature of 10°C; each instar requires a known number of day-degrees for completion of development.

Death-point, thermal Temperature above or below which living organisms die.

Death rate Number of individuals of a population to die over a given period of time.

Deet The chemical diethyltoluamide used as a repellent for biting flies and ticks.

Deme Assemblage of taxonomically closely related individuals.

Demersal Living on or near the bottom of lakes or seas.

Density Degree of crowding of a population.

Density-dependent factors Population-controlling factors that are influenced by the size of the population, such as food, disease susceptibility, etc.

Density-independent factors Population-controlling factors that are not influenced by the size of the population, such as weather conditions.

Depressed Insects with body flattened dorso-ventrally (opp. compressed).

Dermaptera Earwigs; an order of insects, rather primitive, with hind wings concealed under short hemi-elytra and with terminal claspers.

Desiccation Drying; often extreme drying which leads to death.

Detritus Organic material produced by the disintegration of plant and animal bodies after death and decomposition; also used geologically.

Detrivore Animal that feeds on organic detritus and debris.

Diapause Spontaneous state of dormancy in many insects, usually physiologically induced and broken, to avoid a period of inclement weather, usually winter.

Diel Photoperiod, usually diurnal; daily; usually refers to a 24-hour period.

Dimorphic Having two different life forms; two different types of leaves, sexes, etc.

Dioecious Having sexes separate; dioic; male and female flowers on different plants (opp. monoecious).

Diplura Primitively wingless and entognathous order of insects, placed in the subclass Apterygota.

Diptera Flies; a large order of endopterygote insects, the adults having one pair of wings and the hind wings modified into clubbed halteres.

Disease Morbid condition of plant or animal body, or some part of it, resulting from the attack of pathogenic organisms.

Disinfect To free from infection by destruction of pathogens or pests.

Disinfest To kill or inactivate pests present upon the surface of plants or animals in their immediate vicinity (e.g. in soil).

Dispersal Movement of individuals out of a population (emigration) or into a population (immigration); the movement of plant diaspores away from the plants.

Distribution Range of an organism or group in biogeographical divisions of the world; spatial arrangement or pattern of the members of a group.

Diurnal Describes animals which are active during the daytime; flowers that only open during the day (opp. nocturnal; cf. crepuscular).

Diversity Differences in appearance, habits, etc. in a group of animals or plants; also species richness.

Dormant Resting or quiescent; hibernation or aestivation; alive but not growing; ungerminated seeds.

Dose; dosage Quantity of pesticide applied per individual or per unit area, etc.

Drift Lateral displacement of insects, plant diaspores, pesticides, etc. as a result of wind or water currents; transported soils; also genetic drift.

Drift fauna Freshwater fauna gradually swept downstream in a stream or a river.

Drop spectrum Distribution by number or volume of pesticide spray drops into different categories of droplet size.

Droplet size Precise sizes or the various droplets that make up a pesticide spray, with their different properties.

Dulosis Slave-making, as practised by some ants (Hym.; Formicidae).

Ecdemic Out of an area; not native (opp. endemic).

Ecdysis The moulting or skin-shedding of larval Arthropoda between one stage of development and another.

Ecesis Invasion and establishment of colonizing plants; the pioneer stage of dispersal to a new habitat.

Ecoclimate Climate within a plant community, usually a microclimate.

Ecology The study of all the living organisms in an area, including their physical environment and interactions.

Economic damage The injury done to a crop which will justify the cost of artificial control measures.

Economic injury level The lowest pest population density that will cause economic damage.

Economic pest Often defined as a pest causing a crop loss of about 5–10%.

Economic threshold The pest population level at which control measures should be started so as to prevent the pest population from reaching the economic injury level.

Ecosystem The ecological system formed by the interaction of living organisms and their environment.

Ecotope A particular type of habitat within a region.

Ecotype Biotype resulting from selection in a particular habitat; habitat type.

Ectoparasite A parasite that lives on the exterior of an organism (host) (opp. endoparasite).

Edaphic Of or pertaining to soil; influenced by conditions of soil or substratum.

Effectiveness of a pest control measure This is shown by the number of pests remaining after a control treatment.

Efficiency of a pest control measure The more or less fixed reduction of a pest population regardless of the number of pests involved.

Elateriform larva Typically a wireworm (Col.; Elateridae) or one of a few unrelated types (e.g. Mealworm) with thin, hard, cylindrical body with few setae.

Elytron The thickened, hard forewing of the Coleoptera, used for protecting the large, folded hind wing.

Embioptera An order of primitive insects, known as 'web-spinners'.

Emergence The adult insect leaving the pupal case; eclosion.

Emigration Movement of organisms out of a particular area; movement of insect colony to a new location (opp. immigration).

Endemic Native; restricted to a certain part of a region (opp. exotic, foreign, etc.).

Endobiotic Living within a substratum or within another organism.

Endocrine glands Specialized glands that produce hormones which are released into the haemolymph.

Endoparasite An organism living parasitically within another (opp. ectoparasite).

Endopterygota An alternative name for the Holometabola.

Entomology The branch of zoology that deals with the study of insects (class Insecta).

Entomophagous An animal or plant that feeds on insects; insectivorous.

Entomophilous Plants that are pollinated by insects; insect-loving.

Environment The sum total of external influences acting upon an organism including physical, chemical and biotic factors.

Ephemeral Literally means living for a day; a short-lived species (cf. Ephemeroptera).

Ephemeroptera Mayflies; an order of primitive insects, whose larvae live in freshwater; the adults are very short-lived and do not feed.

Epicuticle The outer layer of an insect cuticle.

Epidemic Spread of a disease through a human population (cf. epizoötic, epiphytotic).

Epidemiology The study of development and spread of a disease throughout a host population.

Epidermis Outermost layer of cells covering the body of an organism, usually with a thin superficial cuticle.

Epifauna Animals living above ground, usually applied to beach (littoral) fauna living on sand surface or rocks.

Epigeal Living on or near the ground.

Epiphytotic An epidemic disease among plants.

Epizoötic An epidemic disease spreading through an animal population (cf. epiphytotic, epidemic).

Eradicate Eliminate; completely remove.

Eremic Organisms living in a desert.

Erineum A growth of hair-like bristles; a dense patch on a plant leaf resulting from the attack of certain gall mites (Acarina; Eriophyidae).

Ethiopian African; a zoögeographical region including Africa south of the Sahara, southern Arabia and Madagascar.

Euhaline Living only in saline inland waters (cf. hyperhaline).

Euryhaline Organisms adaptable to a wide range of salinity (opp. stenohaline).

Euryhygric Organisms adaptable to a wide range of atmospheric humidity (opp. stenohygric).

Eurythermic Organisms adaptable to a wide range of temperatures (opp. stenothermic).

Eurytopic Many habitats; having a wide range of geographical distribution (opp. stenotopic).

Evolution The gradual development of organisms from pre-existing organisms since the dawn of life.

Exarate pupa A pupa in which the appendages are free and not fixed to the insect body.

Exclusion Control of a pest organism by excluding it from an area or country, often by use of phytosanitary legislation.

Exopterygota Hemimetabola; insects having exopterygote wing development and only a gradual metamorphosis (opp. endoptygota).

Exotic Introduced; foreign; non-endemic; a foreign animal or plant not acclimatized (opp. native, etc.).

Exponential growth Growth of an organism or population where increase in size is directly related to time; maximum growth rate.

Extinction Annihilation; the death or extinguishing of a species or a population.

Extrinsic Factors or influences that originate from the outside of an organism (opp. intrinsic).

Exuvium The cast skin (cuticle) of an arthropod after moulting.

Facies General appearance; the face of an organism; most organisms have a characteristic appearance.

Facultative Optional; not obligatory; able to choose.

Fallow The agricultural practice of leaving land uncropped for a time.

Family 1. A taxonomic category containing one or more genera of common phylogenetic origin; a closely related group, such as the Vespidae – the social wasps. 2. A group comprising parents and offspring and other close relatives.

Fauna All the animals peculiar to an area, habitat or period.

Fecundity Reproductive capacity; capacity to produce offspring and to multiply rapidly.

Feeding The process of recognizing, manipulating and ingesting food materials.

Feral Wild; an animal which has escaped from domestication and reverted to the wild state.

Filter feeder Animal feeding by filtration of micro-organisms, organic detritus and plankton from water.

Fluviatile Growing in, or near, or inhabiting streams or river deposits.

Food chain Simple sequence of organisms in

which each is food for a later member of the sequence, usually expressed in a ladder-like structure. When complex lateral sequences occur, it is usually designated a food web.

Fossorial Animals with limbs, etc. modified for digging and burrowing; in the habit of digging or burrowing.

Frass The mixture of wood fragments and faeces produced by wood-boring insects.

Fumigant Pesticide exhibiting toxicity in the vapour phase; gaseous pesticide.

Fumigation The application of gases or vapours to infiltrate stored produce or soil to kill pests.

Gall Cecidium; abnormal growth of plant tissues, caused by the stimulus of a pest organism (usually insect or mite); most are of a typical appearance.

Gene An inherited unit determining a character; a definite part of the DNA in each cell.

Generation Period from any given stage in the life cycle (usually adult) to same stage in the offspring.

Genetic engineering The changing of the genetic structure of an organism and genotype by gene splicing, translocation, etc.

Genus A group of closely related species in the classification of plants and animals.

Granule Pesticide impregnated into or on to a coarse particle of inert material, for convenience of application to a crop; a small particle.

Grass Monocotyledenous plant belonging to the family Gramineae (= Poaceae).

Grazing Usually applies to feeding on grass or low vegetation close the ground (cf. browsing).

Grease band Adhesive material applied as a band around a tree trunk to trap or repel ascending wingless moths or ants.

Gregarious Tending to herd together; colonial.

Grub Usually refers to a beetle larva (Coleoptera).

Grub (white) Scarabaeiform larva of the Coleoptera Scarabaeidae; many are pests of root crops or destructive to plant roots.

Grylloblattodea A small order of apterous, rather primitive insects with cursorial limbs.

Gustatory Applies to the sense of taste.

Habit (= habitus) The general appearance of an organism.

Habits The usual behaviour of an animal.

Haematozoa Tiny parasitic animals that live in the blood of the host – usually Protozoa and other unicellular forms.

Hemelytron The partially thickened forewing of heteropteran bugs.

Hemimetabolous Insect having a simple gradual metamorphosis, as in the Orthoptera and Hemiptera.

Hemiptera True bugs. A very large group of exopterygote Insecta with mouthparts modified into a piercing proboscis, sometimes referred to as a rostrum, forewing usually somewhat thickened.

Herbivorous Phytophagous; feeding on plant material; plant-eating.

Hermaphrodite Bisexual; organism with both male and female reproductive organs.

Heterogenesis Alternation of generations; spontaneous generation.

Heterosis Hybrid vigour, resulting from cross-breeding.

Heterotrophic Holozoic; feeding on organic materials, as practised by most animals (opp. autotrophic).

Hibernation Dormancy, or passing the winter in a resting state, often with a greatly reduced metabolism (cf. aestivation, torpor, etc.).

Holarctic Zoögeographical region including northern parts of the Old World (Palaearctic) and the New World (Nearctic).

Holocoenosis The living community considered as a whole, without undue emphasis on any of its parts or causal factors.

Holometabolous Endopterygota; insects

with complete metamorphosis, involving a definite pupal stage, usually regarded as the more advanced orders (Coleoptera, etc.).

Holozoic Heterotrophic; type of feeding practised by animals; feeding on organic materials (plants and other animals).

Homoptera Plant bugs; one of the two large suborders of Hemiptera that contains the plant bugs (Aphididae, etc.).

Homopterous Insects that have both pairs of wings alike, as in suborder Homoptera (order Hemiptera).

Honey-dew Liquid with high sugar content discharged from the anus of certain Homoptera (aphids, scales, etc.).

Horizontal resistance Plant resistance that operates generally against all races of a pest.

Hornworm Larva of Sphingidae (Lepidoptera) with a dorsal spine or horn on the last abdominal segment.

Horticulture Cultivation of plants (crops) which usually receive individual attention (e.g. fruit, vegetables); garden cultivation (opp. agriculture).

Host The organism on or in which a parasite lives; the plant or animal on which an insect feeds.

Humidity The moisture present in the atmosphere as water vapour, usually expressed as relative humidity (or saturation deficiency).

Hyaline Transparent; usually refers to insect wings.

Hydrophobic Water-repellent; intolerant of water or wet conditions (opp. hydrophilic).

Hylophagous Wood-eating; lignivorous; xylophagous.

Hymenoptera A large order of Insecta, now regarded as the most evolutionarily advanced, including wasps, ants and bees, with membraneous wings.

Hypermetamorphosis A form of complete metamorphosis during which there are two different types of larvae.

Hyperparasite A parasite which is parasitic upon another parasite.

Hypersensitivity Reaction to a parasite when both host tissues and parasite are killed.

Imago Adult; imagine; the reproductive stage of an insect.

Immigration The movement of individuals into a population or area (opp. emigration; cf. migration).

Incidence Recruitment; change in the number of cases of disease or parasitism.

Incompatible Not compatible; two organisms that cannot live together; two chemicals incapable of forming a stable mixture.

Indigenous Native; local; not foreign or exotic.

Infauna Loose term describing inhabitants of crevices, holes, soft substrate (beach sand and mud), etc. (opp. epifauna).

Infect To enter and establish a pathogenic relationship with a host organism.

Infest To establish a population of parasites on the outside of a host body (animal or plant); insect pests on a plant.

Inquiline An animal that lives in the home of another and obtains a share of its food.

Insect Hexapod; a member of the class Insecta; an arthropod with three pairs of legs.

Insect pest management (IPM) The careful and deliberate integrated manipulation of an insect pest population; sometimes regarded as synonymous with integrated pest management.

Insecticide A toxin effective against insects.

Insectivorous Insect-eating; entomophagous; can apply to both animals and plants.

Instar Stadium; the form of an insect between successive moults.

Integrated control The very careful use of several different methods of pest control in conjunction with each other to control pests with a minimum of disturbance to the natural environment.

Integrated pest management (IPM) The latest form of pest control based upon pest population management in an integrated manner. It uses many different methods to achieve a total level of population reduction.

Intermediate host Host intervening between two others in the life cycle of various parasites.

Intrinsic Factors or influences produced within an organism (usually of genetic origin).

Irruption Local population explosion, irregular in occurrence.

Isolation Separation of groups within a species due to spatial, topographical, phenological, genetic or other barriers.

Isoptera Termites. An order of hemimetabolous insects renowned for their large social nests and their destructiveness to wood; primarily a tropical group.

Jungle A loose term applied to tropical rain forests, sometimes restricted to second-growth tropical vegetation (after tree-felling), usually dense in its lower layers and scarcely penetrable.

Juvenile Immature stage of animal; larva; nymph.

Larva Immature stage of insects and many other animals; in the Insecta, restricted to young instars of the Endopterygota (cf. nymph).

Larvicide Poison effective against larval insects.

Larviparous Giving birth to larvae instead of eggs (cf. viviparous).

LD 50 Lethal dose of toxicant required to kill 50% of a large group of individuals of the same species.

Leaf area index (LAI) Ratio of leaf surface to soil surface area, in relation to utilization of solar energy for photosynthesis.

Leaf litter Accumulated fallen leaves under trees, especially well developed in forests.

Leaf-miner An insect that lives and feeds in leaf tissues between upper and lower epidermis, including the larvae of some Diptera, Lepidoptera, Hymenoptera and Coleoptera.

Legislation The passing of laws to control certain human activities, such as plant quarantine laws, Colorado Beetle Order, etc.

Lentic Referring to standing water habitats or slow-moving water (opp. lotic).

Lepidoptera Butterflies and moths. A large order of Endopterygote Insecta; larvae are caterpillars and adults have large scaly wings.

Lesion Disruption of host tissue, usually necrosis, caused by the toxic saliva of a feeding insect or pathogen.

Life cycle The various phases through which an individual species passes to maturity.

Life-table Survey technique of dividing up a population into age–class categories, to assess population viability.

Lignivorous Feeding on wood; xylophagous.

Limiting factor Any environmental factor that imposes a restriction on an organism and its development.

Limits of tolerance The upper and lower values of various factors, usually physical, such as temperature and rainfall, which control the existence of plants and animals in an area.

Limnetic Fresh water; living in freshwater marshes or lakes.

Limnology The study of biological, physical and chemical conditions in fresh water.

Litter Accumulated fresh organic debris on the surface of soil under vegetation, usually consisting of leaves, twigs, fruits and some dead animals.

Littoral Living on the seashore; zone between high and low water marks; also marginal part of lakes and ponds.

Localized Confined to a relatively very small area, sometimes a matter of square metres or else up to a kilometre.

Logistic curve Sigmoid-shaped growth curve applicable to both individuals and populations or organisms, plotted graphically.

Longevity The length of life of an individual or a population.

Looper The caterpillar of Geometridae (Lepidoptera), with only one pair of abdominal prolegs in addition to the terminal claspers.

Lotic Flowing (running) water; living in streams or rivers (opp. lentic).

Luminescence Light produced by special organs in various insects and other organisms, or by chemical action; bioluminescence.

Macroclimate General climate prevailing over a large area, considered as a unit.

Macropterous Large or long-winged insect.

Maggot Vermiform larva, legless and usually without a distinct head capsule (Diptera).

Mallophaga Order of endopterygote, wingless insects ectoparasitic on birds and mammals; Bird (Biting) Lice.

Maturation Development beyond the juvenile stage; becoming mature.

Mecoptera A small order of holometabolous insects, adults mandibulate and head extended into a rostrum.

Mediterranean A biogeographical region (i.e. the Mediterranean Basin); the climate of this region.

Meiofauna Animals that pass through the substratum without displacing the particles; in sand usually $50\,\mu m$–3 mm in length.

Mesoclimate Climate of valley or localized area, that differs from the macroclimate (usually warmer).

Metamorphosis Change of body form and structure as an insect develops from egg to adult.

Microbial control Pest control using pathogenic micro-organisms (fungi, bacteria, viruses, etc.).

Microclimate Climate of a microhabitat (e.g. crevice, plant foliage, hole, etc.), typically more constant and less variable than general climate.

Microfauna Fauna composed of microscopic animals.

Microhabitat A small place in the general habitat; a physical niche; possessing a microclimate.

Micro-lepidoptera A general term for a large group of small families of small moths that are regarded as primitive, and whose larvae are often leaf-miners.

Micro-organisms Microbes; microscopic organisms, usually referring to Protozoa, fungi, bacteria, viruses, etc.

Micropredator Small predator, typically smaller than the prey, e.g. mosquito.

Migration Change of habitat according to season, climate, food supply, etc. by certain animals; the strict definition includes the return journey (typical of birds) (cf. emigration, immigration, dispersal).

Mimicry Resemblance in colour, structure, behaviour or sound to another organism or structure as a means of self-preservation; a form of camouflage; the imitation of a poisonous or distasteful species by a harmless and edible species.

Monoculture The extensive cultivation of a single species of plant.

Monoecious Hermaphrodite; ambisexual; monoic; having both male and female organs on the same body.

Monophagous Monotrophic; feeding on one type of food only; insects feeding on one genus of host plant only.

Monoxenous Inhabiting one species of host only, applies to parasites.

Mortality Death rate of organisms in a population; population decrease factor.

Motile Moving; capable of movement.

Moult Ecdysis; periodical shedding of outer cuticle (skin) or body covering of hair, feathers, etc.

Multivoltine Having several generations per year (cf. univoltine, etc.).

Mutation A sudden genetic change, producing a different and inherited character; a difference in genotype.

Mutualism Symbiosis; relationship between two organisms in which both derive benefit; can be complete physiological interdependence.

Myiasis The invasion of living animal tissues

by larvae (maggots) of certain flies (Diptera).

Myrmecophilous Organisms that live in ant nests, or mimicking ants; plants pollinated by ants.

Natality Birth rate; population increase factors.

Natatorial Formed or adapted for swimming.

Native Indigenous; autochthonous; not exotic; originating locally.

Natural selection Processes occurring in nature which result in the survival of the fittest; elimination of individuals less well adapted to their environment.

Naturalized Species introduced into a country from elsewhere and now locally established and breeding normally.

Nearctic Zoögeographical region comprising North America, North Mexico and Greenland.

Necrophagous Feeding on dead bodies.

Necrosis Death of part of a plant or animal.

Nectarivorous Nectar-feeding, as in some insects, etc.

Nectary Nectar-gland; group of modified sub-epidermal cells in flowers, etc. that secrete nectar (for making honey).

Neotropical Zoögeographical region, consisting of South America and Central America to the south of Mexico.

Neuroptera A small order of holometabolous insects, mostly predatory, adults with many extra wing veins.

Neuston Organisms floating or swimming on the water surface, on the film or just under.

Niche The status or place of an organism in its biotic environment; a physical niche can include a hole in a wall or tree, etc.

Nocturnal Active at night (opp. diurnal, cf. crepuscular).

Non-persistent A chemical or disease that is short-lived; only active for a short while after application.

Nozzle Attachment for the spraying of a liquid; several different types are used.

Nymph The immature stage of a hemimetabolous insect that does not have abrupt metamorphosis.

Obligate Compulsory; without choice; not facultative.

Obtect pupa A pupa in which the appendages are fixed to the body surface and most abdominal segments are immovable (most Lepidoptera).

Odonata Dragonflies and Damselflies; a large order of quite primitive insects; nymphs predacious and aquatic, adults predatory and with two pairs of large wings.

Oligophagous Restricted to a few foods; restricted host-specificity; insects restricted to a single family of food plants.

Omnivorous Eating both animal and plant foods (cf. carnivorous, herbivorous).

Onisciform larva A flattened (platyform) larva like a woodlouse (Isopoda) in appearance.

Optimum conditions The most suitable environmental factors for full and rapid development of the organism concerned.

Oriental Far East; zoögraphical region including India through South East Asia to south China.

Orthoptera A large order of primitive exopterygote insects including grasshoppers, locusts, crickets and mole crickets.

Ovicide Poison effective against eggs of insects and mites.

Oviparous Reproduction by the laying of eggs.

Oviposition The actual laying of eggs.

Ovoviviparous Insects that produce eggs with a definite shell but which hatch inside the maternal body.

Paedogenesis Reproduction by young or larval stages, as in certain Diptera.

Palaearctic Zoögeographical region of the Holarctic, including Europe, Asia above the Himalayas and the fringe of North Africa.

Pandemic Cosmopolitan; very widely distributed.

Pantropical Species distributed throughout the tropics (and the subtropics).

Parasite Organism living on or within another, to its own advantage in food or shelter.

Parasitoid Organism alternately parasitic and free-living; applies to many Hymenoptera Parasitica with parasitic larvae and free-living adults.

Parthenogenesis Virgin birth; reproduction without fertilization, usually through eggs but sometimes through viviparity (esp. Aphididae).

Pathogen A parasite causing disease symptoms in its host.

Pellet Seed coated with inert material, usually incorporating pesticide, to ensure uniform size and shape for precision drilling.

Penetrant Oil added to a pesticide spray to enable it to penetrate the waxy insect cuticle; pesticide that can enter the plant body via the epidermis.

Periodicity The rhythmic (periodic) occurrence of an event; cyclical.

Persistence (of a pesticide) Length of time it remains active after application; continued occupation of a site by an organism.

Pest An organism causing damage to humans, livestock, crops and possessions, *sensu lato* including pathogens.

Pest complex The group of pests that attack a host at any one time.

Pest control To reduce damage or pest density to a level at which damage is insignificant, by physical, chemical or biological means (cf. IPM).

Pest density Population level at which a pest species causes economic damage (often 5–10% crop loss).

Pest (economic) Pest population that can be controlled by measures costing less than the expected yield increase subsequent to control.

Pest load The total burden of parasites/pests attacking a host organism at any one time.

Pest management The careful manipulation of a pest situation after consideration of all ecological, as well as economic, aspects.

Pest spectrum The complete range of pest organisms that are recorded attacking a particular crop or host.

Pesticide A toxic chemical used to kill pest organisms; poison; biocide; sometimes specifically defined as insecticide, aphicide, ovicide, etc.

Phasmida A small order of hemimetabolous insects including the stick insects and leaf insects.

Phenology Recording and study of seasonal biotic events in relation to climate, time and other factors.

Pheromone Ectohormone; a chemical substance secreted by an insect to the exterior which causes a specific reaction in the receiving insects.

Phloem Conducting (vascular) tissue in plants which transports organic (food) materials in solution.

Photoperiod Duration of daily exposure to light; length of day favouring optimum functioning of an organism.

Phototaxis Response to the stimulus of light.

Phytoalexin A substance produced by infected host cells (of a plant) toxic to the invading pathogen.

Phytophagous Herbivorous; plant-eating.

Phytosanitation Removal and destruction (roguing) of diseased or infested plant material.

Phytotoxic A chemical liable to damage or kill plants (especially cultivated plants) or plant parts.

Planidum larva A type of first instar larva which undergoes hypermetamorphosis, as in some Diptera and Hymenoptera.

Plankton Tiny freshwater or marine plants and animals floating in the surface layers of the water.

Plecoptera Stoneflies; an order of holometabolous insects whose larvae live in freshwater streams.

Pleioxenous Heteroxenous; parasitic on or in several species of host.

Plesiobiotic Living in close proximity; building contiguous nests, as in some birds, ants and termites.

Poikilotherm 'Cold-blooded'; animal whose temperature varies with that of the surrounding medium (air, water, etc.) (opp. homeotherm).

Poison bait Attractant foodstuff for insect, mollusc or rodent pests, mixed with toxicant.

Pollination Fertilization in flowers; transference of pollen from anthers to stigma and from stigma down to ovule.

Polymorphism Occurrence of different forms of individuals in the same species (as in ants, termites, etc.).

Polyphagous Eating various types of food; animal feeding on a (wide) range of hosts (opp. monophagous).

Polyvoltine Having several broods (generations) in one season (cf. univoltine, etc.).

Pooter Aspirator; glass tube or bottle with tubing and mouthpiece, for sucking up small invertebrates.

Population The collective individuals of a community, either of a single species or of several species.

Potamic Fluviatile; of or pertaining to rivers.

Predation The killing of animals (prey) by other animals (carnivores) for food.

Predator Animal that kills other animals for food; may consume them alive.

Predispose Favour; make more susceptible.

Predisposition Making a host (plant) more susceptible to a pest disease, usually because of environmental, genetic or cultural defects.

Preference The factor by which certain hosts (plants) are more or less attractive to pests (insects) by virtue of colour, smell, taste, etc.

Pre-harvest interval The necessary interval of time between the last application of a pesticide and the safe harvesting of edible crops for immediate consumption.

Pre-oviposition period Maturation period; the period of time between the emergence of an adult female insect and the start of its egg-laying.

Prepupa A quiescent stage between larva and pupa, as in some Diptera and Thysanoptera.

Prevention A pest control measure applied in anticipation of a pest (or disease) attack.

Prey Animal killed or hunted by predators.

Proboscis Any type of extended mouth structure for insect feeding, sometimes purely suctorial (Lepidoptera, some Diptera), but often adapted for piercing and sucking (Hemiptera, some Diptera, lice, fleas, etc.); rostrum; beak, etc.

Progeny Offspring; young; produced either sexually or vegetatively.

Proleg Fleshy abdominal appendage found on caterpillars (Lepidoptera) and larvae of sawflies (Hymenoptera).

Protection The means taken to control a pest organism on a given host or to prevent damage.

Protective clothing Special clothing to protect a spray operator from the toxic effects of crop protection chemicals; clothing to protect apiarist from bee attack.

Protonymph The second instar of mites (first instar is six-legged nymph) (Acarina).

Protura An order of tiny, primitive, wingless insects in the subclass Apterygota.

Pscoptera Bark and Booklice; psocids; a small order of specialized small exopterygote insects, winged or apterous, solitary or gregarious, that feed on algae, lichens or dried organic matter.

Pterostigma A dark spot or thickened opaque area along the costal margin of the wing near the tip (Odonata, Hymenoptera, etc.).

Pupa The non-feeding (and usually inactive) stage between larva and adult in insects with complete metamorphosis.

Puparium The case made of the hardened last larval skin in which the pupa of some Diptera (Cyclorrhapha) is formed.

Pupate To change from larva to adult during the life cycle of a holometabolous insect.

Pupiparous Producing offspring, either larvae ready to pupate (Glossinidae) or already at the pupal stage (Hippoboscidae).

Quadrat Measured area, of any shape or size, used to sample vegetation and sessile animals.

Quarantine Isolation; operations intended to prevent importation or spread of unwanted noxious organisms.

Quiescence Temporary cessation of activity or development due to unfavourable environmental conditions.

Race Subdivision of a species, sometimes synonymous with a geographical subspecies, but often referring to a pathogenic form.

Raptorial Predatory; adapted for predation.

Recruitment The influx of new members into an animal population, either by reproduction or immigration.

Relative humidity The ratio, expressed as a percentage, of the water vapour present in a unit volume of air to that which the air would contain if saturated, at that temperature.

Repellent A chemical which causes avoidance by a particular insect pest.

Reproduction The process of producing young (offspring).

Reproductive capacity Potential reproduction possible for a particular insect population, sometimes expressed as absolute potential or partial potential.

Residue Amount of pesticide remaining on or in the target tissues after a given time.

Resistance Natural or induced capacity of an organism to repel (or avoid) pests or parasites; the ability of a pest to withstand the toxic effects of a pesticide.

Rhythm Periodic occurrence; seasonal variation; regular movement.

Riparian Riverside; living on or frequenting river banks.

Roguing Removal of unhealthy or unwanted plants from a crop; phytosanitation.

Rostrum The piercing and sucking proboscis, or beak, of the Hemiptera.

Ruderals Plants (other than crops) which grow in artificial habitats created by man, such as road sides, gardens, verges, waste ground; secondary pioneers; weeds.

Run-off Surface run-off is the proportion of rainfall that does not penetrate the soil; pesticide run-off is the portion of the spray that is not retained on the target foliage but drips off on to lower leaves or the soil.

Salinity Saltiness; amount of chlorides of sodium, and other inorganic salts, contained in sea water or groundwater.

Saltatorial Jumping; adapted for, or used in, jumping as a mode of locomotion, as in grasshoppers.

Sampling Taking a subset of a population, representative of the whole, to study the whole population.

Sanguivorous Feeding on blood.

Sanitation Removal of dead or diseased material which could act as a focus of infection.

Saprozoic Animal that feeds on dead or decaying organic material, including animal faeces.

Scarab Beetle belonging to the family Scarabaeidae.

Scarabaeiform larva Grub-like beetle larva, with thickened cylindrical body held in a C-shape, well-sclerotized head capsule and large thoracic legs, sluggish in behaviour and usually found in soil.

Scavenger Animal feeding on dead plants or animals or decaying matter.

Seed dressing A coating of protective pesticide applied to seeds before sowing, either a dry or wet preparation.

Semi-looper Caterpillars from the subfamilies Plusiinae, etc. (Lep.; Noctuidae) with a reduced number of prolegs (two to three pairs).

Semiochemicals The chemicals produced by one organism that elicit responses from another organism; hormones, etc.

Senescence Ageing; condition of advancing age, usually in the post-reproductive period.

Sex ratio Number of males per 100 females, or per 100 births; proportion of males to females in a population.

Sexual isolation When, at low population densities, individuals are so scattered that males are unable to locate females at mating time.

Sexual reproduction The fusion of male and female gametes; reproduction involving both males and females.

Shelter Nest; den; lair; area of protection against an inclement environment.

Siblings Offspring of same parents, but not at same birth.

Siphonaptera Fleas; an order of endopterygote insects, small, wingless adults ectoparasitic on warm-blooded vertebrates, larvae saprozoic.

Siphunculata Anoplura; Phthiraptera; sucking lice; a small order of hemimetabolous insects, ectoparasitic on vertebrates.

Siphunculi The paired protruding organs at the posterior end of the abdomen of Aphidoidea; cornicles.

Slave-making Dulosis; the capture of certain ants by nest-raiding parasitic ants, usually as pupae, which are kept as slaves in the host nest.

Smoke Aerial dispersal of minute pesticide particles (or soot) through the use of combustible mixtures.

Social facilitation Social animals where nearby presence of another animal modifies behaviour or physiological processes.

Social organization Development of animal society with co-operation and often polymorphism with division of labour.

Society A group of organisms of the same species having a social structure.

Soil sterilant Toxicant applied to soil to kill pests and pathogens.

Solvent Carrier solution in which the pesticide is dissolved to form the concentrate.

Speciation Evolution of species.

Species Group of interbreeding individuals, producing fertile young, showing common characteristics and included in a genus.

Spine A thorn-like process projecting from the body cuticle, usually solid and immovable (cf. spur).

Spray A pesticide solution applied as fine droplets through a nozzle, expelled by pressure.

Spread The uniformity of a spray deposit over a continuous target surface such as a leaf lamina.

Spreader Wetter; material added to a spray to lower surface tension and to improve spread over the target.

Spur An articulated spine-like process, often on a leg segment.

Stability The ability of a pesticide formulation to resist chemical degradation over a period of time.

Stenoecious Having a narrow range of habitat selection.

Stenomorphic Dwarfed; smaller than the typical form, owing to cramped conditions.

Stenophagous Subsisting on a limited range of food; insects restricted to one genus of host plant (=monophagous).

Stenotopic Having a restricted range of geographical distribution (opp. eurytopic).

Sterilant A substance used to sterilize and kill micro-organisms and other parasites.

Sterile Infertile; devoid of living organisms.

Sterilization To render infertile; castrate; to make produce, etc. sterile by killing micro-organisms.

Sticker Viscous material used to stick powdered seed dressing on to seeds; also can be added to a spray to improve retention on the foliage.

Stimulus A force or substance producing a response from an animal or plant.

Stomach poison A toxicant that kills by

absorption through the alimentary tract after being ingested.

Strepsiptera A small order of holometabolous insects parasitic on other insects and very specialized.

Stylet Rostrum; proboscis; can apply to the whole piercing proboscis, or else just to paired piercing components (Hemiptera).

Subspecies Race; geographically isolated population that differs from the main population in minor characteristics.

Subterranean Underground; living beneath the surface of the earth.

Subtropical A zone of latitude between 23.5° and 34.0° in either hemisphere (after Lincoln, Boxshall and Clark (1982)).

Succession Ecological, seasonal or geological sequence of species; development of plant communities from pioneer to climax.

Supplement spray Adjuvant; additive, to improve either physical or chemical properties of an insecticidal spray.

Surfactant Spreader; wetter, added to pesticide spray.

Swath Width of target area of vegetation sprayed at one pass.

Sweeping Collecting insects and small invertebrates from low vegetation and grass using a special stout (sweep) net.

Sylvan Sylvatic; forest inhabitant; belonging to a forest.

Symbiosis A condition in which two organisms live together in a partnership, beneficial usually to one and sometimes to both.

Sympatric Having the same or overlapping areas of geographical distribution (opp. allopatric).

Symptom Visible sign of insect attack or disease.

Synanthropic Living in, or close to, human habitation, in close association with humans.

Syndrome A group of concomitant symptoms, usually characteristic of a disease or pest attack.

Synecology Ecology of a plant or animal community (opp. autecology).

Synergism Increased virulence of a mixture of pests or pathogens; increased toxicity of a mixture of pesticides.

Systematics Classification of living organisms into a hierarchical series of groups emphasizing phylogenetic relationships; sometimes used synonymously with taxonomy.

Systemic A pesticide absorbed into plant tissues (usually root) and translocated through the plant vascular system.

Taint Unwanted flavour in food from a pesticide used on the growing crop.

Target surface The surface aimed at and intended to receive a spray or dust application.

Taxon Any definite unit in classification of plants and animals; a taxonomic unit (species, genus, tribe, family, etc.).

Taxonomy The theory and practice of naming and classifying animals and plants.

Tegmen The thickened and leathery forewing of Orthoptera and Dictyoptera.

Temperate Climate or region with a moderate climate.

Teneral A newly emerged adult insect, sometimes of a characteristic appearance (some Diptera).

Terrestrial On land; organisms that live on land.

Thelytoky Parthenogenesis when only females are produced, as in Aphidoidea.

Thigmotaxis Organism reaction to the sense of touch, e.g. cockroaches creeping into crevices.

Thysanoptera Thrips; an order of holometabolous insects characterized by thin strap-like fringed wings.

Thysanura Bristletails; a small order of primitive apterygote insects.

Tolerance Resistance of a living organism to excess or deficiency of an element in its environment; plant enduring insect infestation without signs of distress.

Topography Description and mapping of the surface features of an area of land.

Torrent Rushing body of water, usually at the head region of streams or rivers where the water flow is fast and the stream bed incline steep.

Toxic Poisonous.

Toxicant Poison; a chemical exhibiting toxicity.

Toxicity The ability to poison, or to interfere adversely with, any vital bodily function.

Toxin Poison; substance harmful to an organism.

Transect Square, belt or line, to delimit a linear area for studying and charting vegetation or sessile animals; quadrat.

Translaminar A pesticide capable of passing through a leaf lamina into internal tissues.

Translocation The uptake of materials into one part of a plant body and their subsequent dispersal to other parts.

Trap To capture or kill animals using a structure to hold them captive; often using a bait for attraction.

Trap crop A crop grown especially to attract pests or diseases away from a more valuable or vulnerable crop.

Trichoptera Caddisflies; an order of holometabolous insects; adults with hairy wings and larvae in portable cases in fresh waters.

Triungulin larva The active first instar predatory larva of Meloidae (Coleoptera) and Strepsiptera.

Tropical Belonging to the tropical regions of the world, approximately between the Tropics of Cancer and Capricorn geographically.

Tropicopolitan Pantropical; animals and plant species confined to and common in the tropics; throughout the tropics.

Tropism Animal or plant movement in response to stimulus – light, heat, gravity, etc. are the most common stimuli.

Tullgren funnel Metal cylinder three to four inches in diameter, terminating in a funnel, for dry extraction of small invertebrates from soil cores; usually in a bank of six to ten funnels.

Ubiquitous Very widely distributed.

Ultra-low volume spraying Pesticide spraying when the toxicant is often dissolved in oil, or the concentrate itself is used and very small amounts of liquid are applied.

Univoltine One generation per year (cf. bivoltine, multivoltine).

Urban Domestic; the ecological habitat(s) associated with humans.

Urticating Nettling; stinging; usually applies to hollow setae containing formic acid, found on some caterpillars (Lepidoptera).

Vagrant Wandering; unattached; individual(s) blown off course when migrating, or blown to new regions; rare migrant.

Variety Diversity; a group of plants or animals that differ slightly from the typical; less than a subspecies.

Vector A carrier of pathogenic organisms; an agent transferring a parasite from one host to another, usually an insect.

Vegetation Plant formation; total plant cover of an area, consisting of one or more plant communities.

Venomous Poisonous; toxic; noxious; an animal capable of harming another using chemical poison, either a sting or poisonous bite.

Vermiform larva Worm-like; a legless (apodous), headless (acephalic), worm-like larva typical of some Diptera.

Vernacular name Common name; the agreed local common name for a plant or animal other than the formal Latinized scientific name.

Vertical resistance Plant resistance to pest attack only effective against some races of a pathogen; a temporary form of resistance easily broken by new pathotypes (cf. horizontal resistance).

Vestigial Degenerate; non-functional; small and poorly developed.

Viable Alive; alive even if in a dormant state.

Virus Minute intercellular disease agent.

Viviparous Giving birth to living young, as in the Aphidoidea.

Voltinism Number of generations produced per year.

Volunteer Crop plant growing out of place from shed seeds, without intentional cultivation.

Warning coloration Bright, contrasting coloration possessed by some insects indicating their poisonous or distasteful nature, typically yellow/black, etc.

Wetter Surfactant; a substance which lowers the surface tension of water, or a liquid spray.

Wireworm The long, slender, hard-bodied larva of Elateridae (Coleoptera); many species are serious crop pests attacking roots and root crops.

Xenobiosis Hospitality, as in ant colonies; a form of symbiosis where one species lives in the nest of another.

Xenology The study of hosts in relation to the life history of parasites.

Xeric Dry; tolerating or adapted to living in dry conditions.

Xylem The water-conducting transport system in plants whereby water taken up by the roots is translocated to the foliage.

Yield Harvest; biological productivity; number of plants, fruits or weight of biomass produced in a given area of habitat; number of animals reared or trapped in a given area over a given period of time.

Zonation Stratification; layers or strata; arrangement or distribution of plants and animals into distinct zones.

Zoögeography The study of animal distribution on the Earth.

Zoölogy The science dealing with the study of animals.

Zoötic Disease epidemic among animals (cf. epiphytotic, epidemic).

Zoraptera A tiny order of small endopterygote insects, regarded as both primitive and specialized.

SOURCES

Fowler, H.W. and Fowler, F.G. (1958) *The Concise Oxford Dictionary of Current English*, Clarendon Press, Oxford, pp. 1552.

Henderson, J.H. (1960) *A Dictionary of Scientific Terms* (7th edn.), Van Nostrand, Princeton, pp. 595.

Hill, D.S. (1976) *Glossary of Ecological Terms*, Hong Kong University, Zool. Dept., Hong Kong, pp. 75.

Lincoln, R.J., Boxshall, G.A. and Clark, P.F. (1982) *A Dictionary of Ecology, Evolution and Systematics*, Cambridge University Press, Cambridge, pp. 298.

Nichols, S.W. (1989) *The Torre-Bueno Glossary of Entomology*, New York Ent. Soc., New York, pp. 840.

REFERENCES AND FURTHER READING

Alford, D.V. (1984) *A Colour Atlas of Fruit Pests: Their Recognition, Biology and Control*, Wolfe Publishing, London.

Alford, D.V. (1991) *A Colour Atlas of Pests of Ornamental Trees, Shrubs and Flowers*, Wolfe Publishing, London.

Allee, W.C. *et al.* (1955) *Principles of Animal Ecology*, Saunders, Philadelphia.

Andrewartha, H.G. and Birch, L.C. (1961) *The Distribution and Abundance of Animals*, 2nd edn, University of Chicago Press, Chicago.

Annecke, D.P. and Moran, V.C. (1982) *Insects and Mites of Cultivated Plants in South Africa*, Butterworths, London.

Anti-locust Research Centre (1966) *The Locust Handbook*, COPR, London (2nd edn, 1989; available from NRI, Chatham, UK).

Askew, A.A. (1971) *Parasitic Insects*, Heinemann, London.

Avidoz, Z. and Harpas, I. (1969) *Plant Pests of Israel*, Israel University Press, Jerusalem.

Baker, E.W. and Wharton, G.W. (1964) *An Introduction to Acarology*, Macmillan, New York.

Balachowsky, A.S. (ed.) (1966) *Entomologie Appliquée à l'Agriculture*, Vols 1–8, Masson, Paris.

Banziger, H. (1986) Preliminary observations on a skin-piercing bloodsucking moth, *Calyptra eustrigata* (Hamps.) (Lep., Noctuidae) in Malaya. *Bull. Ent. Res.*, **59**, 159–63.

Barbosa, P. and Wagner, M.A. (1989) *Introduction to Forest and Shade Tree Insects*, Academic Press, London.

Batra, S.W.T. (1984) Solitary bees. *Scientific American*, **250**(2), 86–93.

Beaver, R.A. (1977) Bark and ambrosia beetles in tropical forests. *Biotrop. Spec. Pub. No. 2*, 133–47.

Becker, P. (1974) *Pests of Ornamental Plants*, MAFF Bull. No. 97, HMSO, London.

Beirne, B.P. (1967) *Pest Management*, Leonard Hill, London.

Beirne, B.P. (1971) Pest insects of annual crops in Canada. 1. Lepidoptera; 2. Diptera; 3. Coleoptera. *Mem. Ent. Soc. Canada*, No. 78.

Beirne, B.P. (1972) Pest insects of annual crops in Canada. 4. Hemiptera–Homoptera; 5. Orthoptera; 6. Other Groups. *Mem. Ent. Soc. Canada*, No. 85.

Bevan, D. (1987) *Forest Insects*, For. Comm. Hdbk 1, HMSO, London.

Birch, M.C. and Haynes, K.F. (1982) *Insect Pheromones*, Studies in Biology No. 147, Edward Arnold, London.

Blackman, R.L. and Eastop, V.F. (eds) (1984) *Aphids on the World's Crops*, John Wiley, Chichester.

Bodenheimer, F.S. (1951a) *Citrus Entomology in the Middle East*, W. Junk, The Hague.

Bodenheimer, F.S. (1951b) *Insects as Human Food*, W. Junk, The Hague.

Borror, D.J. and DeLong, D.M. (1971) *An Introduction to the Study of Insects*, 3rd edn, Holt, Rinehart and Winston, New York.

Bostanian, N.J., Wilson, L.T. and Dennehy, T.J. (1990) *Monitoring and Integrated Management of Arthropod Pests of Small Fruit Crops*, Intercept, Andover, UK.

Bottrell, D.R. (1979) *Integrated Pest Management*, Council on Environmental Quality, US Govt. Printing Office, Washington, DC.

Box, H.E. (1953) *List of Sugarcane Insects*, CIE, London.

Brader, L. (1979) Integrated pest control in the developing world. *Ann. Rev. Entomol.*, **24**, 154–255.

Brown, A.W.A. and Pal, R. (1971) *Insecticide Resistance in Arthropods*, 2nd edn, WHO, Geneva.

Brown, F.G. (1968) *Pests and Diseases of Forest Plantation Trees: An Annotated List of the Principal Species Occurring in the British Commonwealth*, Clarendon Press, Oxford.

Brown, K.W. (1967) *Forest Insects of Uganda, An Annotated List*, Govt. Printer, Entebbe.

Buczacki, S.T. and Harris, K.M. (1981) *Collins Guide to the Pests, Diseases and Disorders of Garden Plants*, Collins, London.

Burn, A.J., Coaker, T.H. and Jepson, P.C. (1987) *Integrated Pest Management*, Academic Press, London.

Busvine, J.R. (1966) *Insects and Hygiene*, 2nd edn, Methuen, London.

Busvine, J.R. *Recommended Methods for Measurement of Pest Resistance to Pesticides*, FAO PL. Prod. Prot. Paper – 21, FAO, Rome.

Buyckx, E.J.E. (1962) *Précis des Maladies et des Insectes Nuisibles Rencontrées sur les Plantes Cultivées au Congo, au Rwanda et au Burundi*, INEAC.

CAB (1951–1989) *Distribution Maps of Insect Pests*, Series A, Agricultural. Nos 1–510, CIE, London.

CAB (1961–1967) *CIBC Technical Bulletins*, 1–8, CAB, London.

CAB (1980) *Perspective in World Agriculture*, CAB, Slough.

CAB (1981) *List of Research Workers in the Agricultural Sciences in the Commonwealth and in the Republic of Ireland*, 4th edn, CAB, London.

CABI (1992) *Quarantine Pests for Europe*, CABI and EPPO. (CAB is now Commonwealth Bureaux of Agriculture International and is located at Wallingford, Oxon., UK; many specialist publications.)

Caltagirone, L.E. (1981) Landmark examples in classical biological control. *Ann. Rev. Entomol.*, **26**, 213–32.

Cameron, P.J. *et al.* (eds) (1989) *A Review of Biological Control of Invertebrate Pests and Weeds in New Zealand 1874–1987*, CIBC Tech. Comm. No. 10, CABIBC and DSIR, Ascot.

Campion, D.G. (1972) Insect chemosterilants: a review. *Bull. Ent. Res.*, **61**, 577–635.

Caresche, L. *et al.* (1969) *Handbook for Phytosanitary Inspectors in Africa*, Organization of African Unity/STRC, Lagos.

Caswell, G.H. (1962) *Agricultural Entomology in the Tropics*, Edward Arnold, London.

Cavalloro, R. (ed.) (1983) *Fruit Flies of Economic Importance*, Proc. CEC/IOBC Int. Symp.; Athens, Nov. 1982, Balkema, Rotterdam.

Chandler, A.C. and Read, C.P. (1961) *Introduction to Parasitology (With Special Reference to the Parasites of Man)*, 10th edn, John Wiley, New York.

Cheng, T.C. (1967) *The Biology of Animal Parasites*, Saunders, Philadelphia.

Cherrett, J.M. and Peregrine, D.J. (1976) A review of the status of the leaf-cutting ants and their control. *Ann. Appl. Biol.*, **84**, 128–33.

Cherrett, J.M. and Sager, G.R. (1977) *Origins of Pest, Parasite, Disease and Weed Problems*, 18th Symp. Brit. Ecol. Soc., Blackwell, Oxford.

Cherrett, J.M. *et al.* (1971) *The Control of Injurious Animals*, English University Press, London.

Chiang, H.C. (1978) Pest management in corn. *Ann. Rev. Entomol.*, **23**, 101–23.

Chinnery, M. (1986) *Collins Guide to the Insects of Britain and Western Europe*, Collins, London.

CIE Guide to insects of importance to Man.
  (1987) 1. *Lepidoptera*. CAB Int. Inst. Ent. and Brit. Mus. Nat. Hist., London.
  (1989) 2. *Thysanoptera*. CAB Int. Inst. Ent. and Brit. Mus. Nat. Hist., London.
  (1991) *Coleoptera*. Int. Inst. Ent. and The Nat. Hist. Mus., London.

Clark, L.R. *et al.* (1967) *The Ecology of Insect Populations in Theory and Practice*, Methuen, London.

Clausen, C.P. (1940) *Entomophagous Insects*, 1st edn, McGraw-Hill, New York.

Cloudsley-Thompson, J.L. (1986) Cochineal. *Antenna*, **10**(2), 70–72.

Conway, G.R. (1972a) *Pests of Cocoa in Sabah, Malaysia*, Bull. Dept. Agric., Malaysia.

Conway, G.R. (1972b) Ecological aspects of pest control in Malaysia, in *The Careless Technology; Ecological Aspects of International Development*, (ed. J. Milton), Nat. Hist. Press.

COPR (1978) *Pest Control in Tropical Root Crops*, PANS Manual No. 4, COPR, London.

COPR (1981) *Pest Control in Tropical Grain Legumes*, COPR, London.

COPR (1982) *The Locust and Grasshopper Manual*, COPR, London.

COPR (1983) *Pest Control in Tropical Tomatoes*, COPR, London.

Corbet, S. (1987) More bees make better crops. *New Scientist*, 23 July, 40–43.

Coulson, R.N. and Witter, A. (1984) *Forest Entomology: Ecology and Management*, Wiley & Sons, New York.

Cramer, H.H. (1967) *Plant Protection and World Crop*

Production, Bayer Pflanzenschutz, Leverkusen.

Crane, E. and Walker, P. (1983) *The Impact of Pest Management on Bees and Pollination*, Int. Bee Res. Assoc.; Tropical Development and Research Institute, London.

Crowe, T.J., Tadesse, G.M. and Abate, T. (1977) *An Annotated List of Insect Pests in Ethiopia*, Institute of Agricultural Research, Addis Ababa.

CSCPRC (1977) *Insect Control in the People's Republic of China*, National Academy of Science, Washington, DC.

Darlington, A. (1968) *The Pocket Encyclopaedia of Plant Galls in Colour*, Blandford Press, London.

Davidson, R.H. and Peairs, L.M. (1966) *Insect Pests of Farm, Garden and Orchard*, John Wiley, New York.

Davies, R.G. (1988) *Outlines of Entomology*, 7th edn, Chapman & Hall, London.

DeBach, P. (1964) *Biological Control of Insect Pests and Weeds*, Chapman & Hall, London.

DeBach, P. (1974) *Biological Control by Natural Enemies*, Cambridge University Press, Cambridge.

Delucchi, V.L. (ed.) (1976) *Studies in Biological Control* (IBP-9), Cambridge University Press, Cambridge.

Dent, D. (1991) *Insect Pest Management*, CABI, Wallingford, Oxon.

Dobie, P. *et al.* (1984) *Insects and Arachnids of Tropical Stored Products – Their Biology and Identification*, TDRI, Slough.

Dolling, W.R. (1991) *The Hemiptera*, Natural History Museum Publications, Oxford University Press, Oxford.

Duffey, E.A.J. (1957) *African Timber Beetles*, British Museum of Natural History, London.

Eady, R.D. and Quinlan, J. (1963) *Handbooks for the Identification of British Insects. Hymenoptera Cynipoidea*, Vol. 8, Part 1(a). Roy. Ent. Soc. Lond.: London.

Ebbels, D.L. and King, J.E. (1979) *Plant Health*, Blackwell, Oxford.

Ebeling, W. (1959) *Subtropical Fruit Pests*, 2nd edn, University of California Press, Berkeley, California.

Ebeling, W. (1975) *Urban Entomology*, University of California, Division of Agricultural Science, California.

Edwards, C.A. and Heath, G.W. (1964) *Principles of Agricultural Entomology*, Chapman & Hall, London.

Elton, C.S. (1958) *The Ecology of Invasions by Animals and Plants*, Methuen, London.

Emden, H.F. van (ed.) (1972) *Insect/Plant Relationships*, Symp. Roy. Ent. Soc. Lond., No. 6. Blackwell, Oxford.

Emden, H.F. van (1989) *Pest Control*, 2nd edn, IOB New Studies in Biology, Edward Arnold, London.

Entwhistle, P.F. (1972) *Pests of Cocoa*, Longmans, London.

EPA (1976) *List of Insects and Other Organisms*, 3rd edn, Parts 1, 2, 3 and 4, EPA, Washington, DC.

Evans, J. W. (1952) *Injurious Insects of the British Commonwealth*, Commonwealth Institute of Entomology, London.

Evans, G.O., Sheals, J.G. and Macfarlane, D. (1961) *The Terrestrial Acari of the British Isles. Vol. 1, Introduction to Biology*, British Museum of Natrual History, London.

FAO (1966) *Proceedings of the FAO Symposium on Integrated Pest Control (11–15 October 1965)*, 1, 2, 3. FAO, Rome.

FAO (1974) *Proceedings of the FAO Conference on Ecology in Relation to Plant Pest Control. Rome, Italy, 11–15 December 1972*, FAO, Rome.

FAO (1979a) *Guidelines for Integrated Control of Maize Pests*, FAO Pl. Prod. Prot. Paper – 18, FAO, Rome.

FAO (1979b) *Elements of Integrated Control of Sorghum Pests*, FAO Pl. Prod. Prot. Paper – 19, FAO, Rome.

FAO/CABI (1971) *Crop Loss Assessment Methods*, FAO Manual on the Evaluation of Losses by Pests, Diseases and Weeds. Supplement 1. 1973, 2. 1977, 3. 1981.

Feltwell, J. (1990) *The Story of Silk*, Sutton Publishing, Stroud.

Fletcher, W.W. (1974) *The Pest War*, Blackwell, Oxford.

Fox Wilson, G. (1960) *Horticultural Pests – Detection and Control*, 2nd edn, Crosby Lockwood, London.

Free, J.B. (1970) *Insect Pollination of Crops*, Academic Press, London.

Free, J.B. and Williams, I.H. (1977) *The Pollination of Crops by Bees*, Apimondia, Bucharest & Int. Bee Res. Assoc., UK.

Frolich, G. and Rodewald, W. (1970) *General Pests and Diseases of Tropical Crops and Their Control*, Pergamon Press, London.

Gauld, I. and Bolton, B. (1988) *The Hymenoptera*, British Museum of Natural History and Oxford University Press, London.

Gay, F.J. (ed.) (1966) *Scientific and Common Names*

of Insects and Allied Forms Occurring in Australia, Bull. No. 285, CSIRO, Melbourne.

Geier, P.W. (1966) Management of insect pests. *Ann. Rev. Entomol.*, **11**, 471–90.

Gilbert, P. and Hamilton, C.J. (1990) *Entomology – A Guide to Information Sources*, 2nd edn, Mansell, London.

Glass, E.H. (Coordinator) (1975) *Integrated Pest Management: Rationale, Potential, Needs and Implementation*, Ent. Soc. Amer., Special Publ., 75–8.

Good, R. (1953) *The Geography of the Flowering Plants*, 2nd edn, Longmans, London.

Gray, B. (1972) Economic tropical forest entomology, *Ann. Rev. Entomol.*, **17**, 313–54.

Greathead, D.J. (1971) *A Review of Biological Control in the Ethiopian Region*, Tech. Commun. CIBC, No. 5, CAB, London.

Harcourt, D.G. (1969) The development and use of life tables in the study of natural insect populations. *Ann. Rev. Entomol.*, **14**, 175–96.

Harris, W.V. (1969) *Termites as Pests of Crops and Trees*, Commonwealth Institute of Entomology, London.

Harris, W.V. (1971) *Termites, Their Recognition and Control*, 2nd edn, Longmans, London.

Harwood, R.F. and James, M.T. (1979) *Entomology in Human and Animal Health*, Macmillan, New York.

Hassell, M.P. and Southwood, T.R.E. (1978) Foraging strategies of insects. *Ann. Rev. Ecol. Syst.*, **9**, 75–98.

Hill, D.S. (1967) Figs (*Ficus* spp.) and fig-wasps (Chalciodoidea). *Journal of Natural History*, **1**, 413–34.

Hill, D.S. (1983) *Agricultural Insect Pests of the Tropics and Their Control*, 2nd edn, Cambridge University Press, Cambridge.

Hill, D.S. (1987) *Agricultural Insect Pests of Temperate Regions and Their Control*, Cambridge University Press, Cambridge.

Hill, D.S. (1990) *Pests of Stored Products and Their Control*, Belhaven Press, London.

Hill, D.S. (1994) *Agricultural Entomology*, Timber Press, Portland, Oregon.

Hill, D.S. and Waller, J.M. (1982) *Pests and Diseases of Tropical Crops. Vol. 1. Principles and Methods of Control*, Intermediate Tropical Agriculture Series, Longmans, London.

Hill, D.S. and Waller, J.M. (1988) *Pests and Diseases of Tropical Crops. Vol. 2. Field Handbook*, Intermediate Tropical Agriculture Series, Longmans, London.

Hinton, H.E. and Corbet, A.S. (1955) *Common Insect Pests of Stored Food Products*, 3rd edn, Econ. Ser., No. 15, British Museum of Natural History, London. (For 7th edn see Mound, 1989.)

Holt, V.M. (1988) *Why Not Eat Insects?*, 2nd edn.

Huffaker, C.B. (ed.) (1971) *Biological Control*, Plenum, New York.

Hussey, N.W., Read, W.H. and Hesling, J.J. (1969) *The Pests of Protected Cultivation*, Edward Arnold, London.

Imms, A.D. (1960) *A General Textbook of Entomology*, 9th edn, revised by O.W. Richards and R.G. Davies, Methuen, London.

Japan Plant Protection Association (1980) *Major Insect and Other Pests of Economic Plants in Japan*, Japan Plant Protection Association, Tokyo.

Johnson, W.T. and Lyon, H.H. (1976) *Insects that Feed on Trees and Shrubs: An Illustrated Practical Guide*, Cornell University Press, Ithaca, NY.

Jones, F.G.W. and Jones, M. (1974) *Pests of Field Crops*, 2nd edn, Edward Arnold, London.

Kalshoven, L.G.E. (1981) *Pests of Crops in Indonesia*, revised and translated by P.A. van der Laan and G.H.L. Rothschild, Ichtiar Baru, Van Noeve, Jakarta.

Kettle, D.S. (1990) *Medical and Veterinary Entomology*, 2nd edn, CABI, Wallingford, Oxon.

Kettlewell, H.B.D. (1973) *The Evolution of Melanism*, Clarendon Press, Oxford.

Kevan, P.G. and Baker, H.G. (1983) Insects as flower visitors and pollinators. *Ann. Rev. Entomol.*, **28**, 407–53.

Kilgore, W.W. and Doutt, R.L. (1967) *Pest Control – Biological, Physical and Selected Chemical Methods*, Academic Press, New York.

King, A.B.S. and Saunders, J.L. (1984) *The Invertebrate Pests of Annual Food Crops in Central America*, ODA, London.

Kirkpatrick, T.W. (1966) *Insect Life in the Tropics*, Longmans, London.

Kloet, G.S. and Hincks, W.D. (1964–1978) *A Check List of British Insects*, 2nd edn – revised. Roy. Ent. Soc. of Lond., London.

Kranz, J., Schmutterer, H. and Koch, W. (eds) (1979) *Diseases, Pests and Weeds in Tropical Crops*, John Wiley, Chichester.

Laffoon, J.L. (1960) Common names of insects – approved by the Entomological Society of America, *Bull. Ent. Soc. Amer.*, **6**, 175–211.

Lane, R.P. and Crosskey, R.W. (1993) *Medical Insects and Arachnids*, Natural History Museum, London.

Le Pelley, R.H. (1959) *Agricultural Insects of East Africa*, E. Afr. High Comm., Nairobi.

Le Pelley, R.H. (1968) *Pests of Coffee*, Longmans, London.

Lincoln, R.J., Boxshall, G.A. and Clark, P.F. (1982) *A Dictionary of Ecology, Evolution and Systematics*, Cambridge University Press, Cambridge.

Mackerras, I.M. (ed. CSIRO) (1969) *The Insects of Australia*, Melbourne University Press, Victoria.

Martin, E.C. and McGregor, S.E. (1973) Changing trends in insect pollination of commercial crops. *Ann. Rev. Entomol.*, **18**, 207–26.

Massee, A.M. (1954) *The Pests of Fruit and Hops*, 3rd edn, Crosby Lockwood, London.

Matthews, G.A. (1984) *Pest Management*, Longmans, London.

May, R.M. (ed.) (1976) *Theoretical Ecology – Principles and Applications*, Blackwell, Oxford.

Metcalf, C.L., Flint, W.P. and Metcalf, R.L. (1962) *Destructive and Useful Insects*, McGraw-Hill, New York.

Metcalf, R.L. and Luckmann, W.H. (eds) (1975) *Introduction to Insect Pest Management*, John Wiley, New York.

Morton, A. (1989) Thailand's million-dollar moth. *New Scientist*, 25 November, 48–52.

Mound, L.A. (ed.) (1989) *Common Insect Pests of Stored Food Products*, 7th edn, British Museum of Natural History, Econ. Series No. 15, British Museum of Natural History, London.

Mound, L.A. and Halsey, S. H. (1978) *Whiteflies of the World: A Systematic Catalogue of the Aleyrodidae (Homoptera) with Host Plant and Natural Enemy Data*, John Wiley, London.

Munro, J.W. (1966) *Pests of Stored Products*, Hutchinson, London.

Nat. Acad. Sci. U.S. (1969) *Principles of Plant and Animal Pest Control. Vol. 3, Insect-Pest Management and Control*, Nat. Acad. Sci. U.S., No. 1695: Washington, DC.

Nayar, K.K., Ananthakrishnan, T.N. and David, B.V. (1976) *General and Applied Entomology*, Tata McGraw-Hill, New Delhi.

Nichols, S.W. (ed.) (1989) (updated edn, 1992) *The Torre-Bueno Glossary of Entomology*, New York Ent. Soc., New York.

Nobel, E.R. and Nobel, G.E. (1961) *Parasitology (The Biology of Animal Parasites)*, Lea & Febiger, Philadelphia.

Oldroyd, H. (1958) *Preserving and Studying Insects*, Hutchinson, London.

Oldroyd, H. (1964) *The Natural History of Flies*, Weidenfeld & Nicolson, London.

Oldroyd, H. (1968) *Elements of Entomology*, Weidenfeld & Nicolson, London.

Painter, R.H. (1951) *Insect Resistance in Crop Plants*, University Press Kansas, Lawrence.

Pearson, E.O. (1958) *The Insect Pests of Cotton in Tropical Africa*, Commonwealth Institute of Entomology, London.

PESTDOC (1974) *Organism Thesaurus. Vol. 1, Animal Organisms*, Ciba-Geigy, Basle & Derwent Pub., London.

Peterson, A. (1953) *Entomological Techniques*, Edwards Bros 1, Ann Arbor, Michigan.

Pfadt, R.E. (1962) *Fundamentals of Applied Entomology*, Macmillan, New York.

Pirone, P.P. (1978) *Diseases and Pests of Ornamental Plants*, 5th edn, John Wiley, New York.

Price Jones, D. and Solomon, M.E. (1974) *Biology in Pest and Disease Control*, 13th Symp. Brit. Ecol. Soc., Blackwell, Oxford.

Proctor, M. and Yeo, P. (1973) *The Pollination of Flowers*, The New Naturalist, No. 54, Collins, London.

Pschorn-Walcher, H. (1977) Biological control of forest insects. *Ann. Rev. Entomol.*, **22**, 1–22.

Purseglove, J.W. (1968) *Tropical Crops. Dicotyledons*, Vols 1 and 2, Longmans, London.

Purseglove, J.W. (1972) *Tropical Crops. Monocotyledons*, Vols 1 and 2, Longmans, London.

Rabb, R.L. and Guthrie, F.E. (1970) *Concepts of Pest Management*. Proceeds of a conference held at North Carolina State University, Raleigh, North Carolina, 25–27 March 1970, North Carolina State University, Raleigh.

Richards, O.W. and Davies, R.G. (1977) *Imms' General Textbook of Entomology*, 10th edn. Vol. 1. *Structure, Physiology and Development*; Vol. 2. *Classification and Biology*, Chapman & Hall, London.

Roelofs, W.L. (ed.) (1979) *Establishing Efficacy of Sex Attractants and Disruptants for Insect Control*, Ent. Soc. Amer., Washington, DC.

Russell, G.E. (1978) *Plant Breeding for Pest and Disease Resistance*, Butterworths, London.

Schreck, C.E. (1977) Techniques for the evaluation of insect repellents: a critical review. *Ann. Rev. Entomol.*, **22**, 101–19.

Scopes, N. and Ledieu, M. (eds) (1983) *Pest and Disease Control Handbook*, 2nd edn, British Crop Protection Council, London.

Seymour, P.R. (1979) *Invertebrates of Economic Importance in Britain* (formerly MAFF *Tech. Bull.* No. 6), HMSO, London.

Smith, K.G.V. (ed.) (1973) *Insects and Other Arthropods of Medical Importance*, British Museum of Natural History, London.

Smith, K.G.V. (1986) *A Manual of Forensic Entomology*, Intercept, Andover.

Soulsby, E.J.L. (1968) *Helminths, Arthropods and Protozoa of Domesticated Animals*, 6th edn, Baillière, Tindall & Cassell, London.

Southwood, T.R.E. (ed.) (1968) *Insect Abundance, a Symposium*, Roy. Ent. Soc. Lond., London.

Southwood, T.R.E. (1978) *Ecological Methods (With Particular Reference to the Study of Insect Populations)*, 2nd edn, Chapman & Hall, London.

Spencer, K.A. (1990) *Host Specialization in the World Agromyzidae (Diptera)*, Series Ent., Vol. 45, Kluwer Academic Publishers, Dordrecht, Netherlands.

Stern, V.M. *et al.* (1959) The integrated control concept. *Hilgardia*, **29**, 81–101.

Strong, D.R., Lawton, J.H. and Southwood, T.R.E. (1984) *Insects on Plants: Community Patterns and Mechanisms*, Blackwell, Oxford.

Torre-Bueno, J.R. de la (1937) *A Glossary of Entomology*, New York Entomology Society, New York.

University of California Statewise Integrated Pest Management Project.
  - (1983) IPM for rice (Pub. No. 3280).
  - (1984a) IPM for cotton (Pub. No. 3305).
  - (1984b) IPM for citrus (Pub. No. 3303).
  - (1986) IPM for potatoes (Pub. No. 3316).

USDA (1952) *The Yearbook of Agriculture (1952) Insects* (USDA), US Govt. Printing Off., Washington, DC.

USDA (1966) *The Yearbook of Agriculture (1966) Protecting Our Food* (USDA), US Govt. Printing Off., Washington, DC.

Uvarov, B. (1966) *Grasshoppers and Locusts*, Cambridge University Press, Cambridge.

Waters, W.E. and Stark, R.W. (1980) Forest pest management: concept and reality. *Ann. Rev. Entomol.*, **25**, 479–509.

Watson, T.F., Moore, L. and Ware, G.W. (1975) *Practical Insect Pest Management*, WH Freeman & Co., San Francisco.

Watt, A.D. (ed.) (1990) *Population Dynamics of Forest Insects*, Intercept, Andover, UK.

Werner, F.G. (1982) *Common Names of Insects and Related Organisms*, Ent. Soc. Amer., Washington, DC.

Wilson, F. (1960) *A Review of the Biological Control of Insects and Weeds in Australia and Australian New Guinea*, Techn. Comm. No. 1, CIBC, Ottawa.

Wood, A.M. (1989) *Insects of Economic Importance: A Checklist of Preferred Names*, CABI, Wallingford, Oxon.

Wyniger, R. (1962) *Pests of Crops in Warm Climates and Their Control*, Acta Tropica. Supp. 7.

Wyniger, R. (1968) *Control Measures*, 2nd edn, Verlag Recht Ges., Basel, Switzerland.

INDEX